An Introduction to Celestial Mechanics

This accessible text on classical celestial mechanics—the principles governing the motions of bodies in the solar system—provides a clear and concise treatment of virtually all the major features of solar system dynamics. Building on advanced topics in classical mechanics, such as rigid body rotation, Lagrangian mechanics, and orbital perturbation theory, this text has been written for well-prepared undergraduates and beginning graduate students in astronomy, physics, mathematics, and related fields. Specific topics covered include Keplerian orbits; the perihelion precession of the planets; tidal interactions among the Earth, Moon, and Sun; the Roche radius; the stability of Lagrange points in the three-body problem; and lunar motion. More than 100 exercises allow students to gauge their understanding; a solutions manual is available to instructors. Suitable for a first course in celestial mechanics, this text is the ideal bridge to higher-level treatments.

Richard Fitzpatrick is professor of physics at the University of Texas at Austin, where he has been a faculty member since 1994. He earned his MA in physics at the University of Cambridge and his DPhil in astronomy at the University of Sussex. He is a longstanding Fellow of the Royal Astronomical Society and the American Physical Society and author of *Maxwell's Equations and the Principles of Electromagnetism* (2008).

An Introduction to Celestial Mechanics

RICHARD FITZPATRICK

University of Texas at Austin

CAMBRIDGE
UNIVERSITY PRESS

Shaftesbury Road, Cambridge CB2 8EA, United Kingdom

One Liberty Plaza, 20th Floor, New York, NY 10006, USA

477 Williamstown Road, Port Melbourne, VIC 3207, Australia

314–321, 3rd Floor, Plot 3, Splendor Forum, Jasola District Centre, New Delhi – 110025, India

103 Penang Road, #05–06/07, Visioncrest Commercial, Singapore 238467

Cambridge University Press is part of Cambridge University Press & Assessment,
a department of the University of Cambridge.

We share the University's mission to contribute to society through the pursuit of
education, learning and research at the highest international levels of excellence.

www.cambridge.org
Information on this title: www.cambridge.org/9781107023819

First published 2012
4th printing 2022

Printed in the United Kingdom by Print on Demand, World Wide

A catalogue record for this publication is available from the British Library

Library of Congress Cataloging-in-Publication data

Fitzpatrick, Richard, 1963–
An introduction to celestial mechanics / Richard Fitzpatrick.
p.   cm.
Includes bibliographical references and index.
ISBN 978-1-107-02381-9 (hardback)
1. Celestial mechanics. I. Title.
QB351.F565   2012
521–dc23   2012000780

ISBN 978-1-107-02381-9 Hardback

Contents

Preface

The aim of this book is to bridge the considerable gap that exists between standard undergraduate mechanics texts, which rarely cover topics in celestial mechanics more advanced than two-body orbit theory, and graduate-level celestial mechanics texts, such as the well-known books by Moulton (1914), Brouwer and Clemence (1961), Danby (1992), Murray and Dermott (1999), and Roy (2005). The material presented here is intended to be intelligible to an advanced undergraduate or beginning graduate student with a firm grasp of multivariate integral and differential calculus, linear algebra, vector algebra, and vector calculus.

The book starts with a discussion of the fundamental concepts of Newtonian mechanics, as these are also the fundamental concepts of celestial mechanics. A number of more advanced topics in Newtonian mechanics that are needed to investigate the motions of celestial bodies (e.g., gravitational potential theory, motion in rotating reference frames, Lagrangian mechanics, Eulerian rigid body rotation theory) are also described in detail in the text. However, any discussion of the application of Hamiltonian mechanics, Hamilton-Jacobi theory, canonical variables, and action-angle variables to problems in celestial mechanics is left to more advanced texts (see, for instance, Goldstein, Poole, and Safko 2001).

Celestial mechanics (a term coined by Laplace in 1799) is the branch of astronomy that is concerned with the motions of celestial objects—in particular, the objects that make up the solar system—under the influence of gravity. The aim of celestial mechanics is to reconcile these motions with the predictions of Newtonian mechanics. Modern analytic celestial mechanics started in 1687 with the publication of the *Principia* by Isaac Newton (1643–1727) and was subsequently developed into a mature science by celebrated scientists such as Euler (1707–1783), Clairaut (1713–1765), D'Alembert (1717–1783), Lagrange (1736–1813), Laplace (1749–1827), and Gauss (1777–1855). This book is largely devoted to the study of the "classical" problems of celestial mechanics that were investigated by these scientists. These problems include the figure of the Earth; tidal interactions among the Earth, Moon, and Sun; the free and forced precession and nutation of the Earth; the three-body problem; the secular evolution of the solar system; the orbit of the Moon; and the axial rotation of the Moon. However, any discussion of the highly complex problems that arise in modern celestial mechanics, such as the mutual gravitational interaction between the various satellites of Jupiter and Saturn, the formation of the Kirkwood gaps, the dynamics of planetary rings, and the ultimate stability of the solar system, is again left to more advanced texts (see, in particular, Murray and Dermott 1999).

A number of topics, closely related to classical celestial mechanics, are not discussed in this book for the sake of brevity. The first of these is positional astronomy—the branch of astronomy that is concerned with finding the positions of celestial objects in the Earth's sky at a particular instance in time. Interested readers are directed to Smart (1977). The second excluded topic is the development of numerical methods for the solution of problems in celestial mechanics. Interested readers are directed to Danby (1992). The third (mostly) excluded topic is astrodynamics: the application of Newtonian dynamics to the design and analysis of orbits for artificial satellites and space probes. Interested readers are directed to Bate, Mueller, and White (1977). The final excluded topic is the determination of the orbits of celestial objects from observational data. Interested readers are again directed to Danby (1992).

1 Newtonian mechanics

1.1 Introduction

Newtonian mechanics is a mathematical model whose purpose is to account for the motions of the various objects in the universe. The general principles of this model were first enunciated by Sir Isaac Newton in a work titled *Philosophiae Naturalis Principia Mathematica* (Mathematical Principles of Natural Philosophy). This work, which was published in 1687, is nowadays more commonly referred to as the *Principia*.[1]

Until the beginning of the twentieth century, Newtonian mechanics was thought to constitute a *complete* description of all types of motion occurring in the universe. We now know that this is not the case. The modern view is that Newton's model is only an *approximation* that is valid under certain circumstances. The model breaks down when the velocities of the objects under investigation approach the speed of light in a vacuum, and must be modified in accordance with Einstein's *special theory of relativity*. The model also fails in regions of space that are sufficiently curved that the propositions of Euclidean geometry do not hold to a good approximation, and must be augmented by Einstein's *general theory of relativity*. Finally, the model breaks down on atomic and subatomic length scales, and must be replaced by *quantum mechanics*. In this book, we shall (almost entirely) neglect relativistic and quantum effects. It follows that we must restrict our investigations to the motions of *large* (compared with an atom), *slow* (compared with the speed of light) objects moving in *Euclidean* space. Fortunately, virtually all the motions encountered in conventional celestial mechanics fall into this category.

Newton very deliberately modeled his approach in the *Principia* on that taken in Euclid's *Elements*. Indeed, Newton's theory of motion has much in common with a conventional *axiomatic system*, such as Euclidean geometry. Like all axiomatic systems, Newtonian mechanics starts from a set of terms that are *undefined* within the system. In this case, the fundamental terms are *mass*, *position*, *time*, and *force*. It is taken for granted that we understand what these terms mean, and, furthermore, that they correspond to *measurable* quantities that can be ascribed to, or associated with, objects in the world around us. In particular, it is assumed that the ideas of position in space, distance in space, and position as a function of time in space are correctly described by conventional Euclidean vector algebra and vector calculus. The next component of an axiomatic system is a set of *axioms*. These are a set of *unproven* propositions,

---

[1] An excellent discussion of the historical development of Newtonian mechanics, as well as the physical and philosophical assumptions that underpin this theory, is given in Barbour 2001.

involving the undefined terms, from which all other propositions in the system can be derived via logic and mathematical analysis. In the present case, the axioms are called *Newton's laws of motion* and can be justified only via experimental observation. Note, incidentally, that Newton's laws, in their primitive form, are applicable only to *point objects* (i.e., objects of negligible spatial extent). However, these laws can be applied to extended objects by treating them as collections of point objects.

One difference between an axiomatic system and a physical theory is that, in the latter case, even if a given prediction has been shown to follow necessarily from the axioms of the theory, it is still incumbent on us to test the prediction against experimental observations. Lack of agreement might indicate faulty experimental data, faulty application of the theory (for instance, in the case of Newtonian mechanics, there might be forces at work that we have not identified), or, as a last resort, incorrectness of the theory. Fortunately, Newtonian mechanics has been found to give predictions that are in excellent agreement with experimental observations in all situations in which it would be expected to hold.

In the following, it is assumed that we know how to set up a rigid Cartesian frame of reference and how to measure the positions of point objects as functions of time within that frame. It is also taken for granted that we have some basic familiarity with the laws of mechanics.

1.2   Newton's laws of motion

Newton's laws of motion, in the rather obscure language of the *Principia*, take the following form:

1. Every body continues in its state of rest, or uniform motion in a straight line, unless compelled to change that state by forces impressed on it.
2. The change of motion (i.e., momentum) of an object is proportional to the force impressed on it, and is made in the direction of the straight line in which the force is impressed.
3. To every action there is always opposed an equal reaction; or, the mutual actions of two bodies on each other are always equal, and directed to contrary parts.

Let us now examine how these laws can be applied to a system of point objects.

1.3   Newton's first law of motion

Newton's first law of motion essentially states that a point object subject to zero net external force moves in a straight line with a constant speed (i.e., it does not accelerate). However, this is true only in special frames of reference called *inertial frames*. Indeed, we can think of Newton's first law as the definition of an inertial frame: an inertial frame of reference is one in which a point object subject to zero net external force moves in a straight line with constant speed.

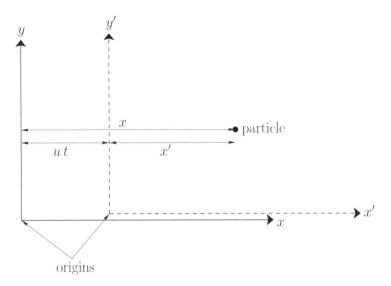

Fig. 1.1 A Galilean coordinate transformation.

Suppose that we have found an inertial frame of reference. Let us set up a Cartesian coordinate system in this frame. The motion of a point object can now be specified by giving its position vector, $\mathbf{r} \equiv (x, y, z)$, with respect to the origin of the coordinate system, as a function of time, t. Consider a second frame of reference moving with some constant velocity $\mathbf{u}$ with respect to the first frame. Without loss of generality, we can suppose that the Cartesian axes in the second frame are parallel to the corresponding axes in the first frame, that $\mathbf{u} \equiv (u, 0, 0)$, and, finally, that the origins of the two frames instantaneously coincide at $t = 0$. (See Figure 1.1.) Suppose that the position vector of our point object is $\mathbf{r}' \equiv (x', y', z')$ in the second frame of reference. It is evident, from Figure 1.1, that at any given time, t, the coordinates of the object in the two reference frames satisfy

$$x' = x - u\,t, \tag{1.1}$$

$$y' = y, \tag{1.2}$$

and

$$z' = z. \tag{1.3}$$

This coordinate transformation was first discovered by Galileo Galilei (1564–1642), and is nowadays known as a *Galilean transformation* in his honor.

By definition, the instantaneous velocity of the object in our first reference frame is given by $\mathbf{v} = d\mathbf{r}/dt \equiv (dx/dt, dy/dt, dz/dt)$, with an analogous expression for the velocity, $\mathbf{v}'$, in the second frame. It follows, from differentiation of Equations (1.1)–(1.3) with respect to time, that the velocity components in the two frames satisfy

$$v'_x = v_x - u, \tag{1.4}$$

$$v'_y = v_y, \tag{1.5}$$

and

$$v_z' = v_z. \tag{1.6}$$

These equations can be written more succinctly as

$$\mathbf{v}' = \mathbf{v} - \mathbf{u}. \tag{1.7}$$

Finally, by definition, the instantaneous acceleration of the object in our first reference frame is given by $\mathbf{a} = d\mathbf{v}/dt \equiv (dv_x/dt, dv_y/dt, dv_z/dt)$, with an analogous expression for the acceleration, $\mathbf{a}'$, in the second frame. It follows, from differentiation of Equations (1.4)–(1.6) with respect to time, that the acceleration components in the two frames satisfy

$$a_x' = a_x, \tag{1.8}$$

$$a_y' = a_y, \tag{1.9}$$

and

$$a_z' = a_z. \tag{1.10}$$

These equations can be written more succinctly as

$$\mathbf{a}' = \mathbf{a}. \tag{1.11}$$

According to Equations (1.7) and (1.11), if an object is moving in a straight line with a constant speed in our original inertial frame (i.e., if $\mathbf{a} = \mathbf{0}$), then it also moves in a (different) straight line with a (different) constant speed in the second frame of reference (i.e., $\mathbf{a}' = \mathbf{0}$). Hence, we conclude that the second frame of reference is also an inertial frame.

A simple extension of the preceding argument allows us to conclude that there is an *infinite* number of different inertial frames moving with constant velocities with respect to one another. Newton thought that one of these inertial frames was special and defined an absolute standard of rest: that is, a static object in this frame was in a state of absolute rest. However, Einstein showed that this is not the case. In fact, there is no absolute standard of rest: in other words, all motion is relative—hence, the name *relativity* for Einstein's theory. Consequently, one inertial frame is just as good as another as far as Newtonian mechanics is concerned.

But what happens if the second frame of reference *accelerates* with respect to the first? In this case, it is not hard to see that Equation (1.11) generalizes to

$$\mathbf{a}' = \mathbf{a} - \frac{d\mathbf{u}}{dt}, \tag{1.12}$$

where $\mathbf{u}(t)$ is the instantaneous velocity of the second frame with respect to the first. According to this formula, if an object is moving in a straight line with a constant speed in the first frame (i.e., if $\mathbf{a} = \mathbf{0}$), then it does not move in a straight line with a constant speed in the second frame (i.e., $\mathbf{a}' \neq \mathbf{0}$). Hence, if the first frame is an inertial frame, then the second is *not*.

A simple extension of the preceding argument allows us to conclude that any frame of reference that accelerates with respect to a given inertial frame is not itself an inertial frame.

For most practical purposes, when studying the motions of objects close to the Earth's surface, a reference frame that is fixed with respect to this surface is approximately inertial. However, if the trajectory of a projectile within such a frame is measured to high precision, then it will be found to deviate slightly from the predictions of Newtonian mechanics. (See Chapter 5.) This deviation is due to the fact that the Earth is rotating, and its surface is therefore *accelerating* toward its axis of rotation. When studying the motions of objects in orbit around the Earth, a reference frame whose origin is the center of the Earth (or, to be more exact, the center of mass of the Earth–Moon system), and whose coordinate axes are fixed with respect to distant stars, is approximately inertial. However, if such orbits are measured to extremely high precision, then they will again be found to deviate very slightly from the predictions of Newtonian mechanics. In this case, the deviation is due to the Earth's orbital motion about the Sun. When studying the orbits of the planets in the solar system, a reference frame whose origin is the center of the Sun (or, to be more exact, the center of mass of the solar system), and whose coordinate axes are fixed with respect to distant stars, is approximately inertial. In this case, any deviations of the orbits from the predictions of Newtonian mechanics due to the orbital motion of the Sun about the galactic center are far too small to be measurable. It should be noted that it is impossible to identify an *absolute* inertial frame—the best approximation to such a frame would be one in which the cosmic microwave background appears to be (approximately) isotropic. However, for a given dynamic problem, it is always possible to identify an *approximate* inertial frame. Furthermore, any deviations of such a frame from a true inertial frame can be incorporated into the framework of Newtonian mechanics via the introduction of so-called fictitious forces. (See Chapter 5.)

1.4 Newton's second law of motion

Newton's second law of motion essentially states that if a point object is subject to an external force, $\mathbf{f}$, then its equation of motion is given by

$$\frac{d\mathbf{p}}{dt} = \mathbf{f}, \tag{1.13}$$

where the momentum, $\mathbf{p}$, is the product of the object's inertial mass, m, and its velocity, $\mathbf{v}$. If m is not a function of time, then Equation (1.13) reduces to the familiar equation

$$m\frac{d\mathbf{v}}{dt} = \mathbf{f}. \tag{1.14}$$

This equation is valid only in an *inertial frame*. Clearly, the inertial mass of an object measures its reluctance to deviate from its preferred state of uniform motion in a straight line (in an inertial frame). Of course, the preceding equation of motion can be solved only if we have an independent expression for the force, $\mathbf{f}$ (i.e., a law of force). Let us suppose that this is the case.

An important corollary of Newton's second law is that force is a *vector quantity*. This must be the case, as the law equates force to the product of a scalar (mass) and a vector (acceleration).[2] Note that acceleration is obviously a vector because it is directly related to displacement, which is the prototype of all vectors. One consequence of force being a vector is that two forces, $\mathbf{f}_1$ and $\mathbf{f}_2$, both acting at a given point, have the same effect as a single force, $\mathbf{f} = \mathbf{f}_1 + \mathbf{f}_2$, acting at the same point, where the summation is performed according to the laws of vector addition. Likewise, a single force, $\mathbf{f}$, acting at a given point, has the same effect as two forces, $\mathbf{f}_1$ and $\mathbf{f}_2$, acting at the same point, provided that $\mathbf{f}_1 + \mathbf{f}_2 = \mathbf{f}$. This method of combining and splitting forces is known as the *resolution of forces*; it lies at the heart of many calculations in Newtonian mechanics.

Taking the scalar product of Equation (1.14) with the velocity, $\mathbf{v}$, we obtain

$$m\,\mathbf{v}\cdot\frac{d\mathbf{v}}{dt} = \frac{m}{2}\frac{d(\mathbf{v}\cdot\mathbf{v})}{dt} = \frac{m}{2}\frac{dv^2}{dt} = \mathbf{f}\cdot\mathbf{v}. \tag{1.15}$$

This can be written

$$\frac{dK}{dt} = \mathbf{f}\cdot\mathbf{v}, \tag{1.16}$$

where

$$K = \frac{1}{2}m\,v^2. \tag{1.17}$$

The right-hand side of Equation (1.16) represents the rate at which the force does work on the object—that is, the rate at which the force transfers energy to the object. The quantity K represents the energy that the object possesses by virtue of its motion. This type of energy is generally known as *kinetic energy*. Thus, Equation (1.16) states that any work done on a point object by an external force goes to increase the object's kinetic energy.

Suppose that under the action of the force, $\mathbf{f}$, our object moves from point P at time t_1 to point Q at time t_2. The net change in the object's kinetic energy is obtained by integrating Equation (1.16):

$$\Delta K = \int_{t_1}^{t_2} \mathbf{f}\cdot\mathbf{v}\,dt = \int_P^Q \mathbf{f}\cdot d\mathbf{r}, \tag{1.18}$$

because $\mathbf{v} = d\mathbf{r}/dt$. Here, $d\mathbf{r}$ is an element of the object's path between points P and Q, and the integral in $\mathbf{r}$ represents the net *work* done by the force as the object moves along the path from P to Q.

As is well known, there are basically two kinds of forces in nature: first, those for which line integrals of the type $\int_P^Q \mathbf{f}\cdot d\mathbf{r}$ depend on the end points but not on the path taken between these points; second, those for which line integrals of the type $\int_P^Q \mathbf{f}\cdot d\mathbf{r}$ depend both on the end points and the path taken between these points. The first kind of force is termed *conservative*, whereas the second kind is termed *non-conservative*. It can be demonstrated that if the line integral $\int_P^Q \mathbf{f}\cdot d\mathbf{r}$ is *path independent*, for all choices of P and Q, then the force $\mathbf{f}$ can be written as the gradient of a scalar field. (See Section A.5.)

---

[2] A *scalar* is a physical quantity that is invariant under rotation of the coordinate axes. A *vector* is a physical quantity that transforms in an analogous manner to a displacement under rotation of the coordinate axes.

In other words, all conservative forces satisfy

$$\mathbf{f}(\mathbf{r}) = -\nabla U \tag{1.19}$$

for some scalar field $U(\mathbf{r})$. [Incidentally, mathematicians, as opposed to physicists and astronomers, usually write $f(\mathbf{r}) = +\nabla U$.] Note that

$$\int_P^Q \nabla U \cdot d\mathbf{r} \equiv \Delta U = U(Q) - U(P), \tag{1.20}$$

irrespective of the path taken between P and Q. Hence, it follows from Equation (1.18) that

$$\Delta K = -\Delta U \tag{1.21}$$

for conservative forces. Another way of writing this is

$$E = K + U = \text{constant}. \tag{1.22}$$

Of course, we recognize Equation (1.22) as an *energy conservation equation*: E is the object's total energy, which is conserved; K is the energy the object has by virtue of its motion, otherwise known as its *kinetic energy*; and U is the energy the object has by virtue of its position, otherwise known as its *potential energy*. Note, however, that we can write energy conservation equations only for conservative forces. Gravity is an obvious example of such a force. Incidentally, potential energy is undefined to an arbitrary additive constant. In fact, it is only the *difference* in potential energy between different points in space that is well defined.

1.5　Newton's third law of motion

Consider a system of N mutually interacting point objects. Let the ith object, whose mass is m_i, be located at position vector $\mathbf{r}_i$. Suppose that this object exerts a force $\mathbf{f}_{ji}$ on the jth object. Likewise, suppose that the jth object exerts a force $\mathbf{f}_{ij}$ on the ith object. Newton's third law of motion essentially states that these two forces are equal and opposite, irrespective of their nature. In other words,

$$\mathbf{f}_{ij} = -\mathbf{f}_{ji}. \tag{1.23}$$

(See Figure 1.2.) One corollary of Newton's third law is that an object cannot exert a force on itself. Another corollary is that all forces in the universe have corresponding reactions. The only exceptions to this rule are the fictitious forces that arise in non-inertial reference frames (e.g., the centrifugal and Coriolis forces that appear in rotating reference frames—see Chapter 5). Fictitious forces do not generally possess reactions.

Newton's third law implies *action at a distance*. In other words, if the force that object i exerts on object j suddenly changes, then Newton's third law demands that there must be an *immediate* change in the force that object j exerts on object i. Moreover, this must be true irrespective of the distance between the two objects. However, we now know that Einstein's special theory of relativity forbids information from traveling through

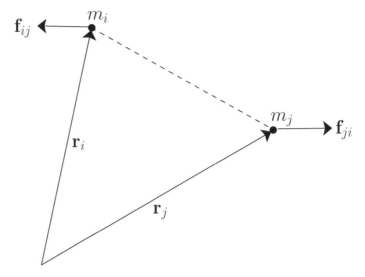

Fig. 1.2   Newton's third law.

the universe faster than the velocity of light in vacuum. Hence, action at a distance is also forbidden. In other words, if the force that object i exerts on object j suddenly changes, then there must be a *time delay*, which is at least as long as it takes a light ray to propagate between the two objects, before the force that object j exerts on object i can respond. Of course, this means that Newton's third law is not, strictly speaking, correct. However, as long as we restrict our investigations to the motions of dynamical systems over timescales that are long compared with the time required for light rays to traverse these systems, Newton's third law can be regarded as being approximately correct.

In an inertial frame, Newton's second law of motion applied to the ith object yields

$$m_i \frac{d^2 \mathbf{r}_i}{dt^2} = \sum_{j=1,N}^{j \neq i} \mathbf{f}_{ij}. \tag{1.24}$$

Note that the summation on the right-hand side of this equation excludes the case $j = i$, as the ith object cannot exert a force on itself. Let us now take this equation and sum it over all objects. We obtain

$$\sum_{i=1,N} m_i \frac{d^2 \mathbf{r}_i}{dt^2} = \sum_{i,j=1,N}^{j \neq i} \mathbf{f}_{ij}. \tag{1.25}$$

Consider the sum over forces on the right-hand side of the preceding equation. Each element of this sum—$\mathbf{f}_{ij}$, say—can be paired with another element—$\mathbf{f}_{ji}$, in this case— that is equal and opposite, according to Newton's third law. In other words, the elements of the sum all cancel out in pairs. Thus, the net value of the sum is *zero*. It follows that Equation (1.25) can be written

$$M \frac{d^2 \mathbf{r}_{cm}}{dt^2} = \mathbf{0}, \tag{1.26}$$

where $M = \sum_{i=1,N} m_i$ is the total mass. The quantity $\mathbf{r}_{cm}$ is the vector displacement of the *center of mass* of the system, which is an imaginary point whose coordinates are the mass weighted averages of the coordinates of the objects that constitute the system:

$$\mathbf{r}_{cm} = \frac{\sum_{i=1,N} m_i \, \mathbf{r}_i}{\sum_{i=1,N} m_i}. \tag{1.27}$$

According to Equation (1.26), the center of mass of the system moves in a uniform straight line, in accordance with Newton's first law of motion, irrespective of the nature of the forces acting between the various components of the system.

Now, if the center of mass moves in a uniform straight line, then the center of mass velocity,

$$\frac{d\mathbf{r}_{cm}}{dt} = \frac{\sum_{i=1,N} m_i \, d\mathbf{r}_i/dt}{\sum_{i=1,N} m_i}, \tag{1.28}$$

is a constant of the motion. However, the momentum of the ith object takes the form $\mathbf{p}_i = m_i \, d\mathbf{r}_i/dt$. Hence, the total momentum of the system is written

$$\mathbf{P} = \sum_{i=1,N} m_i \, \frac{d\mathbf{r}_i}{dt}. \tag{1.29}$$

A comparison of Equations (1.28) and (1.29) suggests that $\mathbf{P}$ is also a constant of the motion. In other words, the total momentum of the system is a *conserved* quantity, irrespective of the nature of the forces acting between the various components of the system. This result (which holds only if there is zero net external force acting on the system) is a direct consequence of Newton's third law of motion.

Taking the vector product of Equation (1.24) with the position vector $\mathbf{r}_i$, we obtain

$$m_i \, \mathbf{r}_i \times \frac{d^2\mathbf{r}_i}{dt^2} = \sum_{j=1,N}^{j \neq i} \mathbf{r}_i \times \mathbf{f}_{ij}. \tag{1.30}$$

The right-hand side of this equation is the net *torque* about the origin that acts on object i as a result of the forces exerted on it by the other objects. It is easily seen that

$$m_i \, \mathbf{r}_i \times \frac{d^2\mathbf{r}_i}{dt^2} = \frac{d}{dt}\left(m_i \, \mathbf{r}_i \times \frac{d\mathbf{r}_i}{dt}\right) = \frac{d\mathbf{l}_i}{dt}, \tag{1.31}$$

where

$$\mathbf{l}_i = m_i \, \mathbf{r}_i \times \frac{d\mathbf{r}_i}{dt} \tag{1.32}$$

is the *angular momentum* of the ith object about the origin of our coordinate system. Moreover, the total angular momentum of the system (about the origin) takes the form

$$\mathbf{L} = \sum_{i=1,N} \mathbf{l}_i. \tag{1.33}$$

Hence, summing Equation (1.30) over all particles, we obtain

$$\frac{d\mathbf{L}}{dt} = \sum_{i,j=1,N}^{i \neq j} \mathbf{r}_i \times \mathbf{f}_{ij}. \tag{1.34}$$

Consider the sum on the right-hand side of Equation (1.34). A general term, $\mathbf{r}_i \times \mathbf{f}_{ij}$, in this sum can always be paired with a matching term, $\mathbf{r}_j \times \mathbf{f}_{ji}$, in which the indices

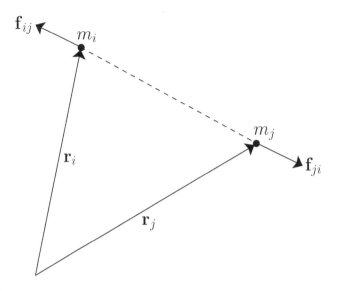

Fig. 1.3   Central forces.

have been swapped. Making use of Equation (1.23), we can write the sum of a general matched pair as

$$\mathbf{r}_i \times \mathbf{f}_{ij} + \mathbf{r}_j \times \mathbf{f}_{ji} = (\mathbf{r}_i - \mathbf{r}_j) \times \mathbf{f}_{ij}. \tag{1.35}$$

Let us assume that the forces acting between the various components of the system are *central* in nature, so that $\mathbf{f}_{ij}$ is parallel to $\mathbf{r}_i - \mathbf{r}_j$. In other words, the force exerted on object j by object i either points directly toward, or directly away from, object i, and vice versa. (See Figure 1.3.) This is a reasonable assumption, as virtually all the forces that we encounter in celestial mechanics are of this type (e.g., gravity). It follows that if the forces are central, then the vector product on the right-hand side of the above expression is zero. We conclude that

$$\mathbf{r}_i \times \mathbf{f}_{ij} + \mathbf{r}_j \times \mathbf{f}_{ji} = \mathbf{0} \tag{1.36}$$

for all values of i and j. Thus, the sum on the right-hand side of Equation (1.34) is zero for any kind of central force. We are left with

$$\frac{d\mathbf{L}}{dt} = \mathbf{0}. \tag{1.37}$$

In other words, the total angular momentum of the system is a *conserved* quantity, provided that the different components of the system interact via *central* forces (and there is zero net external torque acting on the system).

1.6 Nonisolated systems

Up to now, we have considered only *isolated* dynamical systems, in which all the forces acting on the system originate from within the system itself. Let us now generalize our approach to deal with *nonisolated* dynamical systems, in which some of the forces originate outside the system. Consider a system of N mutually interacting point objects. Let m_i and $\mathbf{r}_i$ be the mass and position vector of the ith object, respectively. Suppose that the ith object is subject to two forces: first, an *internal force* that originates from the other objects in the system, and second, an *external force* that originates outside the system. In other words, let the force acting on the ith object take the form

$$\mathbf{f}_i = \sum_{\substack{j=1,N \\ j \neq i}} \mathbf{f}_{ij} + \mathbf{F}_i, \tag{1.38}$$

where $\mathbf{f}_{ij}$ is the internal force exerted by object j on object i, and $\mathbf{F}_i$ the external force acting on object i.

The equation of motion of the ith object is

$$m_i \frac{d^2 \mathbf{r}_i}{dt^2} = \mathbf{f}_i = \sum_{\substack{j=1,N \\ j \neq i}} \mathbf{f}_{ij} + \mathbf{F}_i. \tag{1.39}$$

Summing over all objects, we obtain

$$\sum_{i=1,N} m_i \frac{d^2 \mathbf{r}_i}{dt^2} = \sum_{\substack{i,j=1,N \\ j \neq i}} \mathbf{f}_{ij} + \sum_{i=1,N} \mathbf{F}_i, \tag{1.40}$$

which reduces to

$$\frac{d\mathbf{P}}{dt} = \mathbf{F}, \tag{1.41}$$

where

$$\mathbf{F} = \sum_{i=1,N} \mathbf{F}_i \tag{1.42}$$

is the net external force acting on the system. Here, the sum over the internal forces has canceled out in pairs as a result of Newton's third law of motion. (See Section 1.5.) We conclude that if there is a net external force acting on the system, then the total linear momentum evolves in time according to Equation (1.41) but is completely unaffected by any internal forces. The fact that Equation (1.41) is similar in form to Equation (1.13) suggests that the center of mass of a system consisting of many point objects has analogous dynamics to a single point object whose mass is the total system mass, moving under the action of the net external force.

Taking $\mathbf{r}_i \times$ Equation (1.39), and summing over all objects, we obtain

$$\frac{d\mathbf{L}}{dt} = \boldsymbol{\tau}, \tag{1.43}$$

where

$$\boldsymbol{\tau} = \sum_{i=1,N} \mathbf{r}_i \times \mathbf{F}_i \tag{1.44}$$

is the net external torque (about the origin) acting on the system. Here, the sum over the internal torques has canceled out in pairs, assuming that the internal forces are central in nature. (See Section 1.5.) We conclude that if there is a net external torque acting

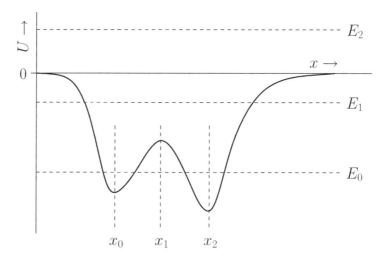

Fig. 1.4   A potential energy curve.

on the system, then the total angular momentum evolves in time according to Equation (1.43) but is completely unaffected by any internal torques.

1.7 Motion in one-dimensional potential

As a simple illustration of the application of Newton's laws of motion, consider a point particle of mass m moving in the x-direction, say, under the action of some x-directed force $f(x)$. Suppose that $f(x)$ is a conservative force, such as gravity. In this case, according to Equation (1.19), we can write

$$f(x) = -\frac{dU(x)}{dx}, \tag{1.45}$$

where $U(x)$ is the potential energy of the particle at position x.

Let the curve $U(x)$ take the form shown in Figure 1.4. For instance, this curve might represent the gravitational potential energy of a cyclist freewheeling in a hilly region. Observe that we have set the potential energy at infinity to zero (which we are generally free to do, as potential energy is undefined to an arbitrary additive constant). This is a fairly common convention. What can we deduce about the motion of the particle in this potential?

Well, we know that the total energy, E—which is the sum of the kinetic energy, K, and the potential energy, U—is a *constant* of the motion [see Equation (1.22)]. Hence, we can write

$$K(x) = E - U(x). \tag{1.46}$$

However, we also know that a kinetic energy can never be negative, because $K = (1/2)\,m\,v^2$, and neither m nor v^2 can be negative. Hence, the preceding expression tells us that the particle's motion is restricted to the region (or regions) in which the potential energy curve $U(x)$ falls below the value E. This idea is illustrated in Figure 1.4. Suppose that the total energy of the system is E_0. It is clear, from the figure, that the particle is trapped inside one or other of the two dips in the potential—these dips are generally referred to as *potential wells*. Suppose that we now raise the energy to E_1. In this case, the particle is free to enter or leave each of the potential wells, but its motion is still *bounded* to some extent, as it clearly cannot move off to infinity. Finally, let us raise the energy to E_2. Now the particle is *unbounded*: that is, it can move off to infinity. In conservative systems in which it makes sense to adopt the convention that the potential energy at infinity is zero, bounded systems are characterized by $E < 0$, whereas unbounded systems are characterized by $E > 0$.

The preceding discussion suggests that the motion of a particle moving in a potential generally becomes less bounded as the total energy E of the system increases. Conversely, we would expect the motion to become more bounded as E decreases. In fact, if the energy becomes sufficiently small, then it appears likely that the system will settle down in some *equilibrium state* in which the particle remains stationary. Let us try to identify any prospective equilibrium states in Figure 1.4. If the particle remains stationary, then it must be subject to zero force (otherwise, it would accelerate). Hence, according to Equation (1.45), an equilibrium state is characterized by

$$\frac{dU}{dx} = 0. \tag{1.47}$$

In other words, an equilibrium state corresponds to either a *maximum* or a *minimum* of the potential energy curve, $U(x)$. It can be seen that the $U(x)$ curve shown in Figure 1.4 has three associated equilibrium states located at $x = x_0$, $x = x_1$, and $x = x_2$.

Let us now make a distinction between *stable* equilibrium points and *unstable* equilibrium points. When the particle is displaced slightly from a stable equilibrium point, then the resultant force acting on it must always be such as to return it to this point. In other words, if $x = x_0$ is an equilibrium point, then we require

$$\left.\frac{df}{dx}\right|_{x_0} < 0 \tag{1.48}$$

for stability: that is, if the particle is displaced to the right, so that $x - x_0 > 0$, then the force must act to the left, so that $f < 0$, and vice versa. Likewise, if

$$\left.\frac{df}{dx}\right|_{x_0} > 0 \tag{1.49}$$

then the equilibrium point $x = x_0$ is unstable. It follows, from Equation (1.45), that stable equilibrium points are characterized by

$$\frac{d^2U}{dx^2} > 0. \tag{1.50}$$

In other words, a stable equilibrium point corresponds to a *minimum* of the potential energy curve, $U(x)$. Likewise, an unstable equilibrium point corresponds to a *maximum* of the $U(x)$ curve. Hence we conclude that, in Figure 1.4, $x = x_0$ and $x = x_2$ are stable equilibrium points, whereas $x = x_1$ is an unstable equilibrium point. Of course, this makes perfect sense if we think of $U(x)$ as a gravitational potential energy curve, so that

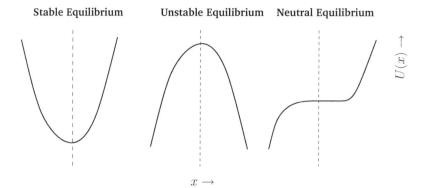

Stable Equilibrium   Unstable Equilibrium   Neutral Equilibrium

Fig. 1.5   Different types of equilibrium points.

U is directly proportional to height. In this case, all we are saying is that it is easy to confine a low energy mass at the bottom of a smooth valley, but very difficult to balance the same mass on the top of a smooth hill (because any slight displacement of the mass will cause it to slide down the hill). Finally, if

$$\frac{dU}{dx} = \frac{d^2U}{dx^2} = 0 \tag{1.51}$$

at any point (or in any region), then we have what is known as a *neutral equilibrium* point. We can move the particle slightly away from such a point and it will still remain in equilibrium (i.e., it will neither attempt to return to its initial state, nor will it continue to move). A neutral equilibrium point corresponds to a *flat spot* in a $U(x)$ curve. See Figure 1.5.

The equation of motion of a particle moving in one dimension under the action of a conservative force is, in principle, integrable. Because $K = (1/2)\,m\,v^2$, the energy conservation equation, Equation (1.46), can be rearranged to give

$$v = \pm \left\{ \frac{2\,[E - U(x)]}{m} \right\}^{1/2}, \tag{1.52}$$

where the $\pm$ signs correspond to motion to the left and to the right, respectively. However, because $v = dx/dt$, this expression can be integrated to give

$$t = \pm \left(\frac{m}{2E} \right)^{1/2} \int_{x_0}^{x} \frac{dx'}{\sqrt{1 - U(x')/E}}, \tag{1.53}$$

where $x(t = 0) = x_0$. For sufficiently simple potential functions, $U(x)$, Equation (1.53) can be solved to give x as a function of t. For instance, if $U = (1/2)\,k\,x^2$, $x_0 = 0$, and the plus sign is chosen, then

$$t = \left(\frac{m}{k} \right)^{1/2} \int_0^{(k/2E)^{1/2}\,x} \frac{dy}{\sqrt{1 - y^2}} = \left(\frac{m}{k} \right)^{1/2} \sin^{-1}\left[\left(\frac{k}{2E} \right)^{1/2} x \right], \tag{1.54}$$

which can be inverted to give

$$x = a\,\sin(\omega\,t), \tag{1.55}$$

where $a = \sqrt{2\,E/k}$ and $\omega = \sqrt{k/m}$. Note that the particle reverses direction each time it reaches one of the so-called *turning points* ($x = \pm a$) at which $U = E$ (and, so, $K = 0$).

1.8 Simple harmonic motion

Consider the motion of a point particle of mass m, moving in one dimension, that is slightly displaced from a *stable* equilibrium point located at $x = 0$. Suppose that the particle is moving in the conservative force field $f(x)$. According to the preceding analysis, for $x = 0$ to correspond to a stable equilibrium point, we require both

$$f(0) = 0 \tag{1.56}$$

and

$$\frac{df(0)}{dx} < 0. \tag{1.57}$$

Our particle obeys Newton's second law of motion,

$$m\,\frac{d^2x}{dt^2} = f(x). \tag{1.58}$$

Let us assume that the particle always stays fairly close to its equilibrium point. In this case, to a good approximation, we can represent $f(x)$ via a truncated Taylor expansion about this point. In other words,

$$f(x) \simeq f(0) + \frac{df(0)}{dx}\,x + \mathcal{O}(x^2). \tag{1.59}$$

However, according to Equations (1.56) and (1.57), the preceding expression can be written

$$f(x) \simeq -m\,\omega_0^2\,x, \tag{1.60}$$

where $df(0)/dx = -m\,\omega_0^2$. Hence, we conclude that our particle satisfies the following approximate equation of motion:

$$\frac{d^2x}{dt^2} + \omega_0^2\,x \simeq 0, \tag{1.61}$$

provided that it does not stray too far from its equilibrium point: in other words, provided $|x|$ does not become too large.

Equation (1.61) is called the *simple harmonic equation*; it governs the motion of all one-dimensional conservative systems that are slightly perturbed from some stable equilibrium state. The solution of Equation (1.61) is well known:

$$x(t) = a\,\sin(\omega_0\,t - \phi_0). \tag{1.62}$$

The pattern of motion described by this expression, which is called *simple harmonic motion*, is *periodic* in time, with repetition period $T_0 = 2\pi/\omega_0$, and oscillates between $x = \pm a$. Here, a is called the *amplitude* of the motion. The parameter ϕ_0, known as the *phase angle*, simply shifts the pattern of motion backward and forward in time.

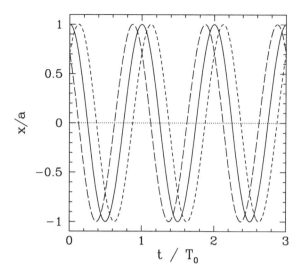

Fig. 1.6 Simple harmonic motion.

Figure 1.6 shows some examples of simple harmonic motion. Here, $\phi_0 = 0$, $+\pi/4$, and $-\pi/4$ correspond to the solid, short-dashed, and long-dashed curves, respectively.

Note that the frequency, ω_0—and, hence, the period, T_0—of simple harmonic motion is determined by the parameters appearing in the simple harmonic equation, Equation (1.61). However, the amplitude, a, and the phase angle, ϕ_0, are the two integration constants of this second-order ordinary differential equation, and are thus determined by the initial conditions: the particle's initial displacement and velocity.

From Equations (1.45) and (1.60), the potential energy of our particle at position x is approximately

$$U(x) \simeq \frac{1}{2} m \omega_0^2 x^2. \tag{1.63}$$

Hence, the total energy is written

$$E = K + U = \frac{1}{2} m \left(\frac{dx}{dt}\right)^2 + \frac{1}{2} m \omega_0^2 x^2, \tag{1.64}$$

giving

$$E = \frac{1}{2} m \omega_0^2 a^2 \cos^2(\omega_0 t - \phi_0) + \frac{1}{2} m \omega_0^2 a^2 \sin^2(\omega_0 t - \phi_0) = \frac{1}{2} m \omega_0^2 a^2, \tag{1.65}$$

where use has been made of Equation (1.62), and the trigonometric identity $\cos^2 \theta + \sin^2 \theta \equiv 1$. Note that the total energy is *constant* in time, as is to be expected for a conservative system, and is proportional to the amplitude squared of the motion.

Consider the motion of a point particle of mass m that is slightly displaced from an *unstable* equilibrium point at $x = 0$. The fact that the equilibrium is unstable implies that

$$f(0) = 0 \tag{1.66}$$

and

$$\frac{df(0)}{dx} > 0. \tag{1.67}$$

As long as $|x|$ remains small, our particle's equation of motion takes the approximate form

$$m\frac{d^2x}{dt^2} \simeq f(0) + \frac{df(0)}{dx} x, \tag{1.68}$$

which reduces to

$$\frac{d^2x}{dt^2} \simeq k^2 x, \tag{1.69}$$

where $df(0)/dx = m k^2$. The most general solution to the preceding equation is

$$x(t) = A\,e^{kt} + B\,e^{-kt}, \tag{1.70}$$

where A and B are arbitrary constants. Thus, unless the initial conditions are such that A is *exactly* zero, the particle's displacement from the unstable equilibrium point grows *exponentially* in time.

1.9 Two-body problem

An isolated dynamical system consisting of two freely moving point objects exerting forces on one another is conventionally termed a *two-body problem*. Suppose that the first object is of mass m_1 and is located at position vector $\mathbf{r}_1$. Likewise, the second object is of mass m_2 and is located at position vector $\mathbf{r}_2$. Let the first object exert a force $\mathbf{f}_{21}$ on the second. By Newton's third law, the second object exerts an equal and opposite force, $\mathbf{f}_{12} = -\mathbf{f}_{21}$, on the first. Suppose that there are no other forces in the problem. The equations of motion of our two objects are thus

$$m_1\frac{d^2\mathbf{r}_1}{dt^2} = -\mathbf{f} \tag{1.71}$$

and

$$m_2\frac{d^2\mathbf{r}_2}{dt^2} = \mathbf{f}, \tag{1.72}$$

where $\mathbf{f} = \mathbf{f}_{21}$.

The center of mass of our system is located at

$$\mathbf{r}_{cm} = \frac{m_1\,\mathbf{r}_1 + m_2\,\mathbf{r}_2}{m_1 + m_2}. \tag{1.73}$$

Hence, we can write

$$\mathbf{r}_1 = \mathbf{r}_{cm} - \frac{m_2}{m_1 + m_2}\,\mathbf{r} \tag{1.74}$$

and

$$\mathbf{r}_2 = \mathbf{r}_{cm} + \frac{m_1}{m_1 + m_2}\,\mathbf{r}, \tag{1.75}$$

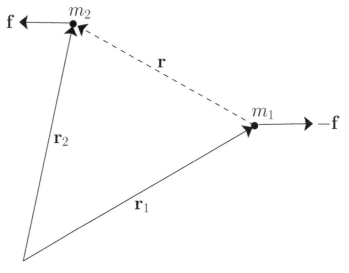

Fig. 1.7   Two-body problem.

where $\mathbf{r} = \mathbf{r}_2 - \mathbf{r}_1$. (See Figure 1.7.) Substituting the preceding two equations into Equations (1.71) and (1.72), and making use of the fact that the center of mass of an isolated system *does not accelerate* (see Section 1.5), we find that both equations yield

$$\mu \frac{d^2\mathbf{r}}{dt^2} = \mathbf{f}, \tag{1.76}$$

where

$$\mu = \frac{m_1\, m_2}{m_1 + m_2} \tag{1.77}$$

is called the *reduced mass*. Hence, we have effectively converted our original two-body problem into an equivalent one-body problem. In the equivalent problem, the force $\mathbf{f}$ is the *same* as that acting on both objects in the original problem (except for a minus sign). However, the mass, μ, is *different*, and it is less than either of m_1 or m_2 (which is why it is called the "reduced" mass). We conclude that the dynamics of an isolated system consisting of two interacting point objects can always be reduced to that of an equivalent system consisting of a single point object moving in a fixed potential.

Exercises

1.1   Derive Equation (1.12).

1.2   Consider a system consisting of N point particles. Let $\mathbf{r}_i$ be the position vector of the ith particle, and let $\mathbf{F}_i$ be the external force acting on this particle. Any internal forces are assumed to be central in nature. The resultant force and torque (about the origin) acting on the system are

$$\mathbf{F} = \sum_{i=1,N} \mathbf{F}_i$$

and

$$\tau = \sum_{i=1,N} \mathbf{r}_i \times \mathbf{F}_i,$$

respectively. A *point of action* of the resultant force is defined as a point whose position vector $\mathbf{r}$ satisfies

$$\mathbf{r} \times \mathbf{F} = \tau.$$

Demonstrate that there are an infinite number of possible points of action lying on the straight line

$$\mathbf{r} = \frac{\mathbf{F} \times \tau}{F^2} + \lambda \frac{\mathbf{F}}{F},$$

where λ is arbitrary. This straight line is known as the *line of action* of the resultant force.

1.3 Consider an isolated system consisting of two extended bodies (which can, of course, be modeled as collections of point particles), A and B. Let $\mathbf{F}_A$ be the resultant force acting on A due to B, and let $\mathbf{F}_B$ be the resultant force acting on B due to A. Demonstrate that $\mathbf{F}_B = -\mathbf{F}_A$, and that both forces have the same line of action.

1.4 An extended body is acted on by two resultant forces, $\mathbf{F}_1$ and $\mathbf{F}_2$. Show that these forces can be only replaced by a single equivalent force, $\mathbf{F} = \mathbf{F}_1 + \mathbf{F}_2$, provided:
a. $\mathbf{F}_1$ and $\mathbf{F}_2$ are parallel (or antiparallel). In this case, the line of action of $\mathbf{F}$ is parallel to those of $\mathbf{F}_1$ and $\mathbf{F}_2$.
b. $\mathbf{F}_1$ and $\mathbf{F}_2$ are not parallel (or antiparallel), but their lines of action cross at a point. In this case, the line of action of $\mathbf{F}$ passes through the crossing point.

1.5 Deduce that if an isolated system consists of three extended bodies, A, B, and C, where $\mathbf{F}_A$ is the resultant force acting on A (due to B and C), $\mathbf{F}_B$ is the resultant force acting on B, and $\mathbf{F}_C$ is the resultant force acting on C; then $\mathbf{F}_A + \mathbf{F}_B + \mathbf{F}_C = \mathbf{0}$, and the forces either all have parallel lines of action or have lines of action that cross at a common point.

1.6 A particle of mass m moves in one dimension and has an instantaneous displacement x. The particle is released at rest from $x = a$, subject to the force $f(x) = -c\,x^{-2}$, where $a, c > 0$. Demonstrate that the time needed for the particle to reach $x = 0$ is

$$\pi \left(\frac{m\,a^3}{8\,c} \right)^{1/2}.$$

(Modified from Fowles and Cassiday 2005.)

1.7 A particle of mass m moves in one dimension and has an instantaneous displacement x. The particle is released at rest from $x = a$, subject to the force

$f(x) = -m\mu (a^5/x^2)^{1/3}$, where $a, \mu > 0$. Show that the particle will reach the origin with a speed $a\sqrt{6\mu}$ after a time $(8/15)(6/\mu)^{1/2}$ has elapsed. (Modified from Smart 1951.)

1.8 A particle moves in one dimension and has an instantaneous displacement x. The particle is released at rest from $x = a$ and accelerates such that $\ddot{x} = \mu (x + a^4/x^3)$, where $a > 0$. Show that the particle will reach the origin after a time $\pi/(4\sqrt{\mu})$ has elapsed, and that its speed is then infinite. (Modified from Smart 1951.)

1.9 A particle of mass m, moving in one dimension with an initial (i.e., at $t = -\infty$) velocity v_0, is subject to a force

$$f(t) = \frac{p_0 \, \delta t}{\pi} \frac{1}{(t - t_0)^2 + (\delta t)^2}.$$

Find the velocity as a function of time. Show that, as $\delta t \to 0$, the motion approaches motion at constant velocity, with an abrupt change in velocity, by an amount p_0/m, at $t = t_0$.

1.10 A particle of mass m moving in one dimension is subject to a force

$$f(x) = -k\,x + \frac{a}{x^3},$$

where $k, a > 0$. Find the potential energy, $U(x)$. Find the equilibrium points. Are they stable or unstable? Determine the angular frequency of small-amplitude oscillations about any stable equilibrium points.

1.11 A particle moving in one dimension with simple harmonic motion has speeds u and v at displacements a and b, respectively, from its mean position. Show that the period of the motion is

$$T = 2\pi \left(\frac{b^2 - a^2}{u^2 - v^2}\right)^{1/2}.$$

Find the amplitude. (From Smart 1951.)

1.12 The potential energy for the force between two atoms in a diatomic molecule has the approximate form

$$U(x) = -\frac{a}{x^6} + \frac{b}{x^{12}},$$

where x is the distance between the atoms, and a, b are positive constants. Find the force.

a. Assuming that one of the atoms is relatively heavy and remains at rest while the other, whose mass is m, moves in a straight line, find the equilibrium distance and the period of small oscillations about the equilibrium position.

b. Assuming that both atoms have the same mass m and move in a straight line, find the equilibrium distance and the period of small oscillations about the equilibrium position.

1.13 Two light springs have spring constants k_1 and k_2, respectively, and are used in a vertical orientation to support an object of mass m. Show that the angular frequency of oscillation is $[(k_1 + k_2)/m]^{1/2}$ if the springs are connected in parallel, and $[k_1 k_2/(k_1 + k_2) m]^{1/2}$ if the springs are connected in series.

1.14 A body of uniform cross-sectional area A and mass density ρ floats in a liquid of density ρ_0 (where $\rho < \rho_0$), and at equilibrium displaces a volume V. Show that the period of small oscillations about the equilibrium position is

$$T = 2\pi \sqrt{\frac{V}{g A}}.$$

1.15 A particle of mass m executes one-dimensional simple harmonic oscillation under the action of a conservative force such that its instantaneous displacement is

$$x(t) = a \cos(\omega t - \phi).$$

Find the average values of x, x^2, $\dot{x}$, and $\dot{x}^2$ over a single cycle of the oscillation. Here, $\dot{} \equiv d/dt$. Find the average values of the kinetic and potential energies of the particle over a single cycle of the oscillation.

1.16 Using the notation of Section 1.9, show that the total momentum and angular momentum of a two-body system take the form

$$\mathbf{P} = M \dot{\mathbf{r}}_{cm}$$

and

$$\mathbf{L} = M \mathbf{r}_{cm} \times \dot{\mathbf{r}}_{cm} + \mu \mathbf{r} \times \dot{\mathbf{r}},$$

respectively, where $M = m_1 + m_2$, and $\dot{} \equiv d/dt$.

a. If the force acting between the bodies is conservative, such that $\mathbf{f} = -\nabla U$, demonstrate that the total energy of the system is written

$$E = \frac{1}{2} M \dot{r}_{cm}^2 + \frac{1}{2} \mu \dot{r}^2 + U.$$

Show, from the equation of motion, $\mu \ddot{\mathbf{r}} = -\nabla U$, that E is constant in time.

b. If the force acting between the particles is central, so that $\mathbf{f} \propto \mathbf{r}$, demonstrate, from the equation of motion, $\mu \ddot{\mathbf{r}} = \mathbf{f}$, that $\mathbf{L}$ is constant in time.

Newtonian gravity

2.1 Introduction

Classical gravity, which is invariably the dominant force in celestial dynamic systems, was first correctly described in Newton's *Principia*. According to Newton, any two point objects exert a gravitational force of attraction on each other. This force is directed along the straight line joining the objects, is directly proportional to the product of their masses, and is inversely proportional to the square of the distance between them. Consider two point objects of mass m_1 and m_2 that are located at position vectors $\mathbf{r}_1$ and $\mathbf{r}_2$, respectively. The gravitational force $\mathbf{f}_{12}$ that mass m_2 exerts on mass m_1 is written

$$\mathbf{f}_{12} = G\, m_1\, m_2\, \frac{\mathbf{r}_2 - \mathbf{r}_1}{|\mathbf{r}_2 - \mathbf{r}_1|^3}. \tag{2.1}$$

The gravitational force $\mathbf{f}_{21}$ that mass m_1 exerts on mass m_2 is equal and opposite: $\mathbf{f}_{21} = -\mathbf{f}_{12}$. (See Figure 1.3.) Here, the constant of proportionality, G, is called the *universal gravitational constant* and takes the value (Yoder 1995)

$$G = 6.673 \times 10^{-11}\, \mathrm{m^3\, kg^{-1}\, s^{-2}}. \tag{2.2}$$

Incidentally, there is something rather curious about Equation (2.1). According to this law, the gravitational force acting on a given object is directly proportional to that object's inertial mass. Why, though, should inertia be related to the force of gravity? After all, inertia measures the reluctance of a given body to deviate from its preferred state of uniform motion in a straight line, in response to some external force. What does this have to do with gravitational attraction? This question perplexed physicists for many years; it was answered only when Albert Einstein published his general theory of relativity in 1916. According to Einstein, inertial mass acts as a sort of gravitational charge, as it is impossible to distinguish an acceleration produced by a gravitational field from an apparent acceleration generated by observing motion in a noninertial reference frame. The assumption that these two types of acceleration are indistinguishable leads directly to all the strange predictions of general relativity, such that clocks in different gravitational potentials run at different rates, mass bends space, and so on.

2.2 Gravitational potential

Consider two point masses, m and m', located at position vectors $\mathbf{r}$ and $\mathbf{r}'$, respectively. According to the preceding analysis, the acceleration $\mathbf{g}$ of mass m as a result of the gravitational force exerted on it by mass m' takes the form

$$\mathbf{g} = G\,m'\,\frac{\mathbf{r}' - \mathbf{r}}{|\mathbf{r}' - \mathbf{r}|^3}. \tag{2.3}$$

The x-component of this acceleration is written

$$g_x = G\,m'\,\frac{x' - x}{[(x' - x)^2 + (y' - y)^2 + (z' - z)^2]^{3/2}}, \tag{2.4}$$

where $\mathbf{r} = (x, y, z)$ and $\mathbf{r}' = (x', y', z')$. However, as is easily demonstrated,

$$\frac{x' - x}{[(x' - x)^2 + (y' - y)^2 + (z' - z)^2]^{3/2}} \equiv \frac{\partial}{\partial x}\left\{\frac{1}{[(x' - x)^2 + (y' - y)^2 + (z' - z)^2]^{1/2}}\right\}. \tag{2.5}$$

Hence,

$$g_x = G\,m'\,\frac{\partial}{\partial x}\left(\frac{1}{|\mathbf{r}' - \mathbf{r}|}\right), \tag{2.6}$$

with analogous expressions for g_y and g_z. It follows that

$$\mathbf{g} = -\nabla\Phi, \tag{2.7}$$

where

$$\Phi(\mathbf{r}) = -\frac{G\,m'}{|\mathbf{r}' - \mathbf{r}|} \tag{2.8}$$

is termed the *gravitational potential*. Of course, we can write $\mathbf{g}$ in the form of Equation (2.7) only because gravity is a *conservative* force. (See Section 1.4.)

It is well known that gravity is a *superposable* force. In other words, the gravitational force exerted on some point mass by a collection of other point masses is simply the vector sum of the forces exerted on the former mass by each of the latter masses taken in isolation. It follows that the gravitational potential generated by a collection of point masses at a certain location in space is the sum of the potentials generated at that location by each point mass taken in isolation. Hence, using Equation (2.8), if there are N point masses, m_i (for $i = 1, N$), located at position vectors $\mathbf{r}_i$, then the gravitational potential generated at position vector $\mathbf{r}$ is simply

$$\Phi(\mathbf{r}) = -G\sum_{i=1,N}\frac{m_i}{|\mathbf{r}_i - \mathbf{r}|}. \tag{2.9}$$

Suppose, finally, that instead of having a collection of point masses, we have a *continuous* mass distribution. In other words, let the mass at position vector $\mathbf{r}'$ be $\rho(\mathbf{r}')\,d^3\mathbf{r}'$, where $\rho(\mathbf{r}')$ is the local mass density, and $d^3\mathbf{r}'$ a volume element. Summing over all space, and taking the limit $d^3\mathbf{r}' \to 0$, we find Equation (2.9) yields

$$\Phi(\mathbf{r}) = -G\int\frac{\rho(\mathbf{r}')}{|\mathbf{r}' - \mathbf{r}|}\,d^3\mathbf{r}', \tag{2.10}$$

where the integral is taken over all space. This is the general expression for the gravitational potential, $\Phi(\mathbf{r})$, generated by a continuous mass distribution, $\rho(\mathbf{r})$.

2.3 Gravitational potential energy

Consider a collection of N point masses m_i located at position vectors $\mathbf{r}_i$ (where i runs from 1 to N). What is the gravitational potential energy stored in such a collection? In other words, how much work would we have to do to assemble the masses, starting from an initial state in which they are all at rest and very widely separated?

We have seen that a gravitational acceleration field can be expressed in terms of a gravitational potential:

$$\mathbf{g}(\mathbf{r}) = -\nabla\Phi. \tag{2.11}$$

We also know that the gravitational force acting on a mass m located at position $\mathbf{r}$ is written

$$\mathbf{f}(\mathbf{r}) = m\,\mathbf{g}(\mathbf{r}). \tag{2.12}$$

The work we would have to do against the gravitational force to *slowly* move the mass from point P to point Q is simply

$$U = -\int_P^Q \mathbf{f}\cdot d\mathbf{r} = -m\int_P^Q \mathbf{g}\cdot d\mathbf{r} = m\int_P^Q \nabla\Phi\cdot d\mathbf{r} = m\left[\Phi(Q) - \Phi(P)\right]. \tag{2.13}$$

The negative sign in the preceding expression comes about because we would have to exert a force $-\mathbf{f}$ on the mass to counteract the force exerted by the gravitational field. Recall, finally, that the gravitational potential generated by a point mass m' located at position $\mathbf{r}'$ is

$$\Phi(\mathbf{r}) = -\frac{G\,m'}{|\mathbf{r}' - \mathbf{r}|}. \tag{2.14}$$

Let us build up our collection of masses one by one. It takes no work to bring the first mass from infinity, as there is no gravitational field to fight against. Let us clamp this mass in position at $\mathbf{r}_1$. To bring the second mass into position at $\mathbf{r}_2$, we have to do work against the gravitational field generated by the first mass. According to Equations (2.13) and (2.14), this work is given by

$$U_2 = -\frac{G\,m_2\,m_1}{|\mathbf{r}_1 - \mathbf{r}_2|}. \tag{2.15}$$

Let us now bring the third mass into position. Because gravitational fields and gravitational potentials are superposable, the work done while moving the third mass from infinity to $\mathbf{r}_3$ is simply the sum of the works done against the gravitational fields generated by masses 1 and 2 taken in isolation:

$$U_3 = -\frac{G\,m_3\,m_1}{|\mathbf{r}_1 - \mathbf{r}_3|} - \frac{G\,m_3\,m_2}{|\mathbf{r}_2 - \mathbf{r}_3|}. \tag{2.16}$$

Thus, the total work done in assembling the arrangement of three masses is given by

$$U = -\frac{G\,m_2\,m_1}{|\mathbf{r}_1 - \mathbf{r}_2|} - \frac{G\,m_3\,m_1}{|\mathbf{r}_1 - \mathbf{r}_3|} - \frac{G\,m_3\,m_2}{|\mathbf{r}_2 - \mathbf{r}_3|}. \tag{2.17}$$

This result can easily be generalized to an arrangement of N point masses, giving

$$U = -\sum_{i,j=1,N}^{j<i} \frac{G\,m_i\,m_j}{|\mathbf{r}_j - \mathbf{r}_i|}. \tag{2.18}$$

The restriction that j must be less than i makes the preceding summation rather messy. If we were to sum without restriction (other than $j \neq i$), then each pair of masses would be counted twice. It is convenient to do just this, and then to divide the result by two. Thus, we obtain

$$U = -\frac{1}{2}\sum_{i,j=1,N}^{j\neq i} \frac{G\,m_i\,m_j}{|\mathbf{r}_j - \mathbf{r}_i|}. \tag{2.19}$$

This is the *potential energy* of an arrangement of point masses. We can think of this quantity as the work required to bring the masses from infinity and assemble them in the required formation. The fact that the work is negative implies that we would gain energy during this process.

Equation (2.19) can be written

$$U = \frac{1}{2}\sum_{i=1,N} m_i\,\Phi_i, \tag{2.20}$$

where

$$\Phi_i = -G\sum_{j=1,N}^{j\neq i} \frac{m_j}{|\mathbf{r}_j - \mathbf{r}_i|} \tag{2.21}$$

is the gravitational potential experienced by the ith mass due to the other masses in the distribution. For the case of a continuous mass distribution, we can generalize the preceding result to give

$$U = \frac{1}{2}\int \rho(\mathbf{r})\,\Phi(\mathbf{r})\,d^3\mathbf{r}, \tag{2.22}$$

where

$$\Phi(\mathbf{r}) = -G\int \frac{\rho(\mathbf{r}')}{|\mathbf{r}' - \mathbf{r}|}\,d^3\mathbf{r}' \tag{2.23}$$

is the familiar gravitational potential generated by a continuous mass distribution of mass density $\rho(\mathbf{r})$.

2.4 Axially symmetric mass distributions

At this point, it is convenient to adopt standard spherical coordinates, r, θ, ϕ, aligned along the z-axis. These coordinates are related to regular Cartesian coordinates as follows (see Section A.8):

$$x = r\sin\theta\cos\phi, \tag{2.24}$$

$$y = r\sin\theta\sin\phi, \tag{2.25}$$

and

$$z = r \cos \theta. \tag{2.26}$$

Consider an *axially symmetric* mass distribution, that is, a $\rho(\mathbf{r})$ that is *independent* of the azimuthal angle, ϕ. We would expect such a mass distribution to generate an axially symmetric gravitational potential, $\Phi(r, \theta)$. Hence, without loss of generality, we can set $\phi = 0$ when evaluating $\Phi(\mathbf{r})$ from Equation (2.10). In fact, given that $d^3\mathbf{r}' = r'^2 \sin \theta' \, dr' \, d\theta' \, d\phi'$ in spherical coordinates, this equation yields

$$\Phi(r, \theta) = -G \int_0^\infty \int_0^\pi \int_0^{2\pi} \frac{r'^2 \rho(r', \theta') \sin \theta'}{|\mathbf{r} - \mathbf{r}'|} d\phi' \, d\theta' \, dr', \tag{2.27}$$

with the right-hand side evaluated at $\phi = 0$. However, because $\rho(r', \theta')$ is independent of ϕ', Equation (2.27) can also be written

$$\Phi(r, \theta) = -2\pi G \int_0^\infty \int_0^\pi r'^2 \rho(r', \theta') \sin \theta' \langle |\mathbf{r} - \mathbf{r}'|^{-1} \rangle d\theta' \, dr', \tag{2.28}$$

where $\langle \cdots \rangle \equiv \oint (\cdots) \, d\phi'/2\pi$ denotes an average over the azimuthal angle, ϕ'.

Now,

$$|\mathbf{r}' - \mathbf{r}|^{-1} = (r^2 - 2\,\mathbf{r} \cdot \mathbf{r}' + r'^2)^{-1/2} \tag{2.29}$$

and

$$\mathbf{r} \cdot \mathbf{r}' = r\,r'\,F, \tag{2.30}$$

where (at $\phi = 0$)

$$F = \sin \theta \, \sin \theta' \, \cos \phi' + \cos \theta \, \cos \theta'. \tag{2.31}$$

Hence,

$$|\mathbf{r}' - \mathbf{r}|^{-1} = (r^2 - 2\,r\,r'\,F + r'^2)^{-1/2}. \tag{2.32}$$

Suppose that $r > r'$. In this case, we can expand $|\mathbf{r}' - \mathbf{r}|^{-1}$ as a convergent power series in r'/r to give

$$|\mathbf{r}' - \mathbf{r}|^{-1} = \frac{1}{r}\left[1 + \left(\frac{r'}{r}\right)F + \frac{1}{2}\left(\frac{r'}{r}\right)^2 (3\,F^2 - 1) + \mathcal{O}\left(\frac{r'}{r}\right)^3\right]. \tag{2.33}$$

Let us now average this expression over the azimuthal angle, ϕ'. Because $\langle 1 \rangle = 1$, $\langle \cos \phi' \rangle = 0$, and $\langle \cos^2 \phi' \rangle = 1/2$, it is easily seen that

$$\langle F \rangle = \cos \theta \, \cos \theta', \tag{2.34}$$

and

$$\begin{aligned} \langle F^2 \rangle &= \frac{1}{2} \sin^2 \theta \, \sin^2 \theta' + \cos^2 \theta \, \cos^2 \theta' \\ &= \frac{1}{3} + \frac{2}{3}\left(\frac{3}{2}\cos^2 \theta - \frac{1}{2}\right)\left(\frac{3}{2}\cos^2 \theta' - \frac{1}{2}\right). \end{aligned} \tag{2.35}$$

Hence,

$$\left\langle |\mathbf{r}' - \mathbf{r}|^{-1} \right\rangle = \frac{1}{r}\left[1 + \left(\frac{r'}{r}\right)\cos\theta\,\cos\theta' \right.$$
$$\left. + \left(\frac{r'}{r}\right)^2 \left(\frac{3}{2}\cos^2\theta - \frac{1}{2}\right)\left(\frac{3}{2}\cos^2\theta' - \frac{1}{2}\right) + \mathcal{O}\left(\frac{r'}{r}\right)^3 \right]. \quad (2.36)$$

The well-known *Legendre polynomials*, $P_n(x)$, are defined (Abramowitz and Stegun 1965) as

$$P_n(x) = \frac{1}{2^n\,n!}\frac{d^n}{dx^n}\left[(x^2 - 1)^n\right], \quad (2.37)$$

for $n = 0, \infty$. It follows that

$$P_0(x) = 1, \quad (2.38)$$
$$P_1(x) = x, \quad (2.39)$$
$$P_2(x) = \frac{1}{2}(3\,x^2 - 1), \quad (2.40)$$
$$P_3(x) = \frac{1}{2}(5\,x^3 - x), \quad (2.41)$$

and so on. The Legendre polynomials are *mutually orthogonal*:

$$\int_{-1}^{1} P_n(x)\,P_m(x)\,dx = \int_0^\pi P_n(\cos\theta)\,P_m(\cos\theta)\,\sin\theta\,d\theta = \frac{\delta_{nm}}{n + 1/2} \quad (2.42)$$

(Abramowitz and Stegun 1965). Here, δ_{nm} is 1 if $n = m$, and 0 otherwise. The Legendre polynomials also form a *complete set*: any function of x that is well behaved in the interval $-1 \le x \le 1$ can be represented as a weighted sum of $P_n(x)$. Likewise, any function of θ that is well behaved in the interval $0 \le \theta \le \pi$ can be represented as a weighted sum of $P_n(\cos\theta)$.

A comparison of Equation (2.36) and Equations (2.38)–(2.40) makes it reasonably clear that when $r > r'$, the complete expansion of $\langle |\mathbf{r}' - \mathbf{r}|^{-1} \rangle$ is

$$\left\langle |\mathbf{r}' - \mathbf{r}|^{-1} \right\rangle = \frac{1}{r}\sum_{n=0,\infty}\left(\frac{r'}{r}\right)^n P_n(\cos\theta)\,P_n(\cos\theta'). \quad (2.43)$$

Similarly, when $r < r'$, we can expand in powers of r/r' to obtain

$$\left\langle |\mathbf{r}' - \mathbf{r}|^{-1} \right\rangle = \frac{1}{r'}\sum_{n=0,\infty}\left(\frac{r}{r'}\right)^n P_n(\cos\theta)\,P_n(\cos\theta'). \quad (2.44)$$

It follows from Equations (2.28), (2.43), and (2.44) that

$$\Phi(r,\theta) = \sum_{n=0,\infty}\Phi_n(r)\,P_n(\cos\theta), \quad (2.45)$$

where

$$\Phi_n(r) = -\frac{2\pi G}{r^{n+1}}\int_0^r\int_0^\pi r'^{\,n+2}\rho(r',\theta')\,P_n(\cos\theta')\,\sin\theta'\,d\theta'\,dr'$$
$$- 2\pi G\,r^n\int_r^\infty\int_0^\pi r'^{\,1-n}\rho(r',\theta')\,P_n(\cos\theta')\,\sin\theta'\,d\theta'\,dr'. \quad (2.46)$$

Given that $P_n(\cos\theta)$ form a complete set, we can always write

$$\rho(r,\theta) = \sum_{n=0,\infty} \rho_n(r)\, P_n(\cos\theta). \tag{2.47}$$

This expression can be inverted, with the aid of Equation (2.42), to give

$$\rho_n(r) = (n+1/2)\int_0^\pi \rho(r,\theta)\, P_n(\cos\theta)\,\sin\theta\,d\theta. \tag{2.48}$$

Hence, Equation (2.46) reduces to

$$\Phi_n(r) = -\frac{2\pi G}{(n+1/2)\,r^{n+1}}\int_0^r r'^{\,n+2}\rho_n(r')\,dr' - \frac{2\pi G\,r^n}{n+1/2}\int_r^\infty r'^{\,1-n}\rho_n(r')\,dr'. \tag{2.49}$$

Thus, we now have a general expression for the gravitational potential, $\Phi(r,\theta)$, generated by an axially symmetric mass distribution, $\rho(r,\theta)$.

2.5 Potential due to a uniform sphere

Let us calculate the gravitational potential generated by a sphere of *uniform* mass density γ and radius R, whose center coincides with the origin. Expressing $\rho(r,\theta)$ in the form of Equation (2.47), we find it clear that

$$\rho_0(r) = \begin{cases} \gamma & \text{for } r \le R \\ 0 & \text{for } r > R \end{cases} \tag{2.50}$$

with $\rho_n(r) = 0$ for $n > 0$. Thus, from Equation (2.49),

$$\Phi_0(r) = -\frac{4\pi G\,\gamma}{r}\int_0^r r'^{\,2}\,dr' - 4\pi G\,\gamma\int_r^R r'\,dr' \tag{2.51}$$

for $r \le R$, and

$$\Phi_0(r) = -\frac{4\pi G\,\gamma}{r}\int_0^R r'^{\,2}\,dr' \tag{2.52}$$

for $r > R$, with $\Phi_n(r) = 0$ for $n > 0$. Hence,

$$\Phi(r) = -\frac{2\pi G\,\gamma}{3}(3R^2 - r^2) = -G\,M\,\frac{(3R^2 - r^2)}{2R^3} \tag{2.53}$$

for $r \le R$, and

$$\Phi(r) = -\frac{4\pi G\,\gamma}{3}\frac{R^3}{r} = -\frac{G\,M}{r} \tag{2.54}$$

for $r > R$. Here, $M = (4\pi/3)\,R^3\,\gamma$ is the total mass of the sphere.

According to Equation (2.54), the gravitational potential outside a uniform sphere of mass M is the same as that generated by a point mass M located at the sphere's center. It turns out that this is a general result for *any* finite spherically symmetric mass distribution. Indeed, from the preceding analysis, it is clear that $\rho(r,\theta) = \rho_0(r)$ and

$\Phi(r, \theta) = \Phi_0(r)$ for such a distribution. Suppose that the distribution extends out to $r = R$. It immediately follows, from Equation (2.49), that

$$\Phi_0(r) = -\frac{G}{r} \int_0^R 4\pi r'^2 \rho_0(r')\, dr' = -\frac{G\,M}{r} \qquad (2.55)$$

for $r > R$, where M is the total mass of the distribution.

Consider a point mass m that lies a distance r from the center of a spherically symmetric mass distribution of mass M (where r exceeds the outer radius of the distribution). Because the external gravitational potential generated by the distribution is the same as that of a point mass M located at its center, the force exerted on the mass m by the distribution is the same as that due to a point mass M located at the center of the distribution. In other words, the force is of magnitude $G\,M\,m/r^2$ and is directed from the mass toward the center of the distribution. Assuming that the system is isolated, the resultant force that the mass exerts on the distribution is of magnitude $G\,M\,m/r^2$ and has a line of action directed from the center of the distribution toward the mass. (See Exercise 1.3.) However, this is the same as the force that the mass would exert on a point mass M located at the center of the distribution. Because gravitational fields are superposable, we conclude that the resultant gravitational force acting on a spherically symmetric mass distribution of mass M situated in the gravitational field generated by many point masses is the same as that which would act on a point mass M located at the center of the distribution.

The center of mass of a spherically symmetric mass distribution lies at the geometric center of the distribution. Moreover, the translational motion of the center of mass is analogous to that of a point particle whose mass is equal to that of the whole distribution, moving under the action of the resultant external force. (See Section 1.6.) If the external force is due to a gravitational field, then the resultant force is the same as that exerted by the field on a point particle, whose mass is that of the distribution, located at the center of mass. We thus conclude that Newton's laws of motion, in their primitive form, apply not only to point masses, but also to the translational motions of extended spherically symmetric mass distributions interacting via gravity (e.g., the Sun and the planets).

2.6 Potential outside a uniform spheroid

Let us now calculate the gravitational potential generated outside a spheroid of uniform mass density γ and mean radius R. A *spheroid* is the solid body produced by rotating an ellipse about a major or a minor axis. Let the axis of rotation coincide with the z-axis, and let the outer boundary of the spheroid satisfy

$$r = R_\theta(\theta) = R\left[1 - \frac{2}{3}\,\epsilon\,P_2(\cos\theta)\right], \qquad (2.56)$$

where ϵ is termed the *ellipticity*. In fact, the radius of the spheroid at the poles (i.e., along the axis) is $R_p = R\,(1 - 2\,\epsilon/3)$, whereas the radius at the equator (i.e., in the bisecting

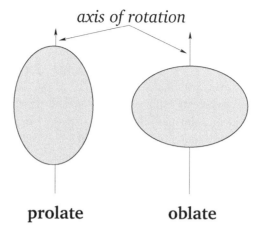

axis of rotation

prolate   **oblate**

Fig. 2.1   Prolate and oblate spheroids.

plane perpendicular to the axis) is $R_e = R\,(1 + \epsilon/3)$. Hence,

$$\epsilon = \frac{R_e - R_p}{R}. \tag{2.57}$$

Let us assume that $|\epsilon| \ll 1$, so that the spheroid is very close to being a sphere. If $\epsilon > 0$, then the spheroid is slightly squashed along its symmetry axis and is termed *oblate*. Likewise, if $\epsilon < 0$, then the spheroid is slightly elongated along its symmetry axis and is termed *prolate*. (See Figure 2.1.) Of course, if $\epsilon = 0$ then the spheroid reduces to a sphere.

Now, according to Equations (2.45) and (2.46), the gravitational potential generated outside an axially symmetric mass distribution can be written

$$\Phi(r, \theta) = \frac{G M}{R} \sum_{n=0,\infty} J_n \left(\frac{R}{r}\right)^{n+1} P_n(\cos\theta), \tag{2.58}$$

where M is the total mass of the distribution, and

$$J_n = -\frac{2\pi R^3}{M} \int \int \left(\frac{r}{R}\right)^{2+n} \rho(r, \theta)\, P_n(\cos\theta)\, \sin\theta\, d\theta\, \frac{dr}{R}. \tag{2.59}$$

Here, the integral is taken over the whole cross section of the distribution in r–θ space.

It follows that for a uniform spheroid, for which $M = (4\pi/3)\,\gamma\,R^3$,

$$J_n = -\frac{3}{2} \int_0^\pi P_n(\cos\theta) \int_0^{R_\theta(\theta)} \frac{r^{2+n}\, dr}{R^{3+n}}\, \sin\theta\, d\theta. \tag{2.60}$$

Hence,

$$J_n = -\frac{3}{2\,(3+n)} \int_0^\pi P_n(\cos\theta) \left[\frac{R_\theta(\theta)}{R}\right]^{3+n} \sin\theta\, d\theta, \tag{2.61}$$

giving

$$J_n \simeq -\frac{3}{2\,(3+n)} \int_0^\pi P_n(\cos\theta) \left[P_0(\cos\theta) - \frac{2}{3}\,(3+n)\,\epsilon\, P_2(\cos\theta)\right] \sin\theta\, d\theta, \tag{2.62}$$

to first order in ϵ. It is thus clear, from Equation (2.42), that, to first order in ϵ, the only nonzero J_n are

$$J_0 = -1 \tag{2.63}$$

and

$$J_2 = \frac{2}{5}\,\epsilon. \tag{2.64}$$

Thus, the gravitational potential outside a uniform spheroid of total mass M, mean radius R, and ellipticity ϵ is

$$\Phi(r,\theta) = -\frac{G\,M}{r} + \frac{2}{5}\,\epsilon\,\frac{G\,M\,R^2}{r^3}\,P_2(\cos\theta) + \mathcal{O}(\epsilon^2). \tag{2.65}$$

By analogy with the preceding analysis, the gravitational potential outside a general (i.e., axisymmetric, but not necessarily uniform) spheroidal mass distribution of mass M, mean radius R, and ellipticity ϵ (where $|\epsilon| \ll 1$) can be written

$$\Phi(r,\theta) = -\frac{G\,M}{r} + J_2\,\frac{G\,M\,R^2}{r^3}\,P_2(\cos\theta) + \mathcal{O}(\epsilon^2), \tag{2.66}$$

where $J_2 \sim \mathcal{O}(\epsilon)$. (See Exercise 2.9.) In particular, the gravitational potential on the surface of the spheroid is

$$\Phi(R_\theta,\theta) = -\frac{G\,M}{R_\theta} + J_2\,\frac{G\,M\,R^2}{R_\theta^3}\,P_2(\cos\theta) + \mathcal{O}(\epsilon^2), \tag{2.67}$$

which yields

$$\Phi(R_\theta,\theta) \simeq -\frac{G\,M}{R}\left[1 + \left(\frac{2}{3}\,\epsilon - J_2\right)P_2(\cos\theta) + \mathcal{O}(\epsilon^2)\right], \tag{2.68}$$

where use has been made of Equation (2.56). For the case of a uniform-density spheroid, for which $J_2 = (2/5)\,\epsilon$ [see Equation (2.64)], the preceding expression simplifies to

$$\Phi(R_\theta,\theta) \simeq -\frac{G\,M}{R}\left[1 + \frac{4}{15}\,\epsilon\,P_2(\cos\theta) + \mathcal{O}(\epsilon^2)\right]. \tag{2.69}$$

Consider a self-gravitating spheroid of mass M, mean radius R, and ellipticity ϵ, such as a star or a planet. Assuming, for the sake of simplicity, that the spheroid is composed of uniform-density incompressible fluid, it follows that the gravitational potential on its surface is given by Equation (2.69). However, the condition for an equilibrium state is that the potential be *constant* over the surface. If this is not the case, then there will be gravitational forces acting *tangential* to the surface. Such forces cannot be balanced by internal fluid pressure, which acts only *normal* to the surface. Hence, from Equation (2.69), it is clear that the condition for equilibrium is $\epsilon = 0$. In other words, the equilibrium configuration of a uniform-density, self-gravitating, fluid, mass distribution is a *sphere*. Deviations from this configuration can be caused only by forces in addition to self-gravity and internal fluid pressure, such as internal tensile forces, centrifugal forces due to rotation, or tidal forces due to orbiting masses. The same is true for a self-gravitating mass distribution of non-uniform density. (See Chapter 5.)

The martian moon Phobos (mean radius 11.1 km). Photograph taken by the European Space Agency's Mars Express spacecraft in 2010. Credit: G. Neukum (ESA/DLR/FU Berlin).

We can estimate how small a rocky asteroid, say, needs to be before its material strength is sufficient to allow it to retain a significantly nonspherical shape. The typical density of rocky asteroids in the solar system is $\gamma \sim 3.5 \times 10^3 \, \mathrm{kg \, m^{-3}}$. Moreover, the critical pressure above which the rock out of which such asteroids are composed ceases to act as a rigid material, and instead deforms and flows like a liquid, is $p_c \sim 2 \times 10^8 \, \mathrm{N \, m^{-2}}$ (de Pater and Lissauer 2010). We must compare this critical pressure with the pressure at the center of the asteroid. Assuming, for the sake of simplicity, that the asteroid is roughly spherical, of radius R, and of uniform density γ, the central pressure is

$$p_0 = \int_0^R \gamma \, g(r) \, dr, \tag{2.70}$$

where $g(r) = (4\pi/3) \, G \, \gamma \, r$ is the gravitational acceleration at radius r [see Equation (2.53)]. This result is a simple generalization of the well-known formula $\rho \, g \, h$ for the pressure a depth h below the surface of a fluid. It follows that

$$p_0 = \frac{2\pi}{3} G \gamma^2 R^2. \tag{2.71}$$

If $p_0 \ll p_c$, then the internal pressure in the asteroid is not sufficiently high to cause its constituent rock to deform like a liquid. Such an asteroid can therefore retain a significantly nonspherical shape. On the other hand, if $p_0 \gg p_c$, then the internal pressure is large enough to render the asteroid fluidlike. Such an asteroid cannot withstand the tendency of self-gravity to make it adopt a spherical shape. The same applies to any rocky body in the solar system. The condition $p_0 \ll p_c$ is equivalent to $R \ll R_c$, where

$$R_c = \left(\frac{3}{2\pi} \frac{p_c}{G \gamma^2} \right)^{1/2} \simeq 230 \, \mathrm{km}. \tag{2.72}$$

It follows that only a rocky body whose radius is significantly less than about 230 km—for instance, the two moons of Mars, Phobos (see Figure 2.2) and Deimos—can retain a highly nonspherical shape. On the other hand, a rocky body whose radius is significantly

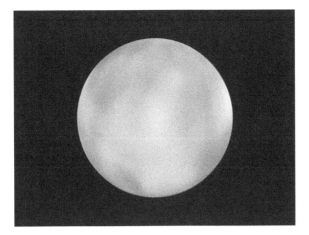

The asteroid Ceres (mean radius 470 km). Photograph taken by the Hubble Space Telescope. Credit: NASA, ESA, J.-Y. Li
(University of Maryland) and G. Bacon (STScl).

greater than about 230 km—for instance, the asteroid Ceres (see Figure 2.3) and the
Earth's Moon—is forced by gravity to be essentially spherical.

According to Equations (2.58) and (2.63), the gravitational potential outside a general
axisymmetric body of mass M and mean radius R can be written

$$\Phi(r,\theta) = -\frac{G\,M}{r}\left[1 - \sum_{n=1,\infty} J_n\,\frac{R^n}{r^n}\,P_n(\cos\theta)\right], \tag{2.73}$$

where the J_n are $\mathcal{O}(1)$ (or smaller) dimensionless parameters that depend on the exact
shape of the body. However, a long way from the body ($r \gg R$), the right-hand side of
the preceding expression is clearly dominated by the first term inside the round brackets,
so that

$$\Phi(r,\theta) \simeq -\frac{G\,M}{r}. \tag{2.74}$$

However, this is just the gravitational potential of a point particle, located at the center
of the body, whose mass is equal to that of the body. This suggests that the gravitational
interaction between two irregularly shaped bodies in the solar system can be approximated as the interaction of two point masses, provided that the distance between the
bodies is much larger than the sum of their radii.

2.7 Potential due to a uniform ring

Consider a uniform ring of mass M, radius a, and negligible cross-sectional area, centered on the origin, and lying in the x–y plane. Let us consider the gravitational potential

$\Phi(r)$ generated by such a ring in the x–y plane (which corresponds to $\theta = 90°$). It follows, from Section 2.4, that for $r > a$,

$$\Phi(r) = -\frac{G M}{a} \sum_{n=0,\infty} [P_n(0)]^2 \left(\frac{a}{r}\right)^{n+1}. \tag{2.75}$$

However, $P_0(0) = 1$, $P_1(0) = 0$, $P_2(0) = -1/2$, $P_3(0) = 0$, $P_4(0) = 3/8$, $P_5(0) = 0$, $P_6(0) = -5/16$, $P(7) = 0$, and $P_8(0) = 35/128$. Hence,

$$\Phi(r) = -\frac{G M}{r} \left[1 + \frac{1}{4}\left(\frac{a}{r}\right)^2 + \frac{9}{64}\left(\frac{a}{r}\right)^4 + \frac{25}{256}\left(\frac{a}{r}\right)^6 + \frac{1225}{16384}\left(\frac{a}{r}\right)^8 + \cdots \right]. \tag{2.76}$$

Likewise, for $r < a$,

$$\Phi(r) = -\frac{G M}{a} \sum_{n=0,\infty} [P_n(0)]^2 \left(\frac{r}{a}\right)^{n}, \tag{2.77}$$

giving

$$\Phi(r) = -\frac{G M}{a} \left[1 + \frac{1}{4}\left(\frac{r}{a}\right)^2 + \frac{9}{64}\left(\frac{r}{a}\right)^4 + \frac{25}{256}\left(\frac{r}{a}\right)^6 + \frac{1225}{16384}\left(\frac{r}{a}\right)^8 + \cdots \right]. \tag{2.78}$$

Exercises

2.1 A particle is projected vertically upward from the Earth's surface with a velocity that would, were gravity uniform, carry it to a height h. Show that if the variation of gravity with height is allowed for, but the resistance of air is neglected, then the height reached will be greater by $h^2/(R - h)$, where R is the Earth's radius. (From Lamb 1923.)

2.2 A particle is projected vertically upward from the Earth's surface with a velocity just sufficient for it to reach infinity (neglecting air resistance). Prove that the time needed to reach a height h is

$$\frac{1}{3}\left(\frac{2R}{g}\right)^{1/2} \left[\left(1 + \frac{h}{R}\right)^{3/2} - 1 \right],$$

where R is the Earth's radius, and g its surface gravitational acceleration. (From Lamb 1923.)

2.3 Assuming that the Earth is a sphere of radius R, and neglecting air resistance, show that a particle that starts from rest a distance R from the Earth's surface will reach the surface with speed $\sqrt{R g}$ after a time $(1 + \pi/2)\sqrt{R/g}$, where g is the surface gravitational acceleration. (Modified from Smart 1951.)

2.4 Demonstrate that if a narrow shaft were drilled though the center of a uniform self-gravitating sphere, then a test mass moving in this shaft executes simple harmonic motion about the center of the sphere with period

$$T = 2\pi \sqrt{\frac{R}{g}},$$

where R is the radius of the sphere, and g the gravitational acceleration at its surface.

2.5 Consider an isolated system consisting of N point objects interacting via gravity. The equation of motion of the ith object is

$$m_i \ddot{\mathbf{r}}_i = \sum_{j=1,N}^{j \neq i} G m_i m_j \frac{\mathbf{r}_j - \mathbf{r}_i}{|\mathbf{r}_j - \mathbf{r}_i|^3},$$

where m_i and $\mathbf{r}_i$ are the mass and position vector of this object, respectively. Moreover, the total potential energy of the system takes the form

$$U = -\frac{1}{2} \sum_{i,j=1,N}^{i \neq j} \frac{G m_i m_j}{|\mathbf{r}_j - \mathbf{r}_i|}.$$

Write an expression for the total kinetic energy, K. Demonstrate from the equations of motion that $K + U$ is constant in time.

2.6 Consider a function of many variables $f(x_1, x_2, \ldots, x_n)$. Such a function that satisfies

$$f(t x_1, t x_2, \ldots, t x_n) = t^a f(x_1, x_2, \ldots, x_n)$$

for all $t > 0$, and all values of x_i, is termed a *homogeneous function of degree a*. Prove the following theorem regarding homogeneous functions:

$$\sum_{i=1,n} x_i \frac{\partial f}{\partial x_i} = a f.$$

2.7 Consider an isolated system consisting of N point particles interacting via attractive central forces. Let the mass and position vector of the ith particle be m_i and $\mathbf{r}_i$, respectively. Suppose that magnitude of the force exerted on particle i by particle j is $k_i k_j |\mathbf{r}_i - \mathbf{r}_j|^{-n}$. Here, k_i measures some constant physical property of the ith particle (e.g., its electric charge). Show that the total potential energy U of the system is written

$$U = -\frac{1}{2} \frac{1}{n-1} \sum_{i,j=1,N}^{j \neq i} \frac{k_i k_j}{|\mathbf{r}_j - \mathbf{r}_i|^{n-1}}.$$

Is this a homogeneous function? If so, what is its degree? Demonstrate that the equation of motion of the ith particle can be written

$$m_i \ddot{\mathbf{r}}_i = -\frac{\partial U}{\partial \mathbf{r}_i}.$$

(This is shorthand for $m_i \ddot{x}_i = -\partial U/\partial x_i$, $m_i \ddot{y}_i = -\partial U/\partial y_i$, etc., where the x_i, y_i, z_i, for $i = 1, N$ are treated as *independent* variables.) Use the mathematical theorem from the previous exercise to show that

$$\frac{1}{2} \frac{d^2 \mathcal{I}}{dt^2} = 2 K + (n-1) U,$$

where $\mathcal{I} = \sum_{i=1,N} m_i r_i^2$, and K is the total kinetic energy. This result is known as the *virial theorem*. Demonstrate that when $n \geq 3$, the system possesses no virial equilbria (i.e., states for which $\mathcal{I}$ does not evolve in time) that are bound.

2.8 Demonstrate that the gravitational potential energy of a spherically symmetric mass distribution of mass density $\rho(r)$ that extends out to $r = R$ can be written

$$U = -16\pi^2 G \int_0^R \int_0^r r'^2 \rho(r') \, r \rho(r) \, dr' \, dr.$$

Hence, show that if the mass distribution is such that

$$\rho(r) = \begin{cases} \rho_0 \, r^{-\alpha} & r \leq R \\ 0 & r > R \end{cases}$$

where $\alpha < 5/2$, then

$$U = -\frac{(3-\alpha)}{(5-2\alpha)} \frac{GM^2}{R},$$

where M is the total mass.

2.9 Consider a spheroidal mass distribution of mass M, mean radius R, and ellipticity ϵ (where $|\epsilon| \ll 1$). Let the density distribution within the spheroid be of the form specified in the previous exercise. Demonstrate that the gravitational potential outside the spheroid takes the form

$$\Phi(r, \theta) = -\frac{GM}{r} + J_2 \frac{GMR^2}{r^3} P_2(\cos\theta) + \mathcal{O}(\epsilon^2),$$

where

$$J_2 = \frac{2}{5}\left(\frac{3-\alpha}{3}\right)\epsilon.$$

Here, r and θ are spherical coordinates whose origin lies at the geometric center of the distribution, and whose symmetry axis coincides with that of the distribution. Let $\mathcal{I}_\parallel$ be the moment of inertia of the distribution about its symmetry axis. Show that

$$\mathcal{I}_\parallel = \frac{2}{3}\left(\frac{3-\alpha}{5-\alpha}\right) MR^2 + \mathcal{O}(\epsilon).$$

2.10 A globular star cluster can be approximated as an isolated self-gravitating virial equilibrium consisting of a great number of equal mass stars. Demonstrate, from the virial theorem, that

$$K = -\frac{1}{2}U$$

for such a cluster. Suppose that the stars in a given cluster are uniformly distributed throughout a spherical volume. Show that

$$\bar{v} = \sqrt{\frac{3}{10}} \, v_{\text{esc}},$$

where $\bar{v}$ is the mean stellar velocity and v_{esc} is the escape speed (i.e., the speed a star at the edge of the cluster would require in order to escape to infinity.) See Exercise 2.7.

2.11 A star can be thought of as a spherical system that consists of a very large number of particles of mass m_i and position vector $\mathbf{r}_i$ interacting via gravity. Show that, for such a system, the virial theorem implies that

$$\frac{d^2 \mathcal{I}}{dt^2} = -2\,U + c,$$

where c is a constant, $\mathcal{I} = \sum m_i\, r_i^2$, and the r_i are measured from the geometric center. Hence, deduce that the angular frequency of small-amplitude radial pulsations of the star (in which the radial displacement is directly proportional to the radial distance from the center) takes the form

$$\omega = \left(\frac{|U_0|}{\mathcal{I}_0}\right)^{1/2},$$

where U_0 and $\mathcal{I}_0$ are the equilibrium values of U and $\mathcal{I}$. Finally, show that if the mass density within the star varies as $r^{-\alpha}$, where r is the radial distance from the geometric center and where $\alpha < 5/2$, then

$$\omega = \left(\frac{5-\alpha}{5-2\alpha}\frac{G\,M}{R^3}\right)^{1/2},$$

where M and R are the stellar mass and radius, respectively. See Exercises 2.7 and 2.8.

Keplerian orbits

3.1 Introduction

Newtonian mechanics was initially developed to account for the motion of the planets around the Sun. Let us now examine this problem. Suppose that the Sun, whose mass is M, is located at the origin of our coordinate system. Consider the motion of a general planet, of mass m, that is located at position vector $\mathbf{r}$. Given that the Sun and all the planets are approximately spherical, the gravitational force exerted on our planet by the Sun can be written (see Chapter 2)

$$\mathbf{f} = -\frac{G\,M\,m}{r^3}\,\mathbf{r}. \tag{3.1}$$

An equal and opposite force to that given in Equation (3.1) acts on the Sun. However, we shall assume that the Sun is so much more massive than the planet that this force does not cause the Sun's position to shift appreciably. Hence, the Sun will always remain at the origin of our coordinate system. Likewise, we shall neglect the gravitational forces exerted on our chosen planet by the other bodies in the solar system, compared with the gravitational force exerted on it by the Sun. This is reasonable because the Sun is much more massive (by a factor of at least 10^3) than any other solar system body. Thus, according to Equation (3.1) and Newton's second law of motion, the equation of motion of our planet (which can effectively be treated as a point object—see Chapter 2) takes the form

$$\frac{d^2\mathbf{r}}{dt^2} = -\frac{G\,M}{r^3}\,\mathbf{r}. \tag{3.2}$$

Note that the planet's mass, m, has canceled out on both sides of this equation.

3.2 Kepler's laws

As is well known, Johannes Kepler (1571–1630) was the first astronomer to correctly describe the motion of the planets (in works published between 1609 and 1619). This motion is summed up in three simple laws:

1. The planetary orbits are all ellipses that are confocal with the Sun (i.e., the Sun lies at one of the foci of each ellipse—see Section A.9).
2. The radius vector connecting each planet to the Sun sweeps out equal areas in equal time intervals.

3. The square of the orbital period of each planet is proportional to the cube of its orbital major radii.

Let us now see whether we can derive Kepler's laws from Equation (3.2).

3.3 Conservation laws

As we have already seen, gravity is a *conservative force*. Hence, the gravitational force in Equation (3.1) can be written (see Section 1.4)

$$\mathbf{f} = -\nabla U, \tag{3.3}$$

where the potential energy, $U(\mathbf{r})$, of our planet in the Sun's gravitational field takes the form

$$U(\mathbf{r}) = -\frac{G\,M\,m}{r}. \tag{3.4}$$

(See Section 2.5.) It follows that the total energy of our planet is a conserved quantity. (See Section 1.4.) In other words,

$$\mathcal{E} = \frac{v^2}{2} - \frac{G\,M}{r} \tag{3.5}$$

is constant in time. Here, $\mathcal{E}$ is actually the planet's total energy per unit mass, and $\mathbf{v} = d\mathbf{r}/dt$.

Gravity is also a *central force*. Hence, the *angular momentum* of our planet is a conserved quantity. (See Section 1.5.) In other words,

$$\mathbf{h} = \mathbf{r} \times \mathbf{v}, \tag{3.6}$$

which is actually the planet's angular momentum per unit mass, is constant in time. Assuming that $|\mathbf{h}| > 0$, and taking the scalar product of the preceding equation with $\mathbf{r}$, we obtain

$$\mathbf{h} \cdot \mathbf{r} = 0. \tag{3.7}$$

This is the equation of a plane that passes through the origin and whose normal is parallel to $\mathbf{h}$. Because $\mathbf{h}$ is a constant vector, it always points in the same direction. We therefore conclude that the orbit of our planet is *two-dimensional*—that is, it is confined to some fixed plane that passes through the origin. Without loss of generality, we can let this plane coincide with the x–y plane.

3.4 Plane polar coordinates

We can determine the instantaneous position of our planet in the x–y plane in terms of standard Cartesian coordinates, x, y, or plane polar coordinates, r, θ, as illustrated in Figure 3.1. Here, $r = (x^2 + y^2)^{1/2}$ and $\theta = \tan^{-1}(y/x)$. It is helpful to define two unit vectors, $\mathbf{e}_r \equiv \mathbf{r}/r$ and $\mathbf{e}_\theta \equiv \mathbf{e}_z \times \mathbf{e}_r$, at the instantaneous position of the planet. The first

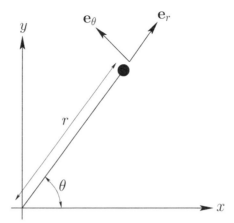

Fig. 3.1 Plane polar coordinates.

always points radially away from the origin, whereas the second is normal to the first, in the direction of increasing θ. As is easily demonstrated, the Cartesian components of $\mathbf{e}_r$ and $\mathbf{e}_\theta$ are

$$\mathbf{e}_r = (\cos\theta,\ \sin\theta) \tag{3.8}$$

and

$$\mathbf{e}_\theta = (-\sin\theta,\ \cos\theta), \tag{3.9}$$

respectively.

We can write the position vector of our planet as

$$\mathbf{r} = r\,\mathbf{e}_r. \tag{3.10}$$

Thus, the planet's velocity becomes

$$\mathbf{v} = \frac{d\mathbf{r}}{dt} = \dot{r}\,\mathbf{e}_r + r\,\dot{\mathbf{e}}_r, \tag{3.11}$$

where $\dot{\ }$ is shorthand for d/dt. Note that $\mathbf{e}_r$ has a nonzero time derivative (unlike a Cartesian unit vector) because its direction changes as the planet moves around. As is easily demonstrated, by differentiating Equation (3.8) with respect to time, we obtain

$$\dot{\mathbf{e}}_r = \dot{\theta}\,(-\sin\theta,\ \cos\theta) = \dot{\theta}\,\mathbf{e}_\theta. \tag{3.12}$$

Thus,

$$\mathbf{v} = \dot{r}\,\mathbf{e}_r + r\,\dot{\theta}\,\mathbf{e}_\theta. \tag{3.13}$$

The planet's acceleration is written

$$\mathbf{a} = \frac{d\mathbf{v}}{dt} = \frac{d^2\mathbf{r}}{dt^2} = \ddot{r}\,\mathbf{e}_r + \dot{r}\,\dot{\mathbf{e}}_r + (\dot{r}\,\dot{\theta} + r\,\ddot{\theta})\,\mathbf{e}_\theta + r\,\dot{\theta}\,\dot{\mathbf{e}}_\theta. \tag{3.14}$$

Again, $\mathbf{e}_\theta$ has a nonzero time derivative because its direction changes as the planet moves around. Differentiation of Equation (3.9) with respect to time yields

$$\dot{\mathbf{e}}_\theta = \dot{\theta}\,(-\cos\theta,\ -\sin\theta) = -\dot{\theta}\,\mathbf{e}_r. \tag{3.15}$$

Hence,

$$\mathbf{a} = (\ddot{r} - r\dot{\theta}^2)\,\mathbf{e}_r + (r\ddot{\theta} + 2\dot{r}\dot{\theta})\,\mathbf{e}_\theta. \tag{3.16}$$

It follows that the equation of motion of our planet, Equation (3.2), can be written

$$\mathbf{a} = (\ddot{r} - r\dot{\theta}^2)\,\mathbf{e}_r + (r\ddot{\theta} + 2\dot{r}\dot{\theta})\,\mathbf{e}_\theta = -\frac{G\,M}{r^2}\,\mathbf{e}_r. \tag{3.17}$$

Because $\mathbf{e}_r$ and $\mathbf{e}_\theta$ are mutually orthogonal, we can separately equate the coefficients of both, in the preceding equation, to give a *radial equation of motion*,

$$\ddot{r} - r\dot{\theta}^2 = -\frac{G\,M}{r^2}, \tag{3.18}$$

and a *tangential equation of motion*,

$$r\ddot{\theta} + 2\dot{r}\dot{\theta} = 0. \tag{3.19}$$

3.5 Kepler's second law

Multiplying our planet's tangential equation of motion, Equation (3.19), by r, we obtain

$$r^2\ddot{\theta} + 2\,r\,\dot{r}\,\dot{\theta} = 0. \tag{3.20}$$

However, this equation can be also written

$$\frac{d(r^2\,\dot{\theta})}{dt} = 0, \tag{3.21}$$

which implies that

$$h = r^2\,\dot{\theta} \tag{3.22}$$

is constant in time. It is easily demonstrated that h is the magnitude of the vector $\mathbf{h}$ defined in Equation (3.6). Thus, the fact that h is constant in time is equivalent to the statement that the angular momentum of our planet is a constant of its motion. As we have already mentioned, this is the case because gravity is a central force.

Suppose that the radius vector connecting our planet to the origin (i.e., the Sun) rotates through an angle $\delta\theta$ between times t and $t + \delta t$. (See Figure 3.2.) The approximately triangular region swept out by the radius vector has the area

$$\delta A \simeq \frac{1}{2}\,r^2\,\delta\theta, \tag{3.23}$$

as the area of a triangle is half its base ($r\,\delta\theta$) times its height (r). Hence, the rate at which the radius vector sweeps out area is

$$\frac{dA}{dt} = \lim_{\delta t \to 0} \frac{r^2\,\delta\theta}{2\,\delta t} = \frac{r^2}{2}\,\frac{d\theta}{dt} = \frac{h}{2}. \tag{3.24}$$

Thus, the radius vector sweeps out area at a constant rate (because h is constant in time)—this is Kepler's second law of planetary motion. We conclude that Kepler's second law is a direct consequence of *angular momentum conservation*.

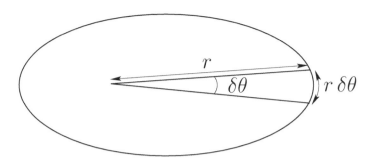

Fig. 3.2   Kepler's second law.

3.6 Kepler's first law

Our planet's radial equation of motion, Equation (3.18), can be combined with Equation (3.22) to give

$$\ddot{r} - \frac{h^2}{r^3} = -\frac{GM}{r^2}.$$ (3.25)

Suppose that $r = u^{-1}$, where $u = u(\theta)$ and $\theta = \theta(t)$. It follows that

$$\dot{r} = -\frac{\dot{u}}{u^2} = -r^2 \frac{du}{d\theta} \frac{d\theta}{dt} = -h \frac{du}{d\theta}.$$ (3.26)

Likewise,

$$\ddot{r} = -h \frac{d^2u}{d\theta^2} \dot{\theta} = -u^2 h^2 \frac{d^2u}{d\theta^2}.$$ (3.27)

Hence, Equation (3.25) can be written in the linear form

$$\frac{d^2u}{d\theta^2} + u = \frac{GM}{h^2}.$$ (3.28)

As can be seen via inspection, the general solution to the preceding equation is

$$u(\theta) = \frac{GM}{h^2} \left[1 + e \cos(\theta - \theta_0) \right],$$ (3.29)

where e and θ_0 are arbitrary constants. Without loss of generality, we can set $\theta_0 = 0$ by rotating our coordinate system about the z-axis. We can also assume that $e \geq 0$. Thus, we obtain

$$r(\theta) = \frac{r_c}{1 + e \cos \theta},$$ (3.30)

where

$$r_c = \frac{h^2}{GM}.$$ (3.31)

Equation (3.30) is the equation of a conic section that is confocal with the origin (i.e., with the Sun). (See Section A.9.) Specifically, for $e < 1$, Equation (3.30) is the equation of an *ellipse*. For $e = 1$, Equation (3.30) is the equation of a *parabola*. Finally, for $e > 1$, Equation (3.30) is the equation of a *hyperbola*. However, a planet cannot have a parabolic or a hyperbolic orbit, because such orbits are appropriate only to objects that are ultimately able to escape from the Sun's gravitational field. Thus, the orbit of our planet is an ellipse that is confocal with the Sun—this is Kepler's first law of planetary motion.

3.7   Kepler's third law

We have seen that the radius vector connecting our planet to the origin sweeps out area at the constant rate $dA/dt = h/2$ [see Equation (3.24)]. We have also seen that the planetary orbit is an ellipse. The major and minor radii of such an ellipse are $a = r_c/(1 - e^2)$ and $b = (1-e^2)^{1/2} a$, respectively. [See Equations (3.34) and (A.108).] The area of the ellipse is $A = \pi a b$. We expect the radius vector to sweep out the whole area of the ellipse in a single orbital period, T. Hence,

$$T = \frac{A}{dA/dt} = \frac{2\pi a b}{h} = \frac{2\pi a^2 (1 - e^2)^{1/2}}{h} = \frac{2\pi a^{3/2} r_c^{1/2}}{h}. \tag{3.32}$$

It follows from Equation (3.31) that

$$T^2 = \frac{4\pi^2 a^3}{G M}. \tag{3.33}$$

In other words, the square of the orbital period of our planet is proportional to the cube of its orbital major radius—this is Kepler's third law of planetary motion.

3.8   Orbital parameters

For an elliptical orbit, the closest distance to the Sun—the *perihelion distance*—is [see Equation (3.30)]

$$r_p = \frac{r_c}{1 + e} = a (1 - e). \tag{3.34}$$

This equation also holds for parabolic and hyperbolic orbits. Likewise, the furthest distance from the Sun—the *aphelion distance*—is

$$r_a = \frac{r_c}{1 - e} = a (1 + e). \tag{3.35}$$

It follows that, for an elliptical orbit, the major radius, a, is simply the mean of the perihelion and aphelion distances,

$$a = \frac{r_p + r_a}{2}. \tag{3.36}$$

The parameter

$$e = \frac{r_a - r_p}{r_a + r_p} \tag{3.37}$$

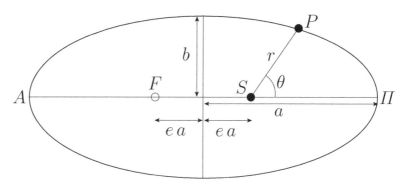

A Keplerian elliptical orbit. S is the Sun, P the planet, F the empty focus, Π the perihelion point, A the aphelion point, a the major radius, b the minor radius, e the eccentricity, r the radial distance, and θ the true anomaly.

is called the *eccentricity*; it measures the deviation of the orbit from circularity. Thus, $e = 0$ corresponds to a circular orbit, whereas $e \to 1$ corresponds to an infinitely elongated elliptical orbit. Note that the Sun is displaced a distance $e\,a$ along the major axis from the geometric center of the orbit. (See Section A.9 and Figure 3.3.)

As is easily demonstrated from the preceding analysis, Kepler's laws of planetary motion can be written in the convenient form

$$r = \frac{a\,(1 - e^2)}{1 + e\,\cos\theta},\tag{3.38}$$

$$h = r^2\,\dot\theta = (1 - e^2)^{1/2}\,n\,a^2,\tag{3.39}$$

and

$$G\,M = n^2\,a^3,\tag{3.40}$$

where a is the mean orbital radius (i.e., the major radius), e the orbital eccentricity, and $n = 2\pi/T$ the mean orbital angular velocity.

3.9 Orbital energies

Let us now generalize our analysis to take into account the orbits of asteroids and comets about the Sun. Such orbits satisfy Equation (3.30) but can be parabolic ($e = 1$) or even hyperbolic ($e > 1$), as well as elliptical ($e < 1$). According to Equations (3.5) and (3.13), the total energy per unit mass of an object in a general orbit around the Sun is given by

$$\mathcal{E} = \frac{\dot r^2 + r^2\,\dot\theta^2}{2} - \frac{G\,M}{r}.\tag{3.41}$$

It follows from Equations (3.22), (3.26), and (3.31) that

$$\mathcal{E} = \frac{h^2}{2}\left[\left(\frac{du}{d\theta}\right)^2 + u^2 - 2\,u\,u_c\right],\tag{3.42}$$

where $u = r^{-1}$ and $u_c = r_c^{-1}$. However, according to Equation (3.30),

$$u(\theta) = u_c (1 + e \cos \theta). \tag{3.43}$$

The previous two equations can be combined with Equations (3.31) and (3.34) to give

$$\mathcal{E} = \frac{u_c^2 h^2}{2} (e^2 - 1) = \frac{G M}{2 r_p} (e - 1). \tag{3.44}$$

We conclude that elliptical orbits ($e < 1$) have *negative* total energies, whereas parabolic orbits ($e = 1$) have *zero* total energies, and hyperbolic orbits ($e > 1$) have *positive* total energies. This makes sense because in a conservative system in which the potential energy at infinity is set to zero [see Equation (3.4)], we expect *bounded* orbits to have negative total energies, and *unbounded* orbits to have positive total energies. (See Section 1.7.) Thus, elliptical orbits, which are clearly bounded, should indeed have negative total energies, whereas hyperbolic orbits, which are clearly unbounded, should indeed have positive total energies. Parabolic orbits are marginally bounded (i.e, an object executing a parabolic orbit only just manages to escape from the Sun's gravitational field), and thus have zero total energy. For the special case of an elliptical orbit, whose major radius a is finite, we can write

$$\mathcal{E} = -\frac{G M}{2 a}. \tag{3.45}$$

It follows that the energy of such an orbit is completely determined by its *major radius*.

3.10 Transfer orbits

Consider an artificial satellite in an elliptical orbit around the Sun (the same considerations also apply to satellites in orbit around the Earth). At perihelion, $\dot{r} = 0$, and Equations (3.41) and (3.44) can be combined to give

$$\frac{v_t}{v_c} = \sqrt{1 + e}. \tag{3.46}$$

Here, $v_t = r \dot{\theta}$ is the satellite's tangential velocity and $v_c = \sqrt{G M / r_p}$ is the tangential velocity that it would need to maintain a circular orbit at the perihelion distance. Likewise, at aphelion,

$$\frac{v_t}{v_c} = \sqrt{1 - e}, \tag{3.47}$$

where $v_c = \sqrt{G M / r_a}$ is now the tangential velocity that the satellite would need to maintain a circular orbit at the aphelion distance.

Suppose that our satellite is initially in a circular orbit of radius r_1 and that we wish to transfer it into a circular orbit of radius r_2, where $r_2 > r_1$. We can achieve this by temporarily placing the satellite in an elliptical orbit whose perihelion distance is r_1 and whose aphelion distance is r_2. It follows, from Equation (3.37), that the required eccentricity of the elliptical orbit is

$$e = \frac{r_2 - r_1}{r_2 + r_1}. \tag{3.48}$$

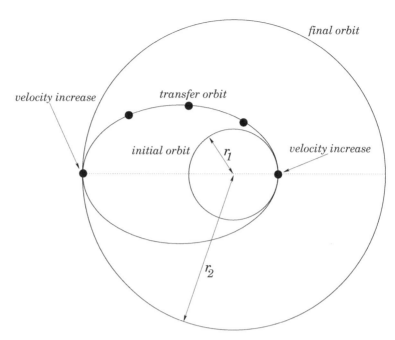

Fig. 3.4 A transfer orbit between two circular orbits.

According to Equation (3.46), we can transfer our satellite from its initial circular orbit into the temporary elliptical orbit by increasing its tangential velocity (by briefly
switching on the satellite's rocket motor) by a factor

$$\alpha_1 = \sqrt{1 + e}. \tag{3.49}$$

We must next allow the satellite to execute half an orbit, so it attains its aphelion distance, and then boost the tangential velocity by a factor [see Equation (3.47)]

$$\alpha_2 = \frac{1}{\sqrt{1 - e}}. \tag{3.50}$$

The satellite will now be in a circular orbit at the aphelion distance, r_2. This process
is illustrated in Figure 3.4. Obviously, we can transfer our satellite from a larger to a
smaller circular orbit by performing the preceding process in reverse. Note, finally, from
Equation (3.46) that if we increase the tangential velocity of a satellite in a circular orbit
about the Sun by a factor greater than $\sqrt{2}$, then we will transfer it into a hyperbolic orbit
($e > 1$), and it will eventually escape from the Sun's gravitational field.

3.11 Elliptical orbits

Let us determine the radial and angular coordinates, r and θ, respectively, of a planet
in an elliptical orbit about the Sun as a function of time. Suppose the planet passes
through its perihelion point, $r = r_p$ and $\theta = 0$, at $t = \tau$. The constant τ is termed the *time*

of perihelion passage. It follows from the previous analysis that

$$r = \frac{r_p(1+e)}{1 + e\,\cos\theta},$$

(3.51)

and

$$\mathcal{E} = \frac{\dot{r}^2}{2} + \frac{h^2}{2\,r^2} - \frac{G\,M}{r},$$

(3.52)

where e, $h = \sqrt{G\,M\,r_p\,(1+e)}$, and $\mathcal{E} = G\,M\,(e-1)/(2\,r_p)$ are the orbital eccentricity, angular momentum per unit mass, and energy per unit mass, respectively. The preceding equation can be rearranged to give

$$\dot{r}^2 = (e-1)\,\frac{G\,M}{r_p} - (e+1)\,\frac{r_p\,G\,M}{r^2} + \frac{2\,G\,M}{r}.$$

(3.53)

Taking the square root and integrating, we obtain

$$\int_{r_p}^{r} \frac{r\,dr}{[2\,r + (e-1)\,r^2/r_p - (e+1)\,r_p]^{1/2}} = \sqrt{G\,M}\,(t-\tau).$$

(3.54)

Consider an elliptical orbit characterized by $0 < e < 1$. Let us write

$$r = \frac{r_p}{1-e}\,(1 - e\,\cos E),$$

(3.55)

where E is termed the *eccentric anomaly*. In fact, E is an angle that varies between $-\pi$ and $+\pi$. Moreover, the perihelion point corresponds to $E = 0$, and the aphelion point to $E = \pi$. Now,

$$dr = \frac{r_p}{1-e}\,e\,\sin E\,dE,$$

(3.56)

whereas

$$2\,r + (e-1)\,\frac{r^2}{r_p} - (e+1)\,r_p = \frac{r_p}{1-e}\,e^2\,(1 - \cos^2 E) = \frac{r_p}{1-e}\,e^2\,\sin^2 E.$$

(3.57)

Thus, Equation (3.54) reduces to

$$\int_0^E (1 - e\,\cos E)\,dE = \left(\frac{G\,M}{a^3}\right)^{1/2}(t-\tau),$$

(3.58)

where $a = r_p/(1-e)$. This equation can immediately be integrated to give

$$E - e\,\sin E = \mathcal{M}.$$

(3.59)

Here,

$$\mathcal{M} = n\,(t-\tau)$$

(3.60)

is termed the *mean anomaly*, $n = 2\pi/T$ is the mean orbital angular velocity, and $T = 2\pi\,(a^3/GM)^{1/2}$ is the orbital period. The mean anomaly is an angle that increases uniformly in time at the rate of 2π radians every orbital period. Moreover, the perihelion point corresponds to $\mathcal{M} = 0$, and the aphelion point to $\mathcal{M} = \pi$. Incidentally, the angle θ, which determines the true angular location of the planet relative to its perihelion point, is called the *true anomaly*. Equation (3.59), which is known as *Kepler's equation*, is

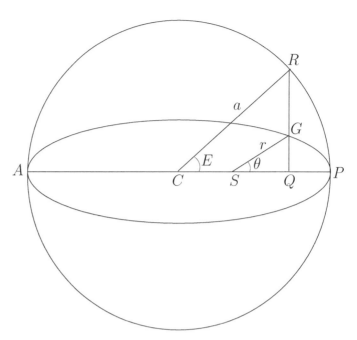

Fig. 3.5   Eccentric anomaly.

a transcendental equation that does not possess a convenient analytic solution. Fortunately, it is fairly straightforward to solve numerically. For instance, when we use an
iterative approach, if E_n is the nth guess, then

$$E_{n+1} = \mathcal{M} + e \sin E_n. \tag{3.61}$$

This iteration scheme converges very rapidly when $0 \leq e \ll 1$ (as is the case for
planetary orbits).

Equations (3.51) and (3.55) can be combined to give

$$r \cos \theta = a (\cos E - e). \tag{3.62}$$

This expression allows us to give a simple geometric interpretation of the eccentric
anomaly, E. Consider Figure 3.5. Let PGA represent the elliptical orbit of a planet,
G, about the Sun, S. Let ACP be the major axis of the orbit, where P is the perihelion
point, A the aphelion point, and C the geometric center. It follows that $CA = CP = a$ and
$CS = e\,a$ (see Section A.9), where a is the orbital major radius and e the eccentricity.
Moreover, the distance SG and the angle GSP correspond to the radial distance, r, and
the true anomaly, θ, respectively. Let PRA be a circle of radius a centered on C. It
follows that AP is a diameter of this circle. Let RGQ be a line, perpendicular to AP,
that passes through G and joins the circle to the diameter. It follows that $CR = a$. Let
us denote the angle RCS as E. Simple trigonometry reveals that $SQ = r \cos \theta$ and
$CQ = a \cos E$. But $CQ = CS + SQ$, or $a \cos E = e\,a + r \cos \theta$, which can be rearranged
to give $r \cos \theta = a (\cos E - e)$, which is identical to Equation (3.62). We thus conclude
that the eccentric anomaly, E, can be identified with the angle RCS in Figure 3.5.

Equations (3.51) and (3.55) can be combined to give

$$\cos\theta = \frac{\cos E - e}{1 - e\,\cos E}.$$ (3.63)

Thus,

$$1 + \cos\theta = 2\,\cos^2(\theta/2) = \frac{2\,(1 - e)\,\cos^2(E/2)}{1 - e\,\cos E},$$ (3.64)

and

$$1 - \cos\theta = 2\,\sin^2(\theta/2) = \frac{2\,(1 + e)\,\sin^2(E/2)}{1 - e\,\cos E}.$$ (3.65)

The previous two equations imply that

$$\tan(\theta/2) = \left(\frac{1 + e}{1 - e}\right)^{1/2}\tan(E/2).$$ (3.66)

The eccentric anomaly, E, and the true anomaly, θ, always lie in the same quadrant (i.e., if $0 \le E \le \pi/2$, then $0 \le \theta \le \pi/2$, etc.) We conclude that in the case of a planet in an elliptical orbit around the Sun, the radial distance, r, and the true anomaly, θ, are specified as functions of time via the solution of the following set of equations:

$$\mathcal{M} = n\,(t - \tau),$$ (3.67)

$$E - e\,\sin E = \mathcal{M},$$ (3.68)

$$r = a\,(1 - e\,\cos E),$$ (3.69)

and

$$\tan(\theta/2) = \left(\frac{1 + e}{1 - e}\right)^{1/2}\tan(E/2).$$ (3.70)

Here, $n = 2\pi/T$, $T = 2\pi\,(a^3/G\,M)^{1/2}$, and $a = r_p/(1 - e)$. Incidentally, it is clear that if $t \to t + T$, then $\mathcal{M} \to \mathcal{M} + 2\pi$, $E \to E + 2\pi$, and $\theta \to \theta + 2\pi$. In other words, the motion is periodic with period T.

3.12 Orbital elements

The previous analysis suffices when considering a single planet orbiting around the Sun. However, it becomes inadequate when dealing with multiple planets whose orbital planes and perihelion directions do not necessarily coincide. Incidentally, for the time being, we are neglecting interplanetary gravitational interactions, which allows us to assume that each planet executes an independent Keplerian elliptical orbit about the Sun.

Let us characterize all planetary orbits using a common Cartesian coordinate system X, Y, Z, centered on the Sun. (See Figure 3.6.) The X–Y plane defines a *reference plane*, which is chosen to be the *ecliptic plane* (i.e., the plane of the Earth's orbit), with the Z-axis pointing toward the ecliptic north pole (i.e., the direction normal to the ecliptic plane in a northward sense). Likewise, the X-axis defines a *reference direction*, which is chosen to point in the direction of the vernal equinox (i.e., the point in the Earth's

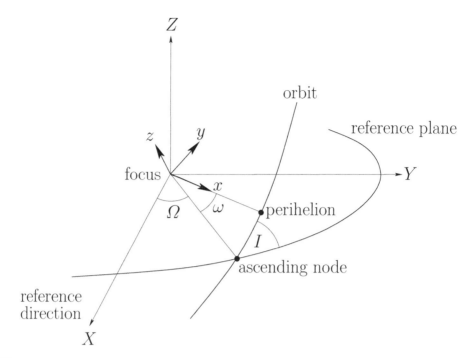

Fig. 3.6 A general planetary orbit.

sky at which the apparent orbit of the Sun passes through the extension of the Earth's
equatorial plane from south to north). Suppose that the plane of a given planetary orbit
is inclined at an angle I to the reference plane. The point at which this orbit crosses the
reference plane in the direction of increasing Z is termed its *ascending node*. The angle
Ω subtended between the reference direction and the direction of the ascending node is
termed the *longitude of the ascending node*. Finally, the angle, ω, subtended between
the direction of the ascending node and the direction of the orbit's perihelion, is termed
the *argument of the perihelion*.

Let us define a second Cartesian coordinate system x, y, z, also centered on the Sun.
Let the x–y plane coincide with the plane of a particular planetary orbit, and let the x-
axis point toward the orbit's perihelion point. Clearly, we can transform from the x, y,
z system to the X, Y, Z system via a series of three rotations of the coordinate system:
first, a rotation through an angle ω about the z-axis (looking down the axis); second, a
rotation through an angle I about the new x-axis; and finally, a rotation through an angle
Ω about the new z-axis. It thus follows from standard coordinate transformation theory
(see Section A.6) that

$$\begin{pmatrix} X \\ Y \\ Z \end{pmatrix} = \begin{pmatrix} \cos\Omega & -\sin\Omega & 0 \\ \sin\Omega & \cos\Omega & 0 \\ 0 & 0 & 1 \end{pmatrix} \begin{pmatrix} 1 & 0 & 0 \\ 0 & \cos I & -\sin I \\ 0 & \sin I & \cos I \end{pmatrix} \begin{pmatrix} \cos\omega & -\sin\omega & 0 \\ \sin\omega & \cos\omega & 0 \\ 0 & 0 & 1 \end{pmatrix} \begin{pmatrix} x \\ y \\ z \end{pmatrix}.$$

$$(3.71)$$

Table 3.1 Planetary data for J2000								
Planet	a(AU)	$\bar{\lambda}_0(°)$	e	$I(°)$	$\varpi(°)$	$\Omega(°)$	T(yr)	m/M
Mercury	0.3871	252.25	0.20564	7.006	77.46	48.34	0.241	1.659×10^{-7}
Venus	0.7233	181.98	0.00676	3.398	131.77	76.67	0.615	2.447×10^{-6}
Earth	1.0000	100.47	0.01673	0.000	102.93	–	1.000	3.039×10^{-6}
Mars	1.5237	355.43	0.09337	1.852	336.08	49.71	1.881	3.226×10^{-7}
Jupiter	5.2025	34.33	0.04854	1.299	14.27	100.29	11.87	9.542×10^{-4}
Saturn	9.5415	50.08	0.05551	2.494	92.86	113.64	29.47	2.857×10^{-4}
Uranus	19.188	314.20	0.04686	0.773	172.43	73.96	84.05	4.353×10^{-5}
Neptune	30.070	304.22	0.00895	1.770	46.68	131.79	164.9	5.165×10^{-5}

a – major radius; $\bar{\lambda}_0$ – mean longitude at epoch; e – eccentricity; I – inclination to ecliptic; ϖ – longitude of perihelion; Ω – longitude of ascending node; T – orbital period; m/M – planetary mass / solar mass.
[a] Source: Standish and Williams 1992.

However, $x = r \cos\theta$, $y = r \sin\theta$, and $z = 0$. Hence,

$$X = r \left[\cos\Omega \cos(\omega + \theta) - \sin\Omega \sin(\omega + \theta) \cos I\right], \tag{3.72}$$

$$Y = r \left[\sin\Omega \cos(\omega + \theta) + \cos\Omega \sin(\omega + \theta) \cos I\right], \tag{3.73}$$

$$Z = r \sin(\omega + \theta) \sin I. \tag{3.74}$$

Thus, a general planetary orbit is determined by Equations (3.67)–(3.70) and (3.72)–(3.74) and is therefore parameterized by six orbital elements: the major radius, a; the time of perihelion passage, τ; the eccentricity, e; the inclination (to the ecliptic plane), I; the argument of the perihelion, ω; and the longitude of the ascending node, Ω. [The mean orbital angular velocity, in radians per year, is $n = 2\pi/a^{3/2}$, where a is measured in astronomical units. Here, an astronomical unit is the mean Earth–Sun distance, and corresponds to 1.496×10^{11} m (Yoder 1995).]

In low-inclination orbits, the argument of the perihelion is usually replaced by

$$\varpi = \Omega + \omega, \tag{3.75}$$

which is termed the *longitude of the perihelion*. Likewise, the time of perihelion passage, τ, is often replaced by the mean longitude at $t = 0$—otherwise known as the *mean longitude at epoch*—where the mean longitude is defined as

$$\bar{\lambda} = \varpi + \mathcal{M} = \varpi + n(t - \tau). \tag{3.76}$$

Thus, if $\bar{\lambda}_0$ denotes the mean longitude at epoch ($t = 0$), then

$$\bar{\lambda} = \bar{\lambda}_0 + n t, \tag{3.77}$$

where $\bar{\lambda}_0 = \varpi - n\tau$. The orbital elements of the major planets at the epoch J2000 (i.e., at 00:00 UT on January 1, 2000) are given in Table 3.1.

The *heliocentric* (i.e., as seen from the Sun) position of a planet is most conveniently expressed in terms of its ecliptic longitude, λ, and ecliptic latitude, β. This type of longitude and latitude is referred to the ecliptic plane, with the Sun as the origin. Moreover, the vernal equinox is defined to be the zero of longitude. It follows that

$$\tan \lambda = \frac{Y}{X} \tag{3.78}$$

and

$$\sin \beta = \frac{Z}{\sqrt{X^2 + Y^2}}, \tag{3.79}$$

where (X, Y, Z) are the heliocentric Cartesian coordinates of the planet.

3.13 Planetary orbits

According to Table 3.1, the planets all have low-inclination orbits characterized by $I \ll 1$ (when I is expressed in radians). In this case, making use of the small angle approximations $\cos I \simeq 1$ and $\sin I \simeq I$, as well as some trigonometric identities (see Section A.3), we find that Equations (3.72)–(3.74) simplify to give

$$X \simeq r \cos(\theta + \varpi), \tag{3.80}$$

$$Y \simeq r \sin(\theta + \varpi), \tag{3.81}$$

and

$$Z \simeq r \, I \, \sin(\theta + \varpi - \Omega), \tag{3.82}$$

where we have made use of Equation (3.75). It thus follows from Equations (3.78) and (3.79) that

$$\lambda \simeq \theta + \varpi \tag{3.83}$$

and

$$\beta \simeq I \sin(\theta + \varpi - \Omega). \tag{3.84}$$

According to Table 3.1, the planets also have *low-eccentricity* orbits, characterized by $0 < e \ll 1$. In this situation, Equations (3.68)–(3.70) can be usefully solved via series expansion in e to give

$$\theta = \mathcal{M} + 2e \sin \mathcal{M} + \frac{5e^2}{4} \sin 2\mathcal{M} + \frac{e^3}{12}(13 \sin 3\mathcal{M} - 3 \sin \mathcal{M}) + \mathcal{O}(e^4), \tag{3.85}$$

$$\frac{r}{a} = 1 - e \cos \mathcal{M} + \frac{e^2}{2}(1 - \cos 2\mathcal{M}) + \frac{3e^3}{8}(\cos \mathcal{M} - \cos 3\mathcal{M}) + \mathcal{O}(e^4). \tag{3.86}$$

(See Section A.10.)

The preceding expressions can be combined with Equations (3.67), (3.77), (3.83), and (3.84) to produce

$$n = \frac{2\pi}{a^{3/2}}, \tag{3.87}$$

$$\bar{\lambda} = \bar{\lambda}_0 + n\,t, \tag{3.88}$$

$$\frac{r}{a} \simeq 1 - e\,\cos(\bar{\lambda} - \varpi), \tag{3.89}$$

$$\lambda \simeq \bar{\lambda} + 2\,e\,\sin(\bar{\lambda} - \varpi), \tag{3.90}$$

and

$$\beta \simeq I\,\sin(\bar{\lambda} - \Omega). \tag{3.91}$$

Here, n is expressed in radians per year, and a in astronomical units. These equations, which are valid up to first order in small quantities (i.e., e and I), illustrate how a planet's six orbital elements—a, $\bar{\lambda}_0$, e, I, ϖ, and Ω—can be used to determine its approximate position relative to the Sun as a function of time. The planet reaches its perihelion point when the mean ecliptic longitude, $\bar{\lambda}$, becomes equal to the longitude of the perihelion, ϖ. Likewise, the planet reaches its aphelion point when $\bar{\lambda} = \varpi + \pi$. Furthermore, the ascending node corresponds to $\bar{\lambda} = \Omega$, and the point of furthest angular distance north of the ecliptic plane (at which $\beta = I$) corresponds to $\bar{\lambda} = \Omega + \pi/2$.

Consider the Earth's orbit about the Sun. As has already been mentioned, ecliptic longitude is measured relative to a point on the ecliptic circle—the circular path that the Sun appears to trace out against the backdrop of the stars—known as the *vernal equinox*. When the Sun reaches the vernal equinox, which it does every year on about March 20, day and night are equally long everywhere on the Earth (because the Sun lies in the Earth's equatorial plane). Likewise, when the Sun reaches the opposite point on the ecliptic circle, known as the *autumnal equinox*, which it does every year on about September 22, day and night are again equally long everywhere on the Earth. The points on the ecliptic circle halfway between the equinoxes are known as the *solstices*. When the Sun reaches the *summer solstice*, which it does every year on about June 21, this marks the longest day in the Earth's northern hemisphere and the shortest day in the southern hemisphere. Likewise, when the Sun reaches the *winter solstice*, which it does every year on about December 21, this marks the shortest day in the Earth's northern hemisphere and the longest day in the southern hemisphere. The period between (the Sun reaching) the vernal equinox and the summer solstice is known as *spring*, that between the summer solstice and the autumnal equinox as *summer*, that between the autumnal equinox and the winter solstice as *autumn*, and that between the winter solstice and the next vernal equinox as *winter*.

Let us calculate the approximate lengths of the seasons. It follows, from the preceding discussion, that the ecliptic longitudes of the Sun, relative to the Earth, at the (times at which the Sun reaches the) vernal equinox, summer solstice, autumnal equinox, and winter solstice are $0°$, $90°$, $180°$, and $270°$, respectively. Hence, the ecliptic longitudes, λ, of the Earth, relative to the Sun, at the same times are $180°$, $270°$, $0°$, and $90°$, respectively. The mean longitude, $\bar{\lambda}$, of the Earth increases *uniformly* in time at the rate of $360°$ per year. Thus, the length of a given season is simply the fraction $\Delta\bar{\lambda}/360°$ of a year, where $\Delta\bar{\lambda}$ is the change in mean longitude associated with the season. Equation (3.90) can be inverted to give

$$\bar{\lambda} \simeq \lambda - 2\,e\,\sin(\lambda - \varpi), \tag{3.92}$$

to first order in e. Hence, the mean longitudes associated with the autumnal equinox, winter solstice, vernal equinox, and summer solstice are

$$\bar{\lambda}_{AE} \simeq 0° + 2\,e\,\sin\varpi = 1.87°, \tag{3.93}$$

$$\bar{\lambda}_{WS} \simeq 90° - 2\,e\,\cos\varpi = 90.43°, \tag{3.94}$$

$$\bar{\lambda}_{VE} \simeq 180° - 2\,e\,\sin\varpi = 178.13°, \tag{3.95}$$

and

$$\bar{\lambda}_{SS} \simeq 270° + 2\,e\,\cos\varpi = 269.57°, \tag{3.96}$$

respectively. (Recall that, according to Table 3.1, $e = 0.01673$ radians and $\varpi = 102.93°$ for the Earth.) Thus,

$$\Delta\bar{\lambda}_{\text{spring}} \simeq 91.44°, \tag{3.97}$$

$$\Delta\bar{\lambda}_{\text{summer}} \simeq 92.30°, \tag{3.98}$$

$$\Delta\bar{\lambda}_{\text{autumn}} \simeq 88.56°, \tag{3.99}$$

and

$$\Delta\bar{\lambda}_{\text{winter}} \simeq 87.70°. \tag{3.100}$$

(See Figure 3.7.) Given that the length of a tropical year (i.e., the mean period between successive vernal equinoxes) is 365.24 days, we deduce that spring, summer, autumn, and winter last 92.8, 93.6, 89.8, and 89.0 days, respectively. Clearly, although the deviations of the Earth's orbit from a uniform circular orbit that is concentric with the Sun seem relatively small, they are still large enough to cause a noticeable difference between the lengths of the various seasons. The preceding calculation was used, in reverse, by ancient Greek astronomers, such as Hipparchus, to determine the eccentricity, and the longitude of the perigee, of the Sun's apparent orbit about the Earth from the observed lengths of the seasons (Evans 1998).

3.14 Parabolic orbits

For the case of a *parabolic* orbit about the Sun, characterized by $e = 1$, similar analysis to that in Section 3.11 yields

$$P + \frac{P^3}{3} = \left(\frac{G\,M}{2\,r_p^3}\right)^{1/2} (t - \tau), \tag{3.101}$$

$$r = r_p\,(1 + P^2), \tag{3.102}$$

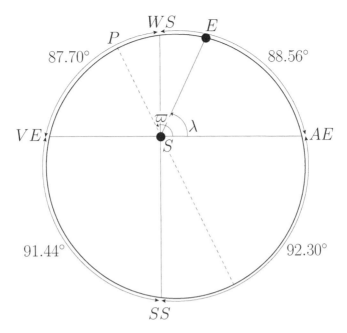

Fig. 3.7 A schematic diagram showing the orbit of the Earth, E, about the Sun, S, as well as the vernal equinox (VE), summer solstice (SS), autumnal equinox (AE), and winter solstice (WS). Here, λ is the Earth's ecliptic longitude, and ϖ the longitude of its perihelion (P).

and

$$\tan(\theta/2) = P. \tag{3.103}$$

Here, P is termed the *parabolic anomaly* and varies between $-\infty$ and $+\infty$, with the perihelion point corresponding to $P = 0$. Note that Equation (3.101) is a cubic equation, possessing a single real root, which can, in principle, be solved analytically. (See Exercise 3.18.) However, a numerical solution is generally more convenient.

3.15 Hyperbolic orbits

For the case of a *hyperbolic* orbit about the Sun, characterized by $e > 1$, similar analysis to that in Section 3.11 gives

$$e \sinh H - H = \left(\frac{G M}{a^3}\right)^{1/2} (t - \tau), \tag{3.104}$$

$$r = a\,(e \cosh H - 1), \tag{3.105}$$

and

$$\tan(\theta/2) = \left(\frac{e + 1}{e - 1}\right)^{1/2} \tanh(H/2). \tag{3.106}$$

Here, H is termed the *hyperbolic anomaly* and varies between $-\infty$ and $+\infty$, with the perihelion point corresponding to $H = 0$. Moreover, $a = r_p/(e - 1)$. As in the elliptical case, Equation (3.104) is a transcendental equation that is most easily solved numerically.

3.16 Binary star systems

Approximately half the stars in our galaxy are members of so-called *binary star systems*. Such systems consist of two stars, of mass m_1 and m_2, and position vectors $\mathbf{r}_1$ and $\mathbf{r}_2$, respectively, orbiting about their common center of mass. The distance separating the stars is generally much less than the distance to the nearest neighbor star. Hence, a binary star system can be treated as a two-body dynamical system to a very good approximation.

In a binary star system, the gravitational force that the first star exerts on the second is

$$\mathbf{f} = -\frac{G\,m_1\,m_2}{r^3}\,\mathbf{r},\tag{3.107}$$

where $\mathbf{r} = \mathbf{r}_2 - \mathbf{r}_1$. As we have seen in Section 1.9, a two-body system can be reduced to an equivalent one-body system whose equation of motion is of the form given in Equation (1.76), where $\mu = m_1\,m_2/(m_1 + m_2)$. Hence, in this particular case, we can write

$$\frac{m_1\,m_2}{m_1 + m_2}\frac{d^2\mathbf{r}}{dt^2} = -\frac{G\,m_1\,m_2}{r^3}\,\mathbf{r},\tag{3.108}$$

which gives

$$\frac{d^2\mathbf{r}}{dt^2} = -\frac{G\,M}{r^3}\,\mathbf{r},\tag{3.109}$$

where

$$M = m_1 + m_2.\tag{3.110}$$

Equation (3.109) is identical to Equation (3.2), which we have already solved. Hence, we can immediately write down the solution:

$$\mathbf{r} = (r\,\cos\theta,\ r\,\sin\theta,\ 0),\tag{3.111}$$

where

$$r = \frac{a\,(1 - e^2)}{1 + e\,\cos\theta}\tag{3.112}$$

and

$$\frac{d\theta}{dt} = \frac{h}{r^2},\tag{3.113}$$

with

$$a = \frac{h^2}{(1 - e^2)\,G\,M}.\tag{3.114}$$

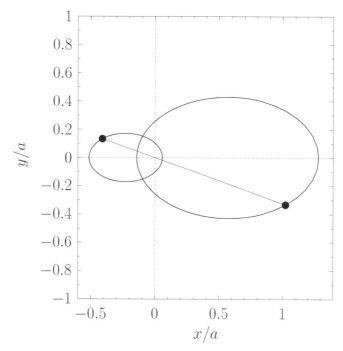

Fig. 3.8 An example binary star orbit.

Here, h is a constant, and we have aligned our Cartesian axes such that the plane of the orbit coincides with the x–y plane. According to this solution, the second star executes a Keplerian elliptical orbit, with major radius a and eccentricity e, relative to the first star, and vice versa. From Equation (3.33), the period of revolution, T, is given by

$$T = \sqrt{\frac{4\pi^2 a^3}{G M}}. \tag{3.115}$$

Moreover, if $n = 2\pi/T$, then

$$n = \frac{\sqrt{G M}}{a^{3/2}}. \tag{3.116}$$

In the *inertial* frame of reference whose origin always coincides with the center of mass—the so-called *center of mass frame*—the position vectors of the two stars are

$$\mathbf{r}_1 = -\frac{m_2}{m_1 + m_2}\,\mathbf{r} \tag{3.117}$$

and

$$\mathbf{r}_2 = \frac{m_1}{m_1 + m_2}\,\mathbf{r}, \tag{3.118}$$

where $\mathbf{r}$ is specified in Equation (3.111). Figure 3.8 shows an example binary star orbit in the center of mass frame, calculated with $m_1/m_2 = 0.4$ and $e = 0.8$. It can be seen that both stars execute elliptical orbits about their common center of mass and, at any point in time, are diagrammatically opposite one another, relative to the origin.

Binary star systems have been very useful to astronomers, as it is possible to determine the masses of both stars in such a system by careful observation. The sum of the masses of the two stars, $M = m_1 + m_2$, can be found from Equation (3.115) after a measurement of the major radius, a (which is the mean of the greatest and smallest distance apart of the two stars during their orbit), and the orbital period, T. The ratio of the masses of the two stars, m_1/m_2, can be determined from Equations (3.117) and (3.118) by observing the fixed ratio of the relative distances of the two stars from the common center of mass about which they both appear to rotate. Obviously, given the sum of the masses and the ratio of the masses, the individual masses themselves can then be calculated.

Exercises

3.1 Demonstrate that if a particle moves in a central force field with zero angular momentum, so that $\mathbf{h} \equiv \mathbf{r} \times \dot{\mathbf{r}} = \mathbf{0}$, then the particle's trajectory lies on a fixed straight line that passes through the origin. [Hint: Show that $d/dt(\mathbf{r}/r) = \mathbf{0}$.]

3.2 Demonstrate that $h = r^2 \dot{\theta}$ is the magnitude of the angular momentum (per unit mass) vector $\mathbf{h} = \mathbf{r} \times \dot{\mathbf{r}}$. Here, r and θ are plane polar coordinates.

3.3 Consider a planet in a Keplerian elliptical orbit about the Sun. Let $\mathbf{r}$ be the planet's position vector, relative to the Sun, and let $\mathbf{h} = \mathbf{r} \times \dot{\mathbf{r}}$ be its angular momentum per unit mass. Demonstrate that the so-called *Laplace-Runge-Lenz* vector,

$$\mathbf{l} = \frac{\dot{\mathbf{r}} \times \mathbf{h}}{GM} - \frac{\mathbf{r}}{r},$$

can be written

$$\mathbf{l} = e \cos\theta \, \mathbf{e}_r - e \sin\theta \, \mathbf{e}_\theta,$$

where M is the solar mass, e the orbital eccentricity, and r, θ are plane polar coordinates in the orbital plane (with the perihelion corresponding to $\theta = 0$). Hence, show that $\mathbf{l}$ is a *constant* vector, of length e, that is directed from the Sun toward the perihelion point.

3.4 Given the Sun's mean apparent radius seen from the Earth (16′), the Earth's mean apparent radius seen from the Moon (57′), and the mean number of lunar revolutions in a year (13.4), show that the ratio of the Sun's mean density to that of the Earth is 0.252. (From Lamb 1923.)

3.5 Prove that the orbital period of a satellite close to the surface of a spherical planet depends on the mean density of the planet, but not on its size. Show that if the mean density is that of water, the period is 3 h. 18 m. (From Lamb 1923.)

3.6 Jupiter's satellite Ganymede has an orbital period of 7 d. 3 h. 43 m. and a mean orbital radius that is 15.3 times the mean radius of the planet. The Moon has an orbital period of 27 d. 7 h. 43 m. and a mean orbital radius that is 60.3 times the Earth's mean radius. Show that the ratio of Jupiter's mean density to that of the Earth is 0.238. (From Lamb 1923.)

3.7   Halley's comet has an orbital eccentricity of 0.967 and a perihelion distance of 55,000,000 miles. Find the orbital period and the comet's speed at perihelion and aphelion.

3.8   Show that the velocity at any point on a Keplerian elliptical orbit can be resolved into two constant components: a velocity $n\,a/\sqrt{1-e^2}$ at right angles to the radius vector, and a velocity $n\,a\,e/\sqrt{1-e^2}$ at right angles to the major axis. Here, n is the mean orbital angular velocity, a the major radius, and e the eccentricity. (From Lamb 1923.)

3.9   The *latus rectum* of a conic section is a chord that passes through a focus; it is perpendicular to the major axis (or the symmetry axis, in the case of a parabola or a hyperbola). Show that, for a body in a Keplerian orbit around the Sun, the maximum value of the radial speed occurs at the points where the latus rectum (associated with the non-empty focus, in the case of an ellipse) intersects the orbit, and that this maximum value is $e\,h/[r_p\,(1+e)]$. Here, h is the angular momentum per unit mass, e the orbital eccentricity, and r_p the perihelion distance.

3.10   A comet is observed a distance R astronomical units from the Sun, traveling at a speed that is V times the Earth's mean orbital speed. Show that the orbit of the comet is hyperbolic, parabolic, or elliptical, depending on whether the quantity $V^2 R$ is greater than, equal to, or less than 2, respectively. (Modified from Fowles and Cassiday 2005.)

3.11   Consider a planet in an elliptical orbit of major radius a and eccentricity e about the Sun. Suppose that the eccentricity of the orbit is small ($0 < e \ll 1$), as is indeed the case for all of the planets. Demonstrate that, to first order in e, the orbit can be approximated as a circle whose center is shifted a distance $e\,a$ from the Sun, and that the planet's angular motion appears uniform when viewed from a point (called the equant) that is shifted a distance $2\,e\,a$ from the Sun, in the same direction as the center of the circle. [This theorem is the basis of Ptolemy's model of planetary motion (Evans 1998).]

3.12   How long (in days) does it take the Sun–Earth radius vector to rotate through $90°$, starting at the perihelion point? How long does it take starting at the aphelion point? The period and eccentricity of the Earth's orbit are $T = 365.24$ days and $e = 0.01673$ radians, respectively.

3.13   If θ is the Sun's ecliptic longitude, measured from the perigee (the point of closest approach to the Earth), show that the Sun's apparent diameter is given by

$$D \simeq D_1 \cos^2(\theta/2) + D_2 \sin^2(\theta/2),$$

where D_1 and D_2 are the greatest and least values of D. (From Lamb 1923.)

3.14   Show that the time-averaged apparent diameter of the Sun, as seen from a planet describing a low-eccentricity elliptical orbit, is approximately equal to the apparent diameter when the planet's distance from the Sun equals the major radius of the orbit. (From Lamb 1923.)

3.15   Consider an asteroid orbiting the Sun. Demonstrate that, at fixed orbital energy, the orbit that maximizes the orbital angular momentum is circular.

3.16   Derive Equations (3.101)–(3.103).

3.17   Derive Equations (3.104)–(3.106).

3.18 A parabolic Keplerian orbit is specified by Equation (3.101), which can be written

$$P + \frac{P^3}{3} = \mathcal{M},$$

where P is the parabolic anomaly and $\mathcal{M} = (G\,M/2\,r_p^3)^{1/2}\,(t - \tau)$ is termed the *parabolic mean anomaly*. Here, M is the solar mass, r_p the perihelion distance, and τ the time of perihelion passage. Demonstrate that the preceding equation has the analytic solution

$$P = \frac{1}{2}\,Q^{1/3} - 2\,Q^{-1/3},$$

where

$$Q = 12\,\mathcal{M} + 4\,\sqrt{4 + 9\,\mathcal{M}^2}.$$

3.19 Consider a comet in an elliptical orbit about the Sun. Let x and y be Cartesian coordinates in the orbital plane, such that $x = y = 0$ corresponds to the Sun and the x-axis is parallel to the orbital major axis. Demonstrate that

$$x = a\,(\cos E - e)$$

and

$$y = a\,(1 - e^2)^{1/2}\,\sin E,$$

where a is the orbital major radius, e the eccentricity, and E the eccentric anomaly.

3.20 Consider a comet in a parabolic orbit about the Sun. Let x and y be Cartesian coordinates in the orbital plane, such that $x = y = 0$ corresponds to the Sun and the x-axis is parallel to the orbital symmetry axis. Demonstrate that

$$x = r_p\,(1 - P^2),$$
$$y = 2\,r_p\,P,$$

where r_p is the perihelion distance and P the parabolic anomaly.

3.21 Consider a comet in an hyperbolic orbit about the Sun. Let x and y be Cartesian coordinates in the orbital plane, such that $x = y = 0$ corresponds to the Sun and the x-axis is parallel to the orbital symmetry axis. Demonstrate that

$$x = a\,(e - \cosh H),$$
$$y = a\,(e^2 - 1)^{1/2}\,\sinh H,$$

where a is the orbital major radius, e the eccentricity, and H the hyperbolic anomaly.

3.22 Consider a comet in an elliptical orbit about the Sun (*Lambert's Theorem*). If r_1 and r_2 are the radial distances from the Sun of two neighboring points, C_1 and C_2, on the orbit, and if s is the length of the straight line joining these two points, prove that the time, t, required for the comet to move from C_1 to C_2 is

$$\Delta t = \frac{T}{2\pi}\,[(\eta - \sin\eta) - (\xi - \sin\xi)],$$

where

$$\sin(\eta/2) = \frac{1}{2}\left(\frac{r_1 + r_2 + s}{a}\right)^{1/2},$$

$$\sin(\xi/2) = \frac{1}{2}\left(\frac{r_1 + r_2 - s}{a}\right)^{1/2}.$$

Here, T and a are the period and the major radius of the orbit, respectively.

3.23 Consider a comet in a parabolic orbit about the Sun (*Euler's Theorem*). If r_1 and r_2 are the radial distances from the Sun of two neighboring points, C_1 and C_2, on the orbit, and if s is the length of the straight line joining these two points, prove that the time required for the comet to move from C_1 to C_2 is

$$\Delta t = \frac{T}{12\pi}\left[\left(\frac{r_1 + r_2 + s}{a}\right)^{3/2} - \left(\frac{r_1 + r_2 - s}{a}\right)^{3/2}\right],$$

where T and a are the period and the major radius of the Earth's orbit, respectively.

3.24 Consider a comet in a hyperbolic orbit about the Sun. If r_1 and r_2 are the radial distances from the Sun of two neighboring points, C_1 and C_2, on the orbit, and if s is the length of the straight line joining these two points, prove that the time, t, required for the comet to move from C_1 to C_2 is

$$\Delta t = \frac{T}{2\pi}\left[(\sinh\eta - \eta) - (\sinh\xi - \xi)\right],$$

where

$$\sinh(\eta/2) = \frac{1}{2}\left(\frac{r_1 + r_2 + s}{a}\right)^{1/2}$$

and

$$\sinh(\xi/2) = \frac{1}{2}\left(\frac{r_1 + r_2 - s}{a}\right)^{1/2}.$$

Here, a is major radius of the orbit and T is the period of an elliptical orbit with the same major radius. (From Smart 1951.)

3.25 A comet is in a parabolic orbit that lies in the plane of the Earth's orbit. Regarding the Earth's orbit as a circle of radius a, show that the points at which the comet intersects the Earth's orbit are given by

$$\cos\theta = -1 + \frac{2r_p}{a},$$

where r_p is the perihelion distance of the comet, defined at $\theta = 0$. Demonstrate that the time interval that the comet remains inside the Earth's orbit is the fraction

$$\frac{2^{1/2}}{3\pi}\left(\frac{2r_p}{a} + 1\right)\left(1 - \frac{r_p}{a}\right)^{1/2}$$

of a year, and that the maximum value of this time interval is $2/3\pi$ year, or about 11 weeks.

3.26 The orbit of a comet around the Sun is a hyperbola of eccentricity e, lying in the ecliptic plane, whose least distance from the Sun is $1/n$ times the radius of the Earth's orbit (which is approximated as a circle). Prove that the time that the comet remains within the Earth's orbit is $(T/\pi)\,(e\,\sinh\phi - \phi)$, where $e\,\cosh\phi = 1 - n\,(1 - e)$, and T is the periodic time of a planet describing an elliptic orbit whose major radius is equal to that of the hyperbolic orbit. (From Smart 1951.)

3.27 Consider a comet in a hyperbolic orbit focused on the Sun. The *impact parameter*, b, is defined as the the distance of closest approach in the absence of any gravitational attraction between the comet and the Sun. Demonstrate that $b = h/(2\,\mathcal{E})^{1/2}$, where h is the comet's angular momentum per unit mass and $\mathcal{E}$ its energy per unit mass. Show that the relationship between the impact parameter, b, and the true distance of closest approach, r_p, is

$$r_p = \frac{2\,b}{\alpha + \sqrt{\alpha^2 + 4}},$$

where $\alpha = G\,M/(\mathcal{E}\,b)$ and M is the solar mass. Hence, deduce that if the comet is to avoid hitting the Sun, then

$$\alpha < \frac{b}{R} - \frac{R}{b}$$

(assuming that $b > R$), where R is the solar radius.

3.28 Spectroscopic analysis has revealed that Spica is a double star whose components revolve around one another with a period of 4.1 days, the greatest relative orbital velocity being 36 miles per second. Show that the mean distance between the components of the star is 2.03×10^6 miles, and that the total mass of the system is 0.083 times that of the Sun. The mean distance of the Earth from the Sun is 92.75 million miles. (From Lamb 1923.)

4   Orbits in central force fields

4.1   Introduction

This chapter examines the motion of a celestial body in a general central potential—that is, a gravitational potential that is a function of the radial coordinate, r, only, but does not necessarily vary as $1/r$.

4.2   Motion in a general central force field

Consider the motion of an object in a general (attractive) central force field characterized by the potential energy per unit mass function $V(r)$. Because the force field is central, it still remains true that

$$h = r^2 \dot{\theta} \tag{4.1}$$

is a constant of the motion. (See Section 3.5.) As is easily demonstrated, Equation (3.28) generalizes to

$$\frac{d^2 u}{d\theta^2} + u = -\frac{1}{h^2} \frac{dV}{du}, \tag{4.2}$$

where $u = r^{-1}$.

Suppose, for instance, that we wish to find the potential $V(r)$ that causes an object to execute the spiral orbit

$$r = r_0 \, \theta^2. \tag{4.3}$$

Substitution of $u = (r_0 \, \theta^2)^{-1}$ into Equation (4.2) yields

$$\frac{dV}{du} = -h^2 \left(6 \, r_0 \, u^2 + u \right). \tag{4.4}$$

Integrating, we obtain

$$V(u) = -h^2 \left(2 \, r_0 \, u^3 + \frac{u^2}{2} \right) \tag{4.5}$$

or

$$V(r) = -h^2 \left(\frac{2 \, r_0}{r^3} + \frac{1}{2 \, r^2} \right). \tag{4.6}$$

In other words, the spiral orbit specified by Equation (4.3) is obtained from a mixture of an inverse-square and inverse-cube potential.

4.3 Motion in a nearly circular orbit

In principle, a circular orbit is a possible orbit for any attractive central force. However, not all such forces result in *stable* circular orbits. Let us now consider the stability of circular orbits in a general central force field. Equation (3.25) generalizes to

$$\ddot{r} - \frac{h^2}{r^3} = f(r), \tag{4.7}$$

where $f(r) = -dV/dr$ is the radial force per unit mass. For a circular orbit, $\ddot{r} = 0$ and the above equation reduces to

$$-\frac{h^2}{r_c^3} = f(r_c), \tag{4.8}$$

where r_c is the radius of the orbit.

Let us now consider *small* departures from circularity. Suppose that

$$x = r - r_c. \tag{4.9}$$

Equation (4.7) can be written

$$\ddot{x} - \frac{h^2}{(r_c + x)^3} = f(r_c + x). \tag{4.10}$$

Expanding the two terms involving $r_c + x$ as power series in x/r_c, and keeping all terms up to first order, we obtain

$$\ddot{x} - \frac{h^2}{r_c^3}\left(1 - 3\,\frac{x}{r_c}\right) = f(r_c) + f'(r_c)\,x, \tag{4.11}$$

where $'$ denotes a derivative. Making use of Equation (4.8), we find that the preceding equation reduces to

$$\ddot{x} + \left[-\frac{3\,f(r_c)}{r_c} - f'(r_c)\right]x = 0. \tag{4.12}$$

If the term in square brackets is positive, then we obtain a simple harmonic equation, which we already know has bounded solutions (see Section 1.8)—that is, the orbit is stable to small perturbations. On the other hand, if the term in square brackets is negative, then we obtain an equation whose solutions grow exponentially in time (see Section 1.8)—that is, the orbit is *unstable* to small perturbations. Thus, the stability criterion for a circular orbit of radius r_c in a central force field characterized by a radial force (per unit mass) function $f(r)$ is

$$f(r_c) + \frac{r_c}{3}\,f'(r_c) < 0. \tag{4.13}$$

For example, consider an attractive power-law force function of the form

$$f(r) = -c\,r^n, \tag{4.14}$$

where $c > 0$. Substituting into the preceding stability criterion, we obtain

$$-c\,r_c^n - \frac{c\,n}{3}\,r_c^n < 0 \tag{4.15}$$

or

$$n > -3. \tag{4.16}$$

We conclude that circular orbits in attractive central force fields that decay faster than r^{-3} are unstable. The case $n = -3$ is special, because the first-order terms in the expansion of Equation (4.10) cancel out exactly, and it is necessary to retain the second-order terms. Doing this, we can easily demonstrate that circular orbits are also unstable for inverse-cube ($n = -3$) forces. (See Exercise 4.9.)

An *apsis* (plural, *apsides*) is a point on an orbit at which the radial distance, r, assumes either a maximum or a minimum value. Thus, the perihelion and aphelion points are the apsides of planetary orbits. The angle through which the radius vector rotates in going between two consecutive apsides is called the *apsidal angle*. Hence, the apsidal angle for elliptical orbits in an inverse-square force field is π.

For the case of stable, nearly circular orbits, we have seen that r oscillates sinusoidally about its mean value, r_c. Indeed, it is clear from Equation (4.12) that the period of the oscillation is

$$T = \frac{2\pi}{[-3\,f(r_c)/r_c - f'(r_c)]^{1/2}}. \tag{4.17}$$

The apsidal angle is the amount by which θ increases in going between a maximum and a minimum of r. The time taken to achieve this is clearly $T/2$. Now, $\dot{\theta} = h/r^2$, where h is a constant of the motion and r is almost constant. Thus, $\dot{\theta}$ is approximately constant. In fact,

$$\dot{\theta} \simeq \frac{h}{r_c^2} = \left[-\frac{f(r_c)}{r_c} \right]^{1/2}, \tag{4.18}$$

where use has been made of Equation (4.8). Thus, the apsidal angle, ψ, is given by

$$\psi = \frac{T}{2}\,\dot{\theta} = \pi \left[3 + r_c\,\frac{f'(r_c)}{f(r_c)} \right]^{-1/2}. \tag{4.19}$$

For the case of attractive power-law central forces of the form $f(r) = -c\,r^n$, where $c > 0$, the apsidal angle becomes

$$\psi = \frac{\pi}{(3 + n)^{1/2}}. \tag{4.20}$$

It should be clear that if an orbit is going to close on itself, the apsidal angle needs to be a *rational* fraction of 2π. There are, in fact, only two small-integer values of the power-law index, n, for which this is the case. As we have seen, for an inverse-square force law (i.e., $n = -2$), the apsidal angle is π. For a linear force law (i.e., $n = 1$), the apsidal angle is $\pi/2$. However, for quadratic (i.e., $n = 2$) or cubic (i.e., $n = 3$) force laws, the apsidal angle is an *irrational* fraction of 2π, which means that non-circular orbits in such force fields never close on themselves.

4.4 Perihelion precession of planets

The solar system consists of eight major planets (Mercury to Neptune) moving around the Sun in slightly elliptical orbits that are approximately coplanar with one another. According to Chapter 3, if we neglect the relatively weak interplanetary gravitational interactions, the perihelia of the various planets (i.e., the points on their orbits at which they are closest to the Sun) remain *fixed* in space. However, once these interactions are taken into account, it turns out that the planetary perihelia all slowly *precess* in a prograde manner (i.e., rotate in the same direction as the orbital motion) around the Sun.[1] We can calculate the approximate rate of perihelion precession of a given planet by treating the other planets as *uniform concentric rings*, centered on the Sun, of mass equal to the planetary mass, and radius equal to the mean orbital radius. This method of calculation, which is due to Gauss, is equivalent to averaging the interplanetary gravitational interactions over the orbits of the other planets. It is reasonable to do this because the precession period in question is very much longer than the orbital period of any planet in the solar system. Thus, by treating the other planets as rings, we can calculate the mean gravitational perturbation due to these planets, and, thereby, determine the desired precession rate. (Actually, Gauss also incorporated the eccentricities, and non-uniform angular velocities, of the planetary orbits into his original calculation.)

We can conveniently index the planets in the solar system by designating Mercury as planet 1, and Neptune planet 8. Let the m_i and the a_i for $i = 1, 8$ be the planetary masses and mean orbital radii, respectively. Furthermore, let M be the mass of the Sun. It follows, from Section 2.7, that the gravitational potential generated in the vicinity of the ith planet by the Sun and the other planets is

$$\Phi_i(r) = -\frac{GM}{r} - \sum_{k=0,\infty} p_k \left[\sum_{j<i} \frac{Gm_j}{a_j} \left(\frac{a_j}{r}\right)^{2k+1} + \sum_{j>i} \frac{Gm_j}{a_j} \left(\frac{r}{a_j}\right)^{2k} \right], \qquad (4.21)$$

where $p_k = [P_{2k}(0)]^2$. Here, $P_n(x)$ is a Legendre polynominal. The radial force per unit mass acting on the ith planet is written $f_i = -d\Phi_i/dr|_{r=R_i}$. Hence, it is easily demonstrated that

$$\left[3 + \frac{r\,df_i/dr}{f_i}\right]^{-1/2}_{r=a_i} \simeq 1 + \sum_{k=1,\infty} k\,(2k+1)\,p_k \left[\sum_{j<i} \frac{m_j}{M} \left(\frac{a_j}{a_i}\right)^{2k} + \sum_{j>i} \frac{m_j}{M} \left(\frac{a_i}{a_j}\right)^{2k+1} \right] \quad (4.22)$$

to first order in the ratio of the planetary masses to the solar mass. Thus, according to Equation (4.19), the apsidal angle for the ith planet is

$$\psi_i \simeq \pi + \pi \sum_{k=1,\infty} k\,(2k+1)\,p_k \left[\sum_{j<i} \frac{m_j}{M} \left(\frac{a_j}{a_i}\right)^{2k} + \sum_{j>i} \frac{m_j}{M} \left(\frac{a_i}{a_j}\right)^{2k+1} \right]. \qquad (4.23)$$

---

[1] Precession can be either *prograde* (in the same sense as orbital motion) or *retrograde* (in the opposite sense). Retrograde precession is often called *regression*.

Table 4.1 Observed and theoretical planetary perihelion precession rates (at J2000)		
Planet	$\dot{\varpi}_{obs}('' \text{ yr}^{-1})^a$	$\dot{\varpi}_{th}('' \text{ yr}^{-1})$
Mercury	5.74	5.54
Venus	2.04	12.07
Earth	11.45	12.79
Mars	16.28	17.75
Jupiter	6.55	7.51
Saturn	19.50	18.59
Uranus	3.34	2.75
Neptune	0.36	0.67

^{a} Source: Standish and Williams 1992.

Hence, the perihelion of the ith planet advances by

$$\delta\varpi_i = 2\,(\psi_i - \pi) \simeq 2\pi \sum_{k=1,\infty} k\,(2k+1)\,p_k \left[\sum_{j<i} \frac{m_j}{M} \left(\frac{a_j}{a_i}\right)^{2k} + \sum_{j>i} \frac{m_j}{M} \left(\frac{a_i}{a_j}\right)^{2k+1} \right] \quad (4.24)$$

radians per revolution around the Sun. Of course, the time required for a single revolution is the orbital period, T_i. Thus, the rate of perihelion precession, in *arc seconds per year*, is given by

$$\dot{\varpi}_i \simeq \frac{1296000}{T_i(\text{yr})} \sum_{k=1,\infty} k\,(2k+1)\,p_k \left[\sum_{j<i} \frac{m_j}{M} \left(\frac{a_j}{a_i}\right)^{2k} + \sum_{j>i} \frac{m_j}{M} \left(\frac{a_i}{a_j}\right)^{2k+1} \right]. \quad (4.25)$$

Table 4.1 and Figure 4.1 compare the observed perihelion precession rates of the planets with the theoretical rates calculated from Equation (4.25) and the planetary data given in Table 3.1. It can be seen that there is excellent agreement between the two, except for the planet Venus. The main reason for this is that Venus has an unusually low eccentricity ($e = 0.0068$), which renders its perihelion point extremely sensitive to small perturbations.

4.5 Perihelion precession of Mercury

If the calculation described in the previous section is carried out more accurately, taking into account the slight eccentricities of the planetary orbits, as well as their small mutual inclinations, then the perihelion precession rate of the planet Mercury is found to be 5.32 arc seconds per year (Stewart 2005). However, the observed precession rate is 5.74 arc seconds per year. It turns out that the cause of this discrepancy is a general relativistic correction to Newtonian gravity.

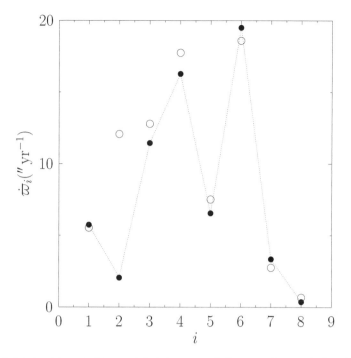

Fig. 4.1 Observed (full circles) and theoretical (empty circles) planetary perihelion precession rates (at J2000). Source (for observational data): Standish and Williams 1992.

General relativity gives rise to a small correction to the force per unit mass exerted by the Sun (mass M) on a planet in a circular orbit of radius r, and angular momentum per unit mass h. This correction is due to the curvature of space in the immediate vicinity of the Sun. In fact, the modified formula for f is (Rindler 1977)

$$f \simeq -\frac{G M}{r^2} - \frac{3 G M h^2}{c^2 r^4}, \tag{4.26}$$

where c is the velocity of light in a vacuum. It follows that

$$\frac{r f'}{f} \simeq -2\left(1 + \frac{3 h^2}{c^2 r^2} + \cdots\right), \tag{4.27}$$

to first order in $h^2/(c^2 r^2) \ll 1$. Hence, from Equation (4.19), the apsidal angle is

$$\psi \simeq \pi\left(1 + \frac{3 h^2}{c^2 r^2}\right). \tag{4.28}$$

Thus, the perihelion advances by

$$\delta\varpi \simeq \frac{6\pi G M}{c^2 r} \tag{4.29}$$

radians per revolution as a consequence of the general relativistic correction to Newtonian gravity. Here, use has been made of $h^2 = G M r$. It follows that the rate of perihelion precession due to the general relativistic correction is

$$\dot{\varpi} \simeq \frac{0.0383}{a\,T} \qquad\qquad (4.30)$$

arc seconds per year, where a is the mean orbital radius in astronomical units, and T the orbital period in years. Hence, from Table 3.1, the general relativistic contribution to $\dot{\varpi}$ for Mercury is 0.41 arc seconds per year. It is easily demonstrated that the corresponding contribution is negligible for the other planets in the solar system. If the preceding calculation is carried out slightly more accurately, taking the eccentricity of Mercury's orbit into account, then the general relativistic contribution to $\delta\varpi$ becomes 0.43 arc seconds per year. (See Exercise 9.2.) It follows that the total perihelion precession rate for Mercury is $5.32 + 0.43 = 5.75$ arc seconds per year. This is in almost exact agreement with the observed precession rate. Indeed, the ability of general relativity to explain the discrepancy between the observed perihelion precession rate of Mercury, and that calculated from Newtonian mechanics, was one of the first major successes of this theory.

Exercises

4.1 Derive Equations (4.2) and (4.7).

4.2 Prove that in the case of a central force varying inversely as the cube of the distance,

$$r^2 = A\,t^2 + B\,t + C,$$

where A, B, C are constants. (From Lamb 1923.)

4.3 The orbit of a particle moving in a central field is a circle that passes through the origin: $r = r_0 \cos\theta$, where $r_0 > 0$. Show that the force law is inverse-fifth power. (Modified from Fowles and Cassiday 2005.)

4.4 The orbit of a particle moving in a central field is the cardoid $r = a\,(1 + \cos\theta)$, where $a > 0$. Show that the force law is inverse fourth power.

4.5 A particle moving in a central field describes a spiral orbit $r = r_0 \exp(k\,\theta)$, where $r_0, k > 0$. Show that the force law is inverse cube, and that θ varies logarithmically with t. Demonstrate that there are two other possible types of orbit in this force field, and give their equations. (Modified from Fowles and Cassiday 2005.)

4.6 A particle moves in the spiral orbit $r = a\,\theta$, where $a > 0$. Suppose that θ increases linearly with t. Is the force acting on the particle central in nature? If not, determine how θ would have to vary with t in order to make the force central. Assuming that the force is central, demonstrate that the particle's potential energy per unit mass is

$$V(r) = -\frac{h^2}{2}\left(\frac{a^2}{r^4} + \frac{1}{r^2}\right),$$

where h is its (constant) angular momentum per unit mass. (Modified from Fowles and Cassiday 2005.)

4.7   A particle moves under the influence of a central force per unit mass of the form

$$f(r) = -\frac{k}{r^2} + \frac{c}{r^3},$$

where k and c are positive constants. Show that the associated orbit can be written

$$r = \frac{a\,(1 - e^2)}{1 + e\,\cos(\alpha\,\theta)},$$

which is a closed ellipse for $e < 1$ and $\alpha = 1$. Discuss the character of the orbit for $\alpha \neq 1$ and $e < 1$. Demonstrate that

$$\alpha = \left(1 - \frac{\gamma}{1 - e^2}\right)^{1/2},$$

where $\gamma = c/(k\,a)$.

4.8   A particle moves in a circular orbit of radius r_0 in an attractive central force field of the form $f(r) = -c\,\exp(-r/a)/r^2$, where $c > 0$ and $a > 0$. Demonstrate that the orbit is stable only provided that $r_0 < a$.

4.9   A particle moves in a circular orbit in an attractive central force field of the form $f(r) = -a\,r^{-3}$, where $a > 0$. Show that the orbit is unstable to small perturbations.

4.10 A particle moves in a nearly circular orbit of radius a under the action of the radial force per unit mass

$$f(r) = -\frac{\mu}{r^2}\,\mathrm{e}^{-k\,r},$$

where $\mu > 0$ and $0 < k\,a \ll 1$. Demonstrate that the so-called apse line, joining successive apse points, rotates in the same direction as the orbital motion through an angle $\pi\,k\,a$ each revolution. (From Lamb 1923.)

4.11 A particle moves in a nearly circular orbit of radius a under the action of the central potential per unit mass

$$V(r) = -\frac{\mu}{r}\,\mathrm{e}^{-k\,r},$$

where $\mu > 0$ and $0 < k\,a \ll 1$. Show that the apse line rotates in the same direction as the orbital motion through an angle $\pi\,k^2\,a^2$ each revolution. (From Lamb 1923.)

4.12 Suppose that the solar system were embedded in a tenuous uniform dust cloud. Demonstrate that the apsidal angle of a planet in a nearly circular orbit around the Sun would be

$$\pi\left(1 - \frac{3}{2}\,\frac{M_0}{M}\right),$$

where M is the mass of the Sun, and M_0 is the mass of dust enclosed by a sphere whose radius matches the major radius of the orbit. It is assumed that $M_0 \ll M$.

4.13 Consider a satellite orbiting around an idealized planet that takes the form of a uniform spheroidal mass distribution of mean radius R and ellipticity ϵ (where $0 < \epsilon \ll 1$). Suppose that the orbit is nearly circular, with a major radius a, and lies in the equatorial plane of the planet. The potential energy per unit mass of the satellite is thus (see Chapter 2)

$$V(r) = -\frac{G\,M}{r}\left(1 + \frac{\epsilon}{5}\,\frac{R^2}{r^2}\right),$$

where r is a radial coordinate in the equatorial plane. Demonstrate that the apse line rotates in the same direction as the orbital motion at the rate

$$\dot{\varpi} = \frac{3}{5}\frac{\epsilon}{}\left(\frac{R}{a}\right)^2 n,$$

where n is the mean orbital angular velocity of the satellite.

Rotating reference frames

5.1 Introduction

As we saw in Chapter 1, Newton's second law of motion is valid only in *inertial* frames of reference. However, it is sometimes convenient to observe motion in noninertial *rotating* reference frames. For instance, it is most convenient for us to observe the motions of objects close to the Earth's surface in a reference frame that is fixed relative to this surface. Such a frame is noninertial in nature, as it accelerates with respect to a standard inertial frame as a result of the Earth's diurnal rotation. (The accelerations of this frame owing to the Earth's orbital motion about the Sun, or the Sun's orbital motion about the galactic center, and so on, are negligible compared with that associated with the Earth's diurnal rotation.) Let us investigate motion observed in a rotating reference frame.

5.2 Rotating reference frames

Suppose that a given object has position vector $\mathbf{r}$ in some inertial (i.e., nonrotating) reference frame. Let us observe the motion of this object in a noninertial reference frame that rotates with constant angular velocity $\mathbf{\Omega}$ about an axis passing through the origin of the inertial frame. Suppose, first of all, that our object appears stationary in the rotating reference frame. Hence, in the nonrotating frame, the object's position vector $\mathbf{r}$ will appear to *precess* about the origin with angular velocity $\mathbf{\Omega}$. It follows from Section A.7 that, in the nonrotating reference frame,

$$\frac{d\mathbf{r}}{dt} = \mathbf{\Omega} \times \mathbf{r}. \tag{5.1}$$

Suppose, now, that our object appears to move in the rotating reference frame with instantaneous velocity $\mathbf{v}'$. It is fairly obvious that the appropriate generalization of the preceding equation is simply

$$\frac{d\mathbf{r}}{dt} = \mathbf{v}' + \mathbf{\Omega} \times \mathbf{r}. \tag{5.2}$$

Let d/dt and d/dt' denote apparent time derivatives in the nonrotating and rotating frames of reference, respectively. Because an object that is stationary in the rotating reference frame appears to move in the nonrotating frame, it is clear that $d/dt \neq d/dt'$. Writing the apparent velocity, $\mathbf{v}'$, of our object in the rotating reference frame as $d\mathbf{r}/dt'$,

Equation (5.2) takes the form

$$\frac{d\mathbf{r}}{dt} = \frac{d\mathbf{r}}{dt'} + \mathbf{\Omega} \times \mathbf{r},\tag{5.3}$$

or

$$\frac{d}{dt} = \frac{d}{dt'} + \mathbf{\Omega} \times,\tag{5.4}$$

because $\mathbf{r}$ is a general position vector. Equation (5.4) expresses the relationship between apparent time derivatives in the nonrotating and rotating reference frames.

Operating on the general position vector $\mathbf{r}$ with the time derivative in Equation (5.4), we get

$$\mathbf{v} = \mathbf{v}' + \mathbf{\Omega} \times \mathbf{r}.\tag{5.5}$$

This equation relates the apparent velocity, $\mathbf{v} = d\mathbf{r}/dt$, of an object with position vector $\mathbf{r}$ in the nonrotating reference frame to its apparent velocity, $\mathbf{v}' = d\mathbf{r}/dt'$, in the rotating reference frame.

Operating twice on the position vector $\mathbf{r}$ with the time derivative in Equation (5.4), we obtain

$$\mathbf{a} = \left(\frac{d}{dt'} + \mathbf{\Omega} \times\right)(\mathbf{v}' + \mathbf{\Omega} \times \mathbf{r}),\tag{5.6}$$

or

$$\mathbf{a} = \mathbf{a}' + \mathbf{\Omega} \times (\mathbf{\Omega} \times \mathbf{r}) + 2\,\mathbf{\Omega} \times \mathbf{v}'.\tag{5.7}$$

This equation relates the apparent acceleration, $\mathbf{a} = d^2\mathbf{r}/dt^2$, of an object with position vector $\mathbf{r}$ in the nonrotating reference frame to its apparent acceleration, $\mathbf{a}' = d^2\mathbf{r}/dt'^2$, in the rotating reference frame.

Applying Newton's second law of motion in the inertial (i.e., nonrotating) reference frame, we obtain

$$m\,\mathbf{a} = \mathbf{f}.\tag{5.8}$$

Here, m is the mass of our object and $\mathbf{f}$ is the (nonfictitious) force acting on it. These quantities are the same in both reference frames. Making use of Equation (5.7), the apparent equation of motion of our object in the rotating reference frame takes the form

$$m\,\mathbf{a}' = \mathbf{f} - m\,\mathbf{\Omega} \times (\mathbf{\Omega} \times \mathbf{r}) - 2\,m\,\mathbf{\Omega} \times \mathbf{v}'.\tag{5.9}$$

The last two terms in this equation are so-called *fictitious forces*. Such forces are always needed to account for motion observed in noninertial reference frames. Fictitious forces can always be distinguished from nonfictitious forces in Newtonian mechanics because the former have no associated reactions. Let us now investigate the two fictitious forces appearing in Equation (5.9).

5.3 Centrifugal acceleration

Let our nonrotating inertial frame be one whose origin lies at the center of the Earth, and let our rotating frame be one whose origin is fixed with respect to some point, of

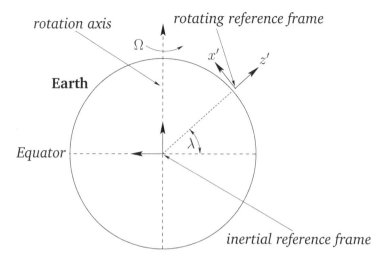

Fig. 5.1
Inertial and noninertial reference frames.

latitude λ, on the Earth's surface. (See Figure 5.1.) The latter reference frame thus rotates with respect to the former (about an axis passing through the Earth's center) with an angular velocity vector, $\boldsymbol{\Omega}$, which points from the center of the Earth toward its north pole and is of magnitude

$$\Omega = \frac{2\pi}{23^{\mathrm{h}}\,56^{\mathrm{m}}\,04^{\mathrm{s}}} = 7.2921 \times 10^{-5}\,\mathrm{rad.\,s^{-1}}. \tag{5.10}$$

Here, $23^{\mathrm{h}}\,56^{\mathrm{m}}\,04^{\mathrm{s}}$ is the length of a *sidereal day*, that is, the Earth's rotation period relative to the distant stars (Yoder 1995).

Consider an object that appears stationary in our rotating reference frame, that is, an object that is stationary with respect to the Earth's surface. According to Equation (5.9), the object's apparent equation of motion in the rotating frame takes the form

$$m\,\mathbf{a}' = \mathbf{f} - m\,\boldsymbol{\Omega} \times (\boldsymbol{\Omega} \times \mathbf{r}). \tag{5.11}$$

Let the nonfictitious force acting on our object be the force of gravity, $\mathbf{f} = m\,\mathbf{g}$. Here, the local gravitational acceleration, $\mathbf{g}$, points directly toward the center of the Earth. It follows that the apparent gravitational acceleration in the rotating frame is written

$$\mathbf{g}' = \mathbf{g} - \boldsymbol{\Omega} \times (\boldsymbol{\Omega} \times \mathbf{R}), \tag{5.12}$$

where $\mathbf{R}$ is the displacement vector of the origin of the rotating frame (which lies on the Earth's surface) with respect to the center of the Earth. Here, we are assuming that our object is situated relatively close to the Earth's surface (i.e., $\mathbf{r} \simeq \mathbf{R}$).

It can be seen from Equation (5.12) that the apparent gravitational acceleration of a stationary object close to the Earth's surface has two components: first, the true gravitational acceleration, $\mathbf{g}$, of magnitude $g \simeq 9.82\,\mathrm{m\,s^{-2}}$, which always points directly toward the center of the Earth (Yoder 1995); and second, the so-called *centrifugal acceleration*, $-\boldsymbol{\Omega} \times (\boldsymbol{\Omega} \times \mathbf{R})$. This acceleration is normal to the Earth's axis of rotation and always points directly away from this axis. The magnitude of the centrifugal acceleration is

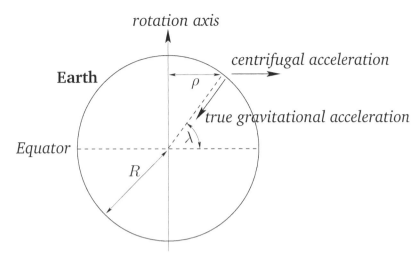

Fig. 5.2 Centrifugal acceleration.

$\Omega^2 \rho = \Omega^2 R \cos \lambda$, where ρ is the perpendicular distance to the Earth's rotation axis, and $R = 6.3710 \times 10^6$ m is the Earth's radius (Yoder 1995). (See Figure 5.2.)

It is convenient to define Cartesian axes in the rotating reference frame such that the z'-axis points vertically upward and the x'- and y'-axes are horizontal, with the x'-axis pointing directly northward and the y'-axis pointing directly westward. (See Figure 5.1.) The Cartesian components of the Earth's angular velocity are thus

$$\mathbf{\Omega} = \Omega \left(\cos \lambda, \, 0, \, \sin \lambda \right), \tag{5.13}$$

while the vectors $\mathbf{R}$ and $\mathbf{g}$ are written

$$\mathbf{R} = (0, 0, R) \tag{5.14}$$

and

$$\mathbf{g} = (0, 0, -g), \tag{5.15}$$

respectively. It follows that the Cartesian coordinates of the apparent gravitational acceleration, from Equation (5.12), are

$$\mathbf{g}' = \left(-\Omega^2 R \cos \lambda \sin \lambda, \, 0, \, -g + \Omega^2 R \cos^2 \lambda \right). \tag{5.16}$$

The magnitude of this acceleration is approximately

$$g' \simeq g - \Omega^2 R \cos^2 \lambda \simeq 9.82 - 0.0338 \cos^2 \lambda \ \mathrm{m\,s^{-2}}. \tag{5.17}$$

According to the preceding equation, the centrifugal acceleration causes the magnitude of the apparent gravitational acceleration on the Earth's surface to vary by about 0.3 percent, being largest at the poles and smallest at the equator. This variation in apparent gravitational acceleration, due (ultimately) to the Earth's rotation, causes the Earth itself to bulge slightly at the equator (see Section 5.5), which has the effect of further intensifying the variation (see Exercise 5.7), as a point on the surface of the Earth at the equator is slightly farther away from the Earth's center than a similar point at one of the poles (and, hence, the true gravitational acceleration is slightly weaker in the former case).

Another consequence of centrifugal acceleration is that the apparent gravitational acceleration on the Earth's surface has a horizontal component aligned in the north–south direction. This horizontal component ensures that the apparent gravitational acceleration does not point directly toward the center of the Earth. In other words, a plumb line on the surface of the Earth does not point vertically downward (toward the center of the Earth), but is deflected slightly away from a true vertical in the north–south direction. The angular deviation from true vertical can easily be calculated from Equation (5.16):

$$\theta_{dev} \simeq -\frac{\Omega^2 R}{2 g} \sin(2\lambda) \simeq -0.1° \sin(2\lambda). \tag{5.18}$$

Here, a positive angle denotes a northward deflection, and vice versa. Thus, the deflection is southward in the northern hemisphere (i.e., $\lambda > 0$) and northward in the southern hemisphere (i.e., $\lambda < 0$). The deflection is zero at the poles and at the equator, and it reaches its maximum magnitude (which is very small) at middle latitudes.

5.4 Coriolis force

We have now accounted for the first fictitious force, $-m\,\boldsymbol{\Omega} \times (\boldsymbol{\Omega} \times \mathbf{r})$, appearing in Equation (5.9). Let us now investigate the second, which takes the form $-2\,m\,\boldsymbol{\Omega} \times \mathbf{v}'$ and is called the *Coriolis force*. Obviously, this force affects only objects that are moving in the rotating reference frame.

Consider a particle of mass m free-falling under gravity in our rotating reference frame. As before, we define Cartesian axes in the rotating frame such that the z'-axis points vertically upward and the x'- and y'-axes are horizontal, with the x'-axis pointing directly northward and the y'-axis pointing directly westward. It follows, from Equation (5.9), that the Cartesian equations of motion of the particle in the rotating reference frame take the form

$$\ddot{x}' = 2\Omega \sin\lambda\,\dot{y}', \tag{5.19}$$

$$\ddot{y}' = -2\Omega \sin\lambda\,\dot{x}' + 2\Omega \cos\lambda\,\dot{z}', \tag{5.20}$$

and

$$\ddot{z}' = -g - 2\Omega \cos\lambda\,\dot{y}'. \tag{5.21}$$

Here, g is the local acceleration due to gravity. In the preceding three equations, we have neglected the centrifugal acceleration for the sake of simplicity. This is reasonable, because the only effect of the centrifugal acceleration is to slightly modify the magnitude and direction of the local gravitational acceleration. We have also neglected air resistance, which is less reasonable.

Consider a particle that is dropped (at $t = 0$) from rest a height h above the Earth's surface. The following solution method exploits the fact that the Coriolis force is much smaller in magnitude that the force of gravity. Hence, Ω can be treated as a *small* parameter. To lowest order (i.e., neglecting Ω), the particle's vertical motion satisfies $\ddot{z}' = -g$, which can be solved, subject to the initial conditions, to give

$$z' = h - \frac{g\,t^2}{2}. \tag{5.22}$$

Substituting this expression into Equations (5.19) and (5.20), neglecting terms involving Ω^2, and solving subject to the initial conditions, we obtain $x' \simeq 0$ and

$$y' \simeq -g\,\Omega\,\cos\lambda\,\frac{t^3}{3}. \tag{5.23}$$

In other words, the particle is deflected eastward (i.e, in the negative y'-direction). The particle hits the ground when $t \simeq \sqrt{2\,h/g}$. Hence, the net eastward deflection of the particle as it strikes the ground is

$$d_{\text{east}} \simeq \frac{\Omega}{3}\,\cos\lambda\,\left(\frac{8\,h^3}{g}\right)^{1/2}. \tag{5.24}$$

This deflection is in the same direction as the Earth's rotation (i.e., west to east) and is greatest at the equator and zero at the poles. A particle dropped from a height of 100 m at the equator is deflected by about 2.2 cm.

Consider a particle launched horizontally with some fairly large velocity,

$$\mathbf{v} = v_0\,(\cos\theta, -\sin\theta, 0). \tag{5.25}$$

Here, θ is the *compass bearing* of the velocity vector (so north is $0°$, east is $90°$, etc.). Neglecting any vertical motion, Equations (5.19) and (5.20) yield

$$\dot{v}_{x'} \simeq -2\,\Omega\,v_0\,\sin\lambda\,\sin\theta \tag{5.26}$$

and

$$\dot{v}_{y'} \simeq -2\,\Omega\,v_0\,\sin\lambda\,\cos\theta, \tag{5.27}$$

which can be integrated to give

$$v_{x'} \simeq v_0\,\cos\theta - 2\,\Omega\,v_0\,\sin\lambda\,\sin\theta\,t \tag{5.28}$$

and

$$v_{y'} \simeq -v_0\,\sin\theta - 2\,\Omega\,v_0\,\sin\lambda\,\cos\theta\,t. \tag{5.29}$$

To lowest order in Ω, the preceding equations are equivalent to

$$v_{x'} \simeq v_0\,\cos(\theta + 2\,\Omega\,\sin\lambda\,t) \tag{5.30}$$

and

$$v_{y'} \simeq -v_0\,\sin(\theta + 2\,\Omega\,\sin\lambda\,t). \tag{5.31}$$

It follows that the Coriolis force causes the compass bearing of the particle's velocity vector to rotate steadily as time progresses. The rotation rate is

$$\frac{d\theta}{dt} \simeq 2\,\Omega\,\sin\lambda. \tag{5.32}$$

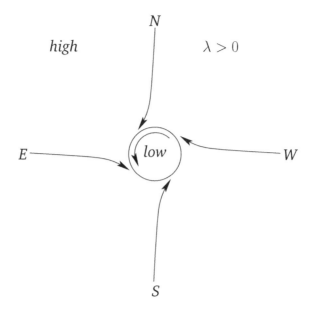

Fig. 5.3 Cyclone in Earth's northern hemisphere.

Hence, the rotation is clockwise (if we look from above) in the northern hemisphere and counterclockwise in the *southern hemisphere*. The rotation rate is zero at the equator and greatest at the poles.

The Coriolis force has a significant effect on terrestrial weather patterns. Near equatorial regions, the Sun's intense heating of the Earth's surface causes hot air to rise. In the northern hemisphere, this causes cooler air to move in a southerly direction toward the equator. The Coriolis force deflects this moving air in a clockwise sense (if we look from above), resulting in the *trade winds*, which blow toward the southwest. In the southern hemisphere, the cooler air moves northward and is deflected by the Coriolis force in a counterclockwise sense, resulting in trade winds that blow toward the northwest. Furthermore, as air flows from high- to low-pressure regions, the Coriolis force deflects the air in a clockwise/counterclockwise manner in the northern/southern hemisphere, producing *cyclonic* rotation. (See Figure 5.3.) It follows that cyclonic rotation is counterclockwise (seen from above) in the northern hemisphere, and clockwise in the southern hemisphere. Thus, this is the direction of rotation of tropical storms (e.g., hurricanes, typhoons) in each hemisphere.

5.5 Rotational flattening

Consider the equilibrium configuration of a self-gravitating celestial body, composed of incompressible fluid, that is rotating steadily and uniformly about some fixed axis passing through its center of mass. Let us assume that the outer boundary of the body is spheroidal. (See Section 2.6.) Let M be the body's total mass, R its mean radius, ϵ its ellipticity, and Ω its angular rotation velocity. Suppose, finally, that the body's axis of rotation coincides with its axis of symmetry, which is assumed to run along the z-axis.

Let us transform to a noninertial frame of reference that co-rotates with the body about the z-axis, and in which the body consequently appears to be stationary. From Section 5.3, the problem is now analogous to that of a nonrotating body, except that the acceleration is written $\mathbf{g} = \mathbf{g}_g + \mathbf{g}_c$, where $\mathbf{g}_g = -\nabla\Phi(r,\theta)$ is the gravitational acceleration, $\mathbf{g}_c$ the centrifugal acceleration, and Φ the gravitational potential. The latter acceleration is of magnitude $r\sin\theta\,\Omega^2$ and is everywhere directed away from the axis of rotation. (See Section 5.2.) Here, r and θ are spherical coordinates whose origin is the body's geometric center and whose symmetry axis coincides with the axis of rotation. The centrifugal acceleration is thus

$$\mathbf{g}_c = r\,\Omega^2\,\sin^2\theta\,\mathbf{e}_r + r\,\Omega^2\,\sin\theta\,\cos\theta\,\mathbf{e}_\theta. \tag{5.33}$$

It follows that $\mathbf{g}_c = -\nabla\chi$, where

$$\chi(r,\theta) = -\frac{\Omega^2\,r^2}{2}\,\sin^2\theta = \frac{\Omega^2\,r^2}{3}\,[P_2(\cos\theta)-1] \tag{5.34}$$

can be thought of as a sort of centrifugal potential. Thus, the total acceleration is

$$\mathbf{g} = -\nabla(\Phi+\chi). \tag{5.35}$$

It is convenient to write the centrifugal potential in the form

$$\chi(r,\theta) = \frac{G\,M}{R}\left(\frac{r}{R}\right)^2\,\zeta\,[P_2(\cos\theta)-1], \tag{5.36}$$

where the dimensionless parameter

$$\zeta = \frac{\Omega^2\,R^3}{3\,G\,M} \tag{5.37}$$

is the typical ratio of the centrifugal acceleration to the gravitational acceleration at $r \sim R$. Let us assume that this ratio is small: $\zeta \ll 1$.

As before (see Section 2.6), the criterion for an equilibrium state is that the total potential be uniform over the body's surface, to eliminate any tangential forces that cannot be balanced by internal pressure. Let us assume that the surface satisfies [see Equation (2.56)]

$$r = R_\theta(\theta) = R\left[1 - \frac{2}{3}\,\epsilon\,P_2(\cos\theta)\right], \tag{5.38}$$

where

$$\epsilon = \frac{R_e - R_p}{R}. \tag{5.39}$$

Here, R is the body's mean radius, $R_p = R(1 - 2\,\epsilon/3)$ the radius at the poles (i.e., along the axis of rotation), and $R_e = R(1 + \epsilon/3)$ the radius at the equator (i.e., perpendicular to the axis of rotation). (See Figure 5.4.) It is assumed that $|\epsilon| \ll 1$, so the body is almost spherical. The external (to the body) gravitational potential can be written [see Equation (2.66)]

$$\Phi(r,\theta) \simeq -\frac{G\,M}{r} + J_2\,\frac{G\,M\,R^2}{r^3}\,P_2(\cos\theta), \tag{5.40}$$

axis of rotation

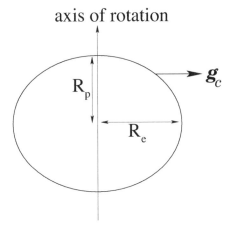

Rotational flattening.

where $J_2 \sim \mathcal{O}(\epsilon)$. The equilibrium configuration is specified by

$$\Phi(R_\theta, \theta) + \chi(R_\theta, \theta) = c, \tag{5.41}$$

where c is a constant. It follows from Equations (5.36), (5.38), and (5.40) that, to first order in ϵ and ζ,

$$-\frac{G M}{R}\left[1 + \left(\frac{2}{3}\epsilon - J_2\right)P_2(\cos\theta)\right] + \frac{G M}{R}\zeta\,[P_2(\cos\theta) - 1] \simeq c, \tag{5.42}$$

which yields

$$\epsilon = \frac{3}{2}(J_2 + \zeta). \tag{5.43}$$

For the special case of a *uniform density* body, we have $J_2 = (2/5)\,\epsilon$ [see Equation (2.64)]. Hence, the previous equation simplifies to

$$\epsilon = \frac{15}{4}\zeta, \tag{5.44}$$

or

$$\frac{R_e - R_p}{R} = \frac{5}{4}\frac{\Omega^2 R^3}{G M}. \tag{5.45}$$

We conclude, from the preceding expression, that the equilibrium configuration of a (relatively slowly) rotating self-gravitating fluid mass is an *oblate spheroid*—a sphere that is slightly flattened along its axis of rotation. The degree of flattening is proportional to the square of the rotation rate.

The result of Equation (5.44) was derived on the assumption that there is zero shear stress at the surface of a uniform-density, rotating, self-gravitating celestial body. This is certainly true for a *fluid* body, as fluids (by definition) are unable to withstand shear stresses. Solids, on the other hand, can withstand such stresses to a limited extent. Hence, it is not necessarily true that there is zero shear stress at the surface of a *solid* rotating body, such as the Earth. Let us investigate whether Equation (5.44) needs to be modified for such a body.

In the presence of the centrifugal potential specified in Equation (5.36), the normal stress at the surface of a spheroidal body, of mean radius R, ellipticity ϵ, and uniform density γ, can be written $\sigma = -X P_2(\cos\theta) + p_0$, where p_0 is a constant,

$$X = \sigma_c \left(\frac{R}{R_c}\right)^2 \left(\zeta - \frac{4}{15}\epsilon\right), \tag{5.46}$$

and σ_c is the *yield stress* of the material from which the body is composed (i.e., the critical shear stress above which the material flows like a liquid; Love 2011). The shear stress is proportional to $\partial\sigma/\partial\theta$. Furthermore,

$$R_c = \left(\frac{3}{4\pi}\frac{\sigma_c}{G\gamma^2}\right)^{1/2}. \tag{5.47}$$

For the rock that makes up the Earth's mantle, $\sigma_c \simeq 2 \times 10^8 \, \mathrm{N\,m^{-2}}$ and $\gamma \simeq 5 \times 10^3 \, \mathrm{kg\,m^{-3}}$, giving $R_c \simeq 169\,\mathrm{km}$ (de Pater and Lissauer 2010). Let us assume that $R \gg R_c$, which implies that, in the absence of the centrifugal potential, the self-gravity of the body in question is sufficiently strong to force it to adopt a spherical shape. (See Section 2.6.) If the surface shear stress is less than the yield stress (i.e., if $|X| < \sigma_c$) then the body responds *elastically* to the stress in such a manner that

$$\epsilon = \frac{15}{38}\frac{X}{\mu}, \tag{5.48}$$

where μ is the *shear modulus*, or *rigidity*, of the body's constituent material (Love 2011). For the rock that makes up the Earth's mantle, $\mu \simeq 1 \times 10^{11} \, \mathrm{N\,m^{-2}}$ (de Pater and Lissauer 2010). It follows that

$$\epsilon = \frac{15}{4}\frac{\zeta}{1+\tilde{\mu}}, \tag{5.49}$$

where

$$\tilde{\mu} = \frac{57}{8\pi}\frac{\mu}{G\gamma^2 R^2} \tag{5.50}$$

is a dimensionless quantity that is termed the body's *effective rigidity*. On the other hand, if the surface shear stress is greater than the yield stress, the body flows like a liquid until the stress becomes zero. So, it follows from Equation (5.46) that

$$\epsilon = \frac{15}{4}\zeta, \tag{5.51}$$

which is identical to Equation (5.44). Hence, we deduce that the rotational flattening of a solid, uniform-density, celestial body is governed by Equation (5.49) if the surface shear stress does not exceed the yield stress, and by Equation (5.44) otherwise. In the former case, the condition $|X| < \sigma_c$ is equivalent to $\zeta < \zeta_c$, where (assuming that $R \ll R_c$)

$$\zeta_c = \frac{2}{19}\frac{\sigma_c}{\mu} + \left(\frac{R_c}{R}\right)^2 \simeq \frac{2}{19}\frac{\sigma_c}{\mu}. \tag{5.52}$$

For the rock that makes up the Earth's mantle [for which $\sigma_c \simeq 2 \times 10^8 \, \mathrm{N\,m^{-2}}$ and $\mu \simeq 1 \times 10^{11} \, \mathrm{N\,m^{-2}}$ (de Pater and Lissauer 2010)], we find that

$$\zeta_c \simeq 2 \times 10^{-4}. \tag{5.53}$$

Thus, if $\zeta < \zeta_c$, the rotational flattening of a uniform body made up of such rock is governed by Equation (5.49), but if $\zeta > \zeta_c$, the flattening is governed by Equation (5.44).

For the case of the Earth itself, $R = 6.37 \times 10^6$ m, $\Omega = 7.29 \times 10^{-5}$ rad. s^{-1}, and $M = 5.97 \times 10^{24}$ kg (Yoder 1995). It follows that

$$\zeta = 1.15 \times 10^{-3}. \tag{5.54}$$

Because $\zeta \gg \zeta_c$, we deduce that the Earth's centrifugal potential is sufficiently strong to force its constituent rock to flow like a liquid. Hence, the rotational flattening is governed by Equation (5.44), which implies that

$$\epsilon = 4.31 \times 10^{-3}. \tag{5.55}$$

This corresponds to a difference between the Earth's equatorial and polar radii of

$$\Delta R = R_e - R_p = \epsilon R = 27.5 \text{ km}. \tag{5.56}$$

In fact, the observed degree of rotational flattening of the Earth is $\epsilon = 3.35 \times 10^{-3}$ (Yoder 1995), corresponding to a difference between equatorial and polar radii of 21.4 km. Our analysis has overestimated the Earth's rotational flattening because, for the sake of simplicity, we assumed that the terrestrial interior is of uniform density. In reality, the Earth's core is much denser than its crust. (See Exercise 5.6.) Incidentally, the observed value of the parameter J_2, which measures the strength of the Earth's quadrupole gravitational field, is 1.08×10^{-3} (Yoder 1995). Hence, $(3/2)(J_2 + \zeta) = 3.35 \times 10^{-3}$. In other words, the Earth's rotational flattening satisfies Equation (5.43) extremely accurately. This confirms that although the Earth is not a uniform density body, its response to the centrifugal potential is indeed fluidlike [because Equation (5.43) was derived on the assumption that the response is fluidlike.]

For the planet Jupiter, $R = 6.92 \times 10^7$ m, $\Omega = 1.76 \times 10^{-4}$ rad. s^{-1}, and $M = 1.90 \times 10^{27}$ kg (Yoder 1995; Seidelmann et al. 2007). Hence,

$$\zeta = 2.70 \times 10^{-2}. \tag{5.57}$$

Because Jupiter is largely composed of liquid, its rotation flattening is governed by Equation (5.44), which yields

$$\epsilon = 0.101. \tag{5.58}$$

This degree of flattening is much larger than that of the Earth, owing to Jupiter's relatively large radius (about ten times that of Earth), combined with its relatively short rotation period (about 0.4 days). In fact, the rotational flattening of Jupiter is clearly apparent from images of this planet. (See Figure 5.5.) The observed degree of rotational flattening of Jupiter is actually $\epsilon = 0.065$ (Yoder 1995). Our estimate for ϵ is slightly too large because Jupiter has a mass distribution that is strongly concentrated at its core. (See Exercise 5.6.) Incidentally, the measured value of J_2 for Jupiter is 1.47×10^{-2} (Yoder 1995). Hence, $(3/2)(J_2 + \zeta) = 0.063$. Thus, Jupiter's rotational flattening also satisfies Equation (5.43) fairly accurately, confirming that its response to the centrifugal potential is fluidlike.

Fig. 5.5 Jupiter. Photograph taken by the Hubble Space Telescope. A circle is superimposed on the image to make the rotational flattening more clearly visible. The axis of rotation is vertical. Credit: NASA.

5.6 Tidal elongation

Consider two point masses, m and m', executing circular orbits about their common center of mass, C, with angular velocity ω. Let a be the distance between the masses and ρ the distance between point C and mass m. (See Figure 5.6.) We know from Section 3.16, that

$$\omega^2 = \frac{G\,M}{a^3},\tag{5.59}$$

and

$$\rho = \frac{m'}{M}\,a,\tag{5.60}$$

where $M = m + m'$.

Let us transform to a noninertial frame of reference that rotates, about an axis perpendicular to the orbital plane and passing through C, at the angular velocity ω. In this

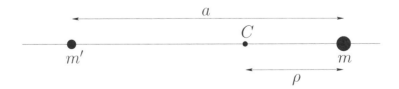

Fig. 5.6 Two orbiting masses.

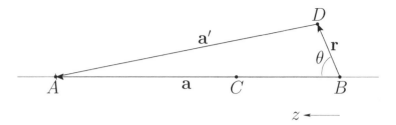

Calculation of tidal forces.

reference frame, both masses appear to be stationary. Consider mass m. In the rotating frame, this mass experiences a gravitational acceleration

$$a_g = \frac{G\,m'}{a^2} \tag{5.61}$$

directed toward the center of mass, and a centrifugal acceleration (see Section 5.3)

$$a_c = \omega^2 \rho \tag{5.62}$$

directed away from the center of mass. However, it is easily demonstrated, using Equations (5.59) and (5.60), that

$$a_c = a_g. \tag{5.63}$$

In other words, the gravitational and centrifugal accelerations balance, as must be the case if mass m is to remain stationary in the rotating frame. Let us investigate how this balance is affected if the masses m and m' have finite spatial extents.

Let the center of the mass distribution m' lie at A, the center of the mass distribution m at B, and the center of mass at C. (See Figure 5.7.) We wish to calculate the centrifugal and gravitational accelerations at some point D in the vicinity of point B. It is convenient to adopt spherical coordinates, centered on point B and aligned such that the z-axis coincides with the line BA.

Let us assume that the mass distribution m is orbiting around C, but is *not* rotating about an axis passing through its center of mass, to exclude rotational flattening from our analysis. If this is the case, it is easily seen that each constituent point of m executes circular motion of angular velocity ω and radius ρ. (See Figure 5.8.) Hence, each point experiences the *same* centrifugal acceleration:

$$\mathbf{g}_c = -\omega^2 \rho\, \mathbf{e}_z. \tag{5.64}$$

It follows that

$$\mathbf{g}_c = -\nabla \chi', \tag{5.65}$$

where

$$\chi' = \omega^2 \rho\, z \tag{5.66}$$

is the centrifugal potential and $z = r \cos\theta$. The centrifugal potential can also be written

$$\chi' = \frac{G\,m'}{a} \frac{r}{a} P_1(\cos\theta). \tag{5.67}$$

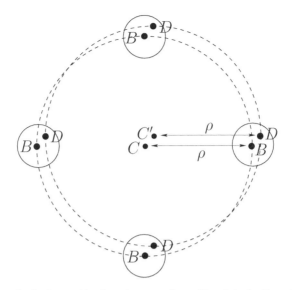

Fig. 5.8 The center B of mass distribution m orbits about the center of mass C in a circle of radius ρ. If m is nonrotating, then a noncentral point D maintains a constant spatial relationship to B, such that D orbits some point C' that has the same spatial relationship to C that D has to B, in a circle of radius ρ.

The gravitational acceleration at point D due to mass m' is given by

$$\mathbf{g}_g = -\nabla \Phi', \tag{5.68}$$

where the gravitational potential takes the form

$$\Phi' = -\frac{G\,m'}{a'}. \tag{5.69}$$

Here, a' is the distance between points A and D. The gravitational potential generated by the mass distribution m' is the same as that generated by an equivalent point mass at A, as long as the distribution is spherically symmetric, which we shall assume to be the case.

Now,

$$\mathbf{a}' = \mathbf{a} - \mathbf{r}, \tag{5.70}$$

where $\mathbf{a}'$ is the vector $\overrightarrow{DA}$, and $\mathbf{a}$ the vector $\overrightarrow{BA}$. (See Figure 5.7.) It follows that

$$a'^{-1} = \left(a^2 - 2\,\mathbf{a} \cdot \mathbf{r} + r^2\right)^{-1/2} = \left(a^2 - 2\,a\,r\cos\theta + r^2\right)^{-1/2}. \tag{5.71}$$

Expanding in powers of r/a, we obtain

$$a'^{-1} = a^{-1} \sum_{n=0,\infty} \left(\frac{r}{a}\right)^n P_n(\cos\theta). \tag{5.72}$$

Hence,

$$\Phi' \simeq -\frac{G\,m'}{a}\left[1 + \frac{r}{a}\,P_1(\cos\theta) + \frac{r^2}{a^2}\,P_2(\cos\theta)\right], \tag{5.73}$$

to second order in r/a, where the $P_n(x)$ are Legendre polynomials.

Adding χ' and Φ', we find that

$$\chi = \chi' + \Phi' \simeq -\frac{G\,m'}{a}\left[1 + \frac{r^2}{a^2}\,P_2(\cos\theta)\right], \qquad (5.74)$$

to second order in r/a. Note that χ is the potential due to the net externally generated force acting on the mass distribution m. This potential is constant up to first order in r/a, because the first-order variations in χ' and Φ' cancel each other. The cancellation is a manifestation of the balance between the centrifugal and gravitational accelerations in the equivalent point mass problem discussed earlier. However, this balance is exact only at the center of the mass distribution m. Away from the center, the centrifugal acceleration remains constant, whereas the gravitational acceleration increases with increasing z. Hence, at positive z, the gravitational acceleration is larger than the centrifugal acceleration, giving rise to a net acceleration in the $+z$ direction. Likewise, at negative z, the centrifugal acceleration is larger than the gravitational giving rise to a net acceleration in the $-z$ direction. It follows that the mass distribution m is subject to a residual acceleration, represented by the second-order variation in Equation (5.74), that acts to elongate it along the z-axis. This effect is known as *tidal elongation*.

Suppose that the mass distribution m is a sphere of radius R and uniform density γ, made up of rock similar to that found in the Earth's mantle. Let us estimate the elongation of this distribution due to the *tidal potential* specified in Equation (5.74), which (neglecting constant terms) can be written

$$\chi(r,\theta) = \frac{G\,m}{R}\left(\frac{r}{R}\right)^2 \zeta\,P_2(\cos\theta). \qquad (5.75)$$

Here, the dimensionless parameter

$$\zeta = -\frac{m'}{m}\left(\frac{R}{a}\right)^3 \qquad (5.76)$$

is (minus) the typical ratio of the tidal acceleration to the gravitational acceleration at $r \sim R$. Let us assume that $|\zeta| \ll 1$. By analogy with the analysis in the previous section, in the presence of the tidal potential, the distribution becomes slightly spheroidal in shape, such that its outer boundary satisfies Equation (5.38). Moreover, the induced ellipticity, ϵ, of the distribution is related to the normalized amplitude, ζ, of the tidal potential according to Equation (5.49) if $|\zeta| < \zeta_c \simeq 2 \times 10^{-4}$, and according to Equation (5.44) if $|\zeta| > \zeta_c$. In the former case, the distribution responds elastically to the tidal potential, whereas in the latter case it responds as a liquid.

Consider the tidal elongation of the Earth due to the Moon. In this case, we have $R = 6.37 \times 10^6$ m, $a = 3.84 \times 10^8$ m, $m = 5.97 \times 10^{24}$ kg, and $m' = 7.35 \times 10^{22}$ kg (Yoder 1995). Hence, we find that

$$\zeta = -5.62 \times 10^{-8}. \qquad (5.77)$$

Note that $|\zeta| \ll \zeta_c$. We conclude that the Earth responds elastically to the tidal potential of the Moon, rather than deforming like a liquid. For the rock that makes up the Earth's mantle, $\mu \simeq 1 \times 10^{11}$ N m^{-2} and $\gamma \simeq 5 \times 10^3$ kg m^{-3} (de Pater and Lissauer 2010). Thus, it follows from Equation (5.50) that

$$\tilde{\mu} \simeq 3.35. \qquad (5.78)$$

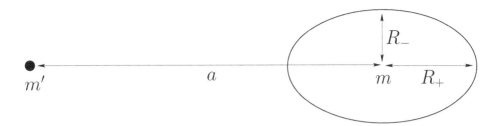

Fig. 5.9 Tidal elongation.

Hence, according to Equation (5.49), the ellipticity of the Earth induced by the tidal effect of the Moon is

$$\epsilon = \frac{15}{4}\left(\frac{\zeta}{1+\bar{\mu}}\right) \simeq -4.8 \times 10^{-8}. \tag{5.79}$$

The fact that ϵ is negative implies that the Earth is elongated along the z-axis, that is, along the axis joining its center to that of the Moon. See Equation (5.38). If R_+ and R_- are the greatest and least radii of the Earth, respectively, due to this elongation (see Figure 5.9), then

$$\Delta R = R_+ - R_- = -\epsilon R = 0.31 \text{ m}. \tag{5.80}$$

Thus, we predict that the tidal effect of the Moon (which is actually due to spatial gradients in the Moon's gravitational field) causes the Earth to elongate along the axis joining its center to that of the Moon by about 31 centimeters. This elongation is only about a quarter of that which would result were the Earth a nonrigid (i.e., liquid) body. The true tidal elongation of the Earth due to the Moon is about 35 centimeters [assuming a Love number $h_2 \simeq 0.6$ (Bertotti et al. 2003)]. We have slightly underestimated this elongation because, for the sake of simplicity, we treated the Earth as a uniform-density body.

Consider the tidal elongation of the Earth due to the Sun. In this case, we have $R = 6.37 \times 10^6$ m, $a = 1.50 \times 10^{11}$ m, $m = 5.97 \times 10^{24}$ kg, and $m' = 1.99 \times 10^{30}$ kg. Hence, we calculate that $\zeta = -2.55 \times 10^{-8}$ and $\epsilon = -2.2 \times 10^{-8}$, or

$$\Delta R = R_+ - R_- = -\epsilon R = 0.14 \text{ m}. \tag{5.81}$$

Thus, the tidal elongation of the Earth due to the Sun is about half that due to the Moon. The true tidal elongation of the Earth due to the Sun is about 16 centimeters [assuming a Love number $h_2 \sim 0.6$ (Bertotti et al. 2003)]. Again, we have slightly underestimated the elongation because we treated the Earth as a uniform-density body.

Because the Earth's oceans are liquid, their tidal elongation is significantly larger than that of the underlying land. (See Exercise 5.10.) Hence, the oceans rise, relative to the land, in the region of the Earth closest to the Moon, and also in the region farthest away. Because the Earth is rotating, whereas the tidal bulge of the oceans remains relatively stationary, the Moon's tidal effect causes the ocean at a given point on the Earth's surface to rise and fall twice daily, giving rise to the phenomenon known as the *tides*. There is also an oceanic tidal bulge due to the Sun that is about half as large as that due to the Moon. Consequently, ocean tides are particularly high when the Sun, the Earth, and the Moon lie approximately in a straight line, so the tidal effects of the Sun and the Moon reinforce one another. This occurs at a new moon, or at a full moon. These types of tides are called *spring tides* (the name has nothing to do with the season). Conversely, ocean tides are particularly low when the Sun, the Earth, and the Moon form a right angle, so that the tidal effects of the Sun and the Moon partially cancel one another. These type of tides are called *neap tides*. Generally, we would expect two spring tides and two neap tides per month.

We can roughly calculate the vertical displacement of the oceans, relative to the underlying land, by treating the oceans as a shallow layer of negligible mass, covering the surface of the Earth. The Earth's external gravitational potential is written [see Equation (2.65)]

$$\Phi(r, \theta) = -\frac{G\,m}{r} + \frac{2}{5}\,\epsilon\,\frac{G\,m\,R^2}{r^3}\,P_2(\cos\theta), \tag{5.82}$$

where ϵ is given by Equation (5.79). Let the ocean surface satisfy

$$r = R'(\theta) = R\left[1 - \frac{2}{3}\,\epsilon'\,P_2(\cos\theta)\right]. \tag{5.83}$$

Because fluids cannot withstand shear stresses, we expect this surface to be an equipotential:

$$\Phi(R'_\theta, \theta) + \chi(R'_\theta, \theta) = c. \tag{5.84}$$

It follows that, to first order in ϵ' and ζ,

$$\epsilon' = \frac{3}{5}\,\epsilon + \frac{3}{2}\,\zeta = \frac{3}{2}\left(\frac{5/2 + \tilde{\mu}}{1 + \tilde{\mu}}\right)\zeta. \tag{5.85}$$

Thus, the maximum vertical displacement of the ocean relative to the underlying land is

$$\Delta R = -(\epsilon' - \epsilon)\,R = -\frac{3}{2}\left(\frac{\tilde{\mu}}{1 + \tilde{\mu}}\right)\zeta\,R. \tag{5.86}$$

As we saw earlier, $\tilde{\mu} \simeq 3.35$ for the Earth. Moreover, the tidal potential due to the Moon is such that $\zeta = -5.62 \times 10^{-8}$. We thus conclude that the Moon causes the oceans to rise a maximum vertical distance of 0.41 m relative to the land. Likewise, the tidal potential due to the Sun is such that $\zeta = -2.55 \times 10^{-8}$. Hence, we predict that the Sun causes the oceans to rise a maximum vertical distance of 0.19 m relative to the land.

In reality, the relationship between ocean tides and the Moon and Sun is much more complicated than that indicated in the previous discussion. This is partly because of the presence of the continents, which impede the flow of the oceanic tidal bulge around the Earth, and partly because of the finite inertia of the oceans.

Note, finally, that as a consequence of friction within the Earth's crust and friction between the oceans and the underlying land, there is a time lag of roughly 12 minutes between the Moon (or Sun) passing directly overhead (or directly below) and the corresponding maximum in the net tidal elongation of the Earth and the oceans (Bertotti et al. 2003).

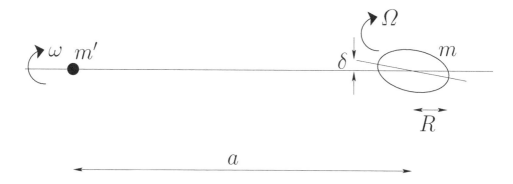

Fig. 5.10 Origin of tidal torque.

5.7 Tidal torques

The fact that there is a time lag between the Moon passing overhead and the corresponding maximum net tidal elongation of the Earth and the oceans suggests the physical scenario illustrated in Figure 5.10. According to this scenario, the Moon, which is of mass m' and which is treated as a point particle, orbits the Earth (it actually orbits the center of mass of the Earth–Moon system, but this amounts to almost the same thing) in an approximately circular orbit of radius a. Moreover, the orbital angular velocity of the Moon is [see Equation (5.59)]

$$\omega = \frac{(G M)^{1/2}}{a^{3/2}}, \tag{5.87}$$

where $M = m + m' \simeq m$ is the total mass of the Earth–Moon system. The Earth (including the oceans) is treated as a uniform sphere of mass m and radius R that rotates daily about its axis (which is approximately normal to the orbital plane of the Moon) at the angular velocity Ω. Note, incidentally, that the Earth rotates in the same sense that the Moon orbits, as indicated in the figure. As we saw in the previous section, spatial gradients in the Moon's gravitational field produce a slight tidal elongation of the Earth. However, because of frictional effects, this elongation does not quite line up along the axis joining the centers of the Earth and Moon. In fact, because $\Omega > \omega$, the tidal elongation is carried ahead (in the sense defined by the Earth's rotation) of this axis by some small angle δ (say), as shown in the figure.

Defining a spherical coordinate system, r, θ, ϕ, whose origin is the center of the Earth, and which is oriented such that the Earth–Moon axis always corresponds to $\theta = 0$ (see Figure 5.7), we find the Earth's external gravitational potential is [cf. Equation (2.65)]

$$\Phi(r,\theta) = -\frac{G m}{r} + \frac{\epsilon}{5} \frac{G m R^2}{r^3} \left[3 \cos^2(\theta - \delta) - 1 \right], \tag{5.88}$$

where ϵ is the ellipticity induced by the tidal field of the Moon. Note that the second term on the right-hand side of this expression is the contribution of the Earth's tidal bulge, which attains its maximum amplitude at $\theta = \delta$, rather than $\theta = 0$, because of the aforementioned misalignment between the bulge and the Earth–Moon axis. Equations (5.76), (5.79), and (5.88) can be combined to give

$$\Phi(r, \theta) = -\frac{G m}{r} - \frac{3}{4} \frac{G m' R^2}{(1 + \tilde{\mu}) r^3} \left(\frac{R}{a}\right)^3 \left[3 \cos^2(\theta - \delta) - 1\right]. \tag{5.89}$$

From Equation (2.7), the torque about the Earth's center that the terrestrial gravitational field exerts on the Moon is

$$\tau = -m' \left.\frac{\partial \Phi}{\partial \theta}\right|_{\theta=0, r=a} \simeq \frac{9}{2} \frac{G m'^2}{R} \left(\frac{R}{a}\right)^6 \frac{\delta}{1 + \tilde{\mu}}, \tag{5.90}$$

where use has been made of Equation (5.89), as well as the fact that δ is a small angle. There is zero torque in the absence of a misalignment between the Earth's tidal bulge and the Earth–Moon axis. The torque τ acts to increase the Moon's orbital angular momentum. By conservation of angular momentum, an equal and opposite torque, $-\tau$, is applied to the Earth; it acts to decrease its rotational angular momentum. Incidentally, if the Moon were sufficiently close to the Earth that its orbital angular velocity exceeded the Earth's rotational angular velocity (i.e., if $\omega > \Omega$), then the phase lag between the Earth's tidal elongation and the Moon's tidal field would cause the tidal bulge to fall slightly behind the Earth–Moon axis (i.e., $\delta < 0$). In this case, the gravitational torque would act to reduce the Moon's orbital angular momentum and to increase the Earth's rotational angular momentum.

The Earth's rotational equation of motion is

$$\mathcal{I}_{\|} \dot{\Omega} = -\tau, \tag{5.91}$$

where $\mathcal{I}_{\|}$ is its moment of inertia about its axis of rotation. Very crudely approximating the Earth as a uniform sphere, we have $\mathcal{I}_{\|} = (2/5) m R^2$. Hence, the previous two equations can be combined to give

$$\frac{\dot{\Omega}}{\Omega} \simeq -\frac{45}{4} \frac{\omega^2}{\Omega} \left(\frac{m'}{m}\right)^2 \left(\frac{R}{a}\right)^3 \frac{\delta}{1 + \tilde{\mu}}, \tag{5.92}$$

where use has been made of Equation (5.87), as well as the fact that $m \gg m'$. A time lag of 12 minutes between the Moon being overhead and a maximum of the Earth's tidal elongation implies a lag angle of $\delta \sim 0.05$ radians (i.e., $\delta \sim 3°$). Hence, employing the observed values $m = 5.97 \times 10^{24}$ kg, $m' = 7.35 \times 10^{22}$ kg, $R = 6.37 \times 10^6$ m, $a = 3.84 \times 10^8$ m, $\Omega = 7.29 \times 10^{-5}$ rad. s^{-1}, and $\omega = 2.67 \times 10^{-6}$ rad. s^{-1} (Yoder 1995), as well as the estimate $\tilde{\mu} \simeq 3.35$ (from the previous section), we find that

$$\frac{\dot{\Omega}}{\Omega} \simeq -8.7 \times 10^{-18} \text{ s}^{-1}. \tag{5.93}$$

It follows that under the influence of the tidal torque, the Earth's axial rotation is gradually decelerating. Indeed, according to this estimate, the length of a day should be increasing at the rate of about 2.3 milliseconds per century. An analysis of ancient and medieval solar and lunar eclipse records indicates that the length of the day is actually increasing at the rate of 1.7 milliseconds per century (Stephenson and Morrison 1995). The timescale for the tidal torque to significantly reduce the Earth's rotational angular velocity is estimated to be

$$T \simeq \frac{\Omega}{|\dot{\Omega}|} \simeq 4 \times 10^9 \text{ years.} \qquad (5.94)$$

This timescale is comparable with the Earth's age, which is thought to be 4.5×10^9 years. Hence, we conclude that, although the Earth is certainly old enough for the tidal torque to have significantly reduced its rotational angular velocity, it is plausible that it is not sufficiently old for the torque to have driven the Earth–Moon system to a final steady state. Such a state, in which the Earth's rotational angular velocity matches the Moon's orbital angular velocity, is termed *synchronous*. In a synchronous state, the Moon would appear stationary to an observer on the Earth's surface, and, hence, there would be no tides (from the observer's perspective), no phase lag, and no tidal torque.

Up to now, we have concentrated on the effect of the tidal torque on the rotation of the Earth. Let us now examine its effect on the orbit of the Moon. The total angular momentum of the Earth–Moon system is

$$L \simeq \frac{2}{5} m R^2 \Omega + m' a^2 \omega, \qquad (5.95)$$

where the first term on the right-hand side is the rotational angular momentum of the Earth and the second the orbital angular momentum of the Moon. Of course, L is a conserved quantity. Moreover, ω and a are related according to Equation (5.87). It follows that

$$\frac{\dot{a}}{a} \simeq -\frac{4}{5} \frac{m}{m'} \left(\frac{R}{a}\right)^2 \frac{\dot{\Omega}}{\omega} \simeq 4.3 \times 10^{-18} \text{ s}^{-1}, \qquad (5.96)$$

where use has been made of Equation (5.93). In other words, the tidal torque causes the radius of the Moon's orbit to gradually increase. According to this estimate, this increase should take place at the rate of about 5 cm a year. The observed rate, which is obtained from lunar laser ranging data, is 3.8 cm a year (Chapront et al. 2002). This suggests that, despite the numerous approximations we have made, our calculation remains reasonably accurate. We also have

$$\frac{\dot{\omega}}{\omega} = -\frac{3}{2} \frac{\dot{a}}{a} \simeq -6.5 \times 10^{-18} \text{ s}^{-1}. \qquad (5.97)$$

In other words, the tidal torque produces a gradual angular deceleration in the Moon's orbital motion. According to the above estimate, this deceleration should take place at the rate of 35 arc seconds per century squared. The measured deceleration is about 26 arc seconds per century squared (Yoder 1995).

The net rate at which the tidal torques acting on the Moon and the Earth do work is

$$\dot{E} = \tau (\omega - \Omega). \qquad (5.98)$$

Note that $\dot{E} < 0$, because $\tau > 0$ and $\Omega > \omega$. This implies that the deceleration of the Earth's rotation and that of the Moon's orbital motion induced by tidal torques are necessarily associated with the dissipation of energy. This dissipation manifests itself as frictional heating of the Earth's crust and the oceans (Bertotti et al. 2003).

Of course, we would expect spatial gradients in the gravitational field of the Earth to generate a tidal bulge in the Moon. We would also expect dissipative effects to produce a phase lag between this bulge and the Earth. This would allow the Earth to exert a gravitational torque that acts to drive the Moon toward a synchronous state in which its rotational angular velocity matches its orbital angular velocity. By analogy with the previous analysis, the de-spinning rate of the Moon is estimated to be

$$\frac{\dot{\Omega}'}{\Omega'} \simeq -\frac{45}{4}\frac{\omega^2}{\Omega'}\frac{m}{m'}\left(\frac{R'}{a}\right)^3\frac{\delta}{1+\tilde{\mu}} \sim -2.2 \times 10^{-14}\,\mathrm{s}^{-1}, \tag{5.99}$$

where

$$\tilde{\mu} = \frac{57}{8\pi}\frac{\mu}{G\,\gamma^2\,R'^{\,2}} \simeq 51.5 \tag{5.100}$$

is the Moon's effective rigidity, Ω' its rotational angular velocity, $R' = 1.74 \times 10^6\,\mathrm{m}$ its radius, $\gamma = 3.3 \times 10^3\,\mathrm{m\,s}^{-1}$ its density (Yoder 1995), $\mu \sim 5 \times 10^{10}\,\mathrm{N\,m}^{-2}$ its shear modulus (Zhang 1992), and δ the lag angle. The above numerical estimate is made with the guesses $\Omega' = 2\,\omega$ and $\delta = 0.01$ radians. Thus, the time required for the Moon to achieve a synchronous state is

$$T' \simeq \frac{\Omega'}{|\dot{\Omega}'|} \simeq 1.5 \times 10^6 \text{ years.} \tag{5.101}$$

This is considerably less than the age of the Moon. Hence, it is not surprising that the Moon has actually achieved a synchronous state, as evidenced by the fact that the same side of the Moon is always visible from the Earth. (See Section 7.11.)

5.8 Roche radius

Consider a spherical moon of mass m and radius R that is in a circular orbit of radius a about a spherical planet of mass m' and radius R'. (Strictly speaking, the moon and the planet execute circular orbits about their common center of mass. However, if the planet is much more massive than the moon, the center of mass lies very close to the planet's center.) According to the analysis in Section 5.6, a constituent element of the moon experiences a force per unit mass, due to the gravitational field of the planet, that takes the form

$$\mathbf{g}' = -\nabla\chi, \tag{5.102}$$

where

$$\chi = -\frac{G\,m'}{a^3}(z^2 - x^2/2 - y^2/2) + \text{const.} \tag{5.103}$$

Here, x, y, z is a Cartesian coordinate system whose origin is the center of the moon, and whose z-axis always points toward the center of the planet. It follows that

$$\mathbf{g}' = \frac{2\,G\,m'}{a^3}\left(-\frac{x}{2}\,\mathbf{e}_x - \frac{y}{2}\,\mathbf{e}_y + z\,\mathbf{e}_z\right). \tag{5.104}$$

This so-called *tidal force* is generated by the spatial variation of the planet's gravitational field over the interior of the moon and acts to elongate the moon along an axis joining its center to that of the planet and to compress it in any direction perpendicular to this axis. Note that the magnitude of the tidal force increases strongly as the radius, a, of the moon's orbit decreases. In fact, if the tidal force becomes sufficiently strong, it can overcome the moon's self-gravity and thereby rip the moon apart. It follows that there is a minimum radius, generally referred to as the *Roche radius*, at which a moon can orbit a planet without being destroyed by tidal forces.

Let us derive an expression for the Roche radius. Consider a small mass element at the point on the surface of the moon that lies closest to the planet, and at which the tidal force is consequently largest (i.e., $x = y = 0$, $z = R$). According to Equation (5.104), the mass experiences an upward (from the moon's surface) tidal acceleration due to the gravitational attraction of the planet of the form

$$\mathbf{g}' = \frac{2\,G\,m'\,R}{a^3}\,\mathbf{e}_z. \tag{5.105}$$

The mass also experiences a downward gravitational acceleration due to the gravitational influence of the moon, which is written

$$\mathbf{g} = -\frac{G\,m}{R^2}\,\mathbf{e}_z. \tag{5.106}$$

Thus, the effective surface gravity at the point in question is

$$g_{\text{eff}} = \frac{G\,m}{R^2}\left(1 - 2\,\frac{m'}{m}\,\frac{R^3}{a^3}\right). \tag{5.107}$$

If $a < a_c$, where

$$a_c = \left(2\,\frac{m'}{m}\right)^{1/3} R, \tag{5.108}$$

then the effective gravity is negative; in other words, the tidal force due to the planet is strong enough to overcome surface gravity and lift objects off the moon's surface. If this is the case, and the tensile strength of the moon is negligible, then it is fairly clear that the tidal force will start to break the moon apart. Hence, a_c is the Roche radius. Now, $m'/m = (\rho'/\rho)\,(R'/R)^3$, where ρ and ρ' are the mean mass densities of the moon and planet, respectively. Thus, the above expression for the Roche radius can also be written

$$a_c = 1.41 \left(\frac{\rho'}{\rho}\right)^{1/3} R'. \tag{5.109}$$

The previous calculation is somewhat inaccurate, as it fails to take into account the inevitable distortion of the moon's shape in the presence of strong tidal forces. (In fact, the calculation assumes that the moon always remains spherical.) A more accurate calculation, which treats the moon as a self-gravitating incompressible fluid body, yields (Chandrasekhar 1969)

$$a_c = 2.44 \left(\frac{\rho'}{\rho}\right)^{1/3} R'. \tag{5.110}$$

It follows that if the planet and the moon have the same mean density, then the Roche radius is 2.44 times the planet's radius. Note that small bodies such as rocks, or even very small moons, can survive intact within the Roche radius because they are held together by internal tensile forces rather than by gravitational attraction. However, this mechanism becomes progressively less effective as the size of the body in question increases. (See Section 2.6.) Not surprisingly, virtually all large planetary moons found in the solar system have orbital radii that exceed the relevant Roche radius, whereas virtually all planetary ring systems (which consist of myriads of small orbiting rocks) have radii that lie inside the relevant Roche radius.

Exercises

5.1 A ball bearing is dropped down an elevator shaft in the Empire State Building ($h = 381$ m, latitude $= 41°$ N). Find the ball bearing's horizontal deflection (magnitude and direction) at the bottom of the shaft due to the Coriolis force. Neglect air resistance. (Modified from Fowles and Cassiday 2005.)

5.2 A projectile is fired due east, at an elevation angle β, from a point on the Earth's surface whose latitude is $+\lambda$. Demonstrate that the projectile strikes the ground with a lateral deflection $4\Omega v_0^3 \sin \lambda \sin^2 \beta \cos \beta / g^2$. Is the deflection northward or southward? Here, Ω is the Earth's angular velocity, v_0 the projectile's initial speed, and g the acceleration due to gravity. Neglect air resistance. (Modified from Thornton and Marion 2004.)

5.3 A particle is thrown vertically upward, reaches some maximum height, and falls back to the ground. Show that the horizontal Coriolis deflection of the particle when it returns to the ground is opposite in direction, and four times greater in magnitude, than the Coriolis deflection when it is dropped at rest from the same maximum height. Neglect air resistance. (From Goldstein et al. 2002.)

5.4 A ball of mass m rolls without friction over a horizontal plane located on the surface of the Earth. Show that in the northern hemisphere it rolls in a clockwise sense (seen from above) around a circle of radius

$$r = \frac{v}{2\Omega \sin \lambda},$$

where v is the speed of the ball, Ω the Earth's angular velocity, and λ the terrestrial latitude.

5.5 A satellite is in a circular orbit of radius a about the Earth. Let us define a set of co-moving Cartesian coordinates, centered on the satellite, such that the x-axis always points toward the center of the Earth, the y-axis in the direction of the satellite's orbital motion, and the z-axis in the direction of the satellite's orbital angular velocity, ω. Demonstrate that the equations of motion of a small mass in orbit about the satellite are

$$\ddot{x} = 3\omega^2 x + 2\omega \dot{y}$$

and

$$\ddot{y} = -2\omega \dot{x},$$

assuming that $|x|/a \ll 1$ and $|y|/a \ll 1$. Neglect the gravitational attraction between the satellite and the mass. Show that the mass executes a retrograde (i.e., in the opposite sense to the satellite's orbital rotation) elliptical orbit about the satellite whose period matches that of the satellite's orbit, and whose major and minor axes are in the ratio 2:1, and are aligned along the y- and x-axes, respectively.

5.6 Show that

$$\epsilon = \frac{5}{2(2+\alpha)}\frac{\Omega^2 R^3}{GM}$$

for a self-gravitating, rotating, fluidlike, spheroid of ellipticity $\epsilon \ll 1$, mass M, mean radius R, and angular velocity Ω, whose mass density varies as $r^{-\alpha}$ (where $\alpha < 3$). Demonstrate that this formula matches the observed rotational flattening of the Earth when $\alpha = 0.575$, and of Jupiter when $\alpha = 1.12$. (See Exercise 2.9.)

5.7 Treating the Earth as a uniform-density, liquid spheroid, and taking rotational flattening into account, show that the variation of the surface acceleration, g, with terrestrial latitude, λ, is

$$g \simeq \frac{GM}{R^2} + \frac{1}{6}\Omega^2 R - \frac{5}{4}\Omega^2 R \cos^2 \lambda.$$

Here, M is the terrestrial mass, R the mean terrestrial radius, and Ω the terrestrial axial angular velocity.

5.8 The Moon's orbital period about the Earth is approximately 27.32 days, and is in the same direction as the Earth's axial rotation (whose period is 24 hours). Use these data to show that high tides at a given point on the Earth occur every 12 hours and 27 minutes.

5.9 Demonstrate that the mean time interval between successive spring tides is 14.76 days.

5.10 Let us model the Earth as a completely rigid sphere that is covered by a shallow ocean of negligible density. Demonstrate that the tidal elongation of the ocean layer due to the Moon is

$$\frac{\Delta R}{R} = \frac{3}{2}\frac{m'}{m}\left(\frac{R}{a}\right)^3,$$

where m is the mass of the Earth, m' the mass of the Moon, R the radius of the Earth, and a the radius of the lunar orbit. Show that $\Delta R = 0.54$ m, and also that the tidal elongation of the ocean layer due to the Sun is such that $\Delta R = 0.24$ m.

5.11 Estimate the tidal heating rate of the Earth due to the Moon. Is this rate significant compared with the net heating rate from solar radiation?

5.12 Estimate the tidal elongation of the Sun due to the Earth.

5.13 An approximately spherical moon of uniform density, mass m, radius R, and effective rigidity $\tilde{\mu}$ is in a circular orbit of major radius a about a spherical planet of mass $m_p \gg m$. The moon rotates about an axis passing through its center of mass that is directed normal to the orbital plane. Suppose that the moon is in a synchronous state such that its rotational angular velocity, Ω, matches its orbital angular velocity, $\omega \simeq (Gm_p/a^3)^{1/2}$. Let x, y, z be a set of Cartesian coordinates, centered on the moon, such that the z-axis is normal to the orbital plane, and the x-axis is directed from the center of the moon to the center of the planet. Assuming that the moon responds elastically to the centrifugal and tidal potentials, show that the changes in radius, parallel to the three coordinate axes, induced by the centrifugal potential are

$$\frac{\delta R_x}{R} = \frac{5}{12} \frac{\omega^2 R^3}{G m} \frac{1}{1 + \tilde{\mu}},$$

$$\frac{\delta R_y}{R} = \frac{5}{12} \frac{\omega^2 R^3}{G m} \frac{1}{1 + \tilde{\mu}},$$

and

$$\frac{\delta R_z}{R} = -\frac{5}{6} \frac{\omega^2 R^3}{G m} \frac{1}{1 + \tilde{\mu}},$$

whereas the corresponding changes induced by the tidal potential are

$$\frac{\delta R_x}{R} = \frac{5}{2} \frac{\omega^2 R^3}{G m} \frac{1}{1 + \tilde{\mu}},$$

$$\frac{\delta R_y}{R} = -\frac{5}{4} \frac{\omega^2 R^3}{G m} \frac{1}{1 + \tilde{\mu}},$$

and

$$\frac{\delta R_z}{R} = -\frac{5}{4} \frac{\omega^2 R^3}{G m} \frac{1}{1 + \tilde{\mu}}.$$

Assuming that these changes are additive, deduce that

$$\frac{\delta R_y - \delta R_z}{\delta R_x - \delta R_z} = \frac{1}{4}.$$

Estimate δR_x, δR_y, and δR_z for the Earth's Moon (which is in a synchronous state).

5.14 An artificial satellite consists of two point objects of mass $m/2$ connected by a light rigid rod of length l. The satellite is placed in a circular orbit of radius $a \gg l$ (measured from the midpoint of the rod) around a planet of mass m'. The rod is oriented such that it always points toward the center of the planet. Demonstrate that the tension in the rod is

$$T = \frac{3}{4} \frac{G m m' l}{a^3} - \frac{1}{4} \frac{G m^2}{l^2}.$$

Lagrangian mechanics

6.1 Introduction

This chapter describes an elegant reformulation of the laws of Newtonian mechanics that is due to the French-Italian scientist Joseph Louis Lagrange (1736–1813). This reformulation is particularly useful for finding the equations of motion of complicated dynamical systems.

6.2 Generalized coordinates

Let the q_i, for $i = 1, \mathcal{F}$, be a set of coordinates that uniquely specifies the instantaneous configuration of some dynamical system. Here, it is assumed that each of the q_i can vary independently. The q_i might be Cartesian coordinates, angles, or some mixture of both types of coordinate, and are therefore termed *generalized coordinates*. A dynamical system whose instantaneous configuration is fully specified by $\mathcal{F}$ independent generalized coordinates is said to have $\mathcal{F}$ *degrees of freedom*. For instance, the instantaneous position of a particle moving freely in three dimensions is completely specified by its three Cartesian coordinates, x, y, and z. Moreover, these coordinates are clearly independent of one another. Hence, a dynamical system consisting of a single particle moving freely in three dimensions has three degrees of freedom. If there are two freely moving particles then the system has six degrees of freedom, and so on.

Suppose that we have a dynamical system consisting of N particles moving freely in three dimensions. This is an $\mathcal{F} = 3N$ degree-of-freedom system whose instantaneous configuration can be specified by $\mathcal{F}$ Cartesian coordinates. Let us denote these coordinates the x_j, for $j = 1, \mathcal{F}$. Thus, x_1, x_2, x_3 are the Cartesian coordinates of the first particle, x_4, x_5, x_6 the Cartesian coordinates of the second particle, and so on. Suppose that the instantaneous configuration of the system can also be specified by $\mathcal{F}$ generalized coordinates, which we shall denote the q_i, for $i = 1, \mathcal{F}$. Thus, the q_i might be the spherical coordinates of the particles. In general, we expect the x_j to be functions of the q_i. In other words,

$$x_j = x_j(q_1, q_2, \ldots, q_\mathcal{F}, t) \tag{6.1}$$

for $j = 1, \mathcal{F}$. Here, for the sake of generality, we have included the possibility that the functional relationship between the x_j and the q_i might depend on the time, t, explicitly. This would be the case if the dynamical system were subject to time-varying constraints—for instance, a system consisting of a particle constrained to move on a surface that is itself moving. Finally, by the chain rule, the variation of the x_j due to a variation of the q_i (at constant t) is given by

$$\delta x_j = \sum_{i=1,\mathcal{F}} \frac{\partial x_j}{\partial q_i} \delta q_i \qquad (6.2)$$

for $j = 1, \mathcal{F}$.

6.3 Generalized forces

The work done on the dynamical system when its Cartesian coordinates change by δx_j is simply

$$\delta W = \sum_{j=1,\mathcal{F}} f_j \, \delta x_j. \qquad (6.3)$$

Here, the f_j are the Cartesian components of the forces acting on the various particles making up the system. Thus, f_1, f_2, f_3 are the components of the force acting on the first particle, f_4, f_5, f_6 the components of the force acting on the second particle, and so on. Using Equation (6.2), we can also write

$$\delta W = \sum_{j=1,\mathcal{F}} f_j \sum_{i=1,\mathcal{F}} \frac{\partial x_j}{\partial q_i} \delta q_i. \qquad (6.4)$$

The preceding expression can be rearranged to give

$$\delta W = \sum_{i=1,\mathcal{F}} Q_i \, \delta q_i, \qquad (6.5)$$

where

$$Q_i = \sum_{j=1,\mathcal{F}} f_j \frac{\partial x_j}{\partial q_i}. \qquad (6.6)$$

Here, the Q_i are termed *generalized forces*. A generalized force does not necessarily have the dimensions of force. However, the product $Q_i q_i$ must have the dimensions of work. Thus, if a particular q_i is a Cartesian coordinate, then the associated Q_i is a force. Conversely, if a particular q_i is an angle, then the associated Q_i is a torque.

Suppose that the dynamical system in question is *conservative*. It follows that

$$f_j = -\frac{\partial U}{\partial x_j} \qquad (6.7)$$

for $j = 1, \mathcal{F}$, where $U(x_1, x_2, \ldots, x_\mathcal{F}, t)$ is the system's potential energy. Hence, according to Equation (6.6),

$$Q_i = -\sum_{j=1,\mathcal{F}} \frac{\partial U}{\partial x_j} \frac{\partial x_j}{\partial q_i} = -\frac{\partial U}{\partial q_i} \qquad (6.8)$$

for $i = 1, \mathcal{F}$.

6.4 Lagrange's equation

The Cartesian equations of motion of our system take the form

$$m_j \ddot{x}_j = f_j \tag{6.9}$$

for $j = 1, \mathcal{F}$, where m_1, m_2, m_3 are each equal to the mass of the first particle; m_4, m_5, m_6 are each equal to the mass of the second particle; and so forth. Furthermore, the kinetic energy of the system can be written

$$K = \frac{1}{2} \sum_{j=1,\mathcal{F}} m_j \dot{x}_j^2. \tag{6.10}$$

Because $x_j = x_j(q_1, q_2, \ldots, q_{\mathcal{F}}, t)$, we can write

$$\dot{x}_j = \sum_{i=1,\mathcal{F}} \frac{\partial x_j}{\partial q_i} \dot{q}_i + \frac{\partial x_j}{\partial t} \tag{6.11}$$

for $j = 1, \mathcal{F}$. Hence, it follows that $\dot{x}_j = \dot{x}_j(\dot{q}_1, \dot{q}_2, \ldots, \dot{q}_{\mathcal{F}}, q_1, q_2, \ldots, q_{\mathcal{F}}, t)$. According to the preceding equation,

$$\frac{\partial \dot{x}_j}{\partial \dot{q}_i} = \frac{\partial x_j}{\partial q_i}, \tag{6.12}$$

where we are treating the $\dot{q}_i$ and the q_i as *independent* variables.

Multiplying Equation (6.12) by $\dot{x}_j$, and then differentiating with respect to time, we obtain

$$\frac{d}{dt}\left(\dot{x}_j \frac{\partial \dot{x}_j}{\partial \dot{q}_i} \right) = \frac{d}{dt}\left(\dot{x}_j \frac{\partial x_j}{\partial q_i} \right) = \ddot{x}_j \frac{\partial x_j}{\partial q_i} + \dot{x}_j \frac{d}{dt}\left(\frac{\partial x_j}{\partial q_i} \right). \tag{6.13}$$

Now,

$$\frac{d}{dt}\left(\frac{\partial x_j}{\partial q_i} \right) = \sum_{k=1,\mathcal{F}} \frac{\partial^2 x_j}{\partial q_i\, \partial q_k} \dot{q}_k + \frac{\partial^2 x_j}{\partial q_i\, \partial t}. \tag{6.14}$$

Furthermore,

$$\frac{1}{2} \frac{\partial \dot{x}_j^2}{\partial \dot{q}_i} = \dot{x}_j \frac{\partial \dot{x}_j}{\partial \dot{q}_i} \tag{6.15}$$

and

$$\frac{1}{2} \frac{\partial \dot{x}_j^2}{\partial q_i} = \dot{x}_j \frac{\partial \dot{x}_j}{\partial q_i} = \dot{x}_j \frac{\partial}{\partial q_i}\left(\sum_{k=1,\mathcal{F}} \frac{\partial x_j}{\partial q_k} \dot{q}_k + \frac{\partial x_j}{\partial t} \right)$$

$$= \dot{x}_j \left(\sum_{k=1,\mathcal{F}} \frac{\partial^2 x_j}{\partial q_i\, \partial q_k} \dot{q}_k + \frac{\partial^2 x_j}{\partial q_i\, \partial t} \right) = \dot{x}_j \frac{d}{dt}\left(\frac{\partial x_j}{\partial q_i} \right), \tag{6.16}$$

where use has been made of Equation (6.14). Thus, it follows from Equations (6.13), (6.15), and (6.16) that

$$\frac{d}{dt}\left(\frac{1}{2} \frac{\partial \dot{x}_j^2}{\partial \dot{q}_i} \right) = \ddot{x}_j \frac{\partial x_j}{\partial q_i} + \frac{1}{2} \frac{\partial \dot{x}_j^2}{\partial q_i}. \tag{6.17}$$

Let us take Equation (6.17), multiply by m_j, and then sum over all j. We obtain

$$\frac{d}{dt}\left(\frac{\partial K}{\partial \dot{q}_i}\right) = \sum_{j=1,\mathcal{F}} f_j \frac{\partial x_j}{\partial q_i} + \frac{\partial K}{\partial q_i}, \tag{6.18}$$

where we have made use of Equations (6.9) and (6.10). Thus, it follows from Equation (6.6) that

$$\frac{d}{dt}\left(\frac{\partial K}{\partial \dot{q}_i}\right) = Q_i + \frac{\partial K}{\partial q_i}. \tag{6.19}$$

Finally, making use of Equation (6.8), we get

$$\frac{d}{dt}\left(\frac{\partial K}{\partial \dot{q}_i}\right) = -\frac{\partial U}{\partial q_i} + \frac{\partial K}{\partial q_i}. \tag{6.20}$$

It is helpful to introduce a function $\mathcal{L}$, called the *Lagrangian*, which is defined as the difference between the kinetic and potential energies of the dynamical system under investigation:

$$\mathcal{L} = K - U. \tag{6.21}$$

Becuase the potential energy U is clearly independent of the $\dot{q}_i$, it follows from Equation (6.20) that

$$\frac{d}{dt}\left(\frac{\partial \mathcal{L}}{\partial \dot{q}_i}\right) - \frac{\partial \mathcal{L}}{\partial q_i} = 0 \tag{6.22}$$

for $i = 1, \mathcal{F}$. This equation is known as *Lagrange's equation*.

According to the preceding analysis, if we can express the kinetic and potential energies of our dynamical system solely in terms of our generalized coordinates and their time derivatives, then we can immediately write down the equations of motion of the system, expressed in terms of the generalized coordinates, using Lagrange's equation, Equation (6.22). Unfortunately, this scheme works only for conservative systems.

As an example, consider a particle of mass m moving in two dimensions in the central potential $U(r)$. This is clearly a two-degree-of-freedom dynamical system. As described in Section 3.4, the particle's instantaneous position is most conveniently specified in terms of the plane polar coordinates r and θ. These are our two generalized coordinates. According to Equation (3.13), the square of the particle's velocity can be written

$$v^2 = \dot{r}^2 + (r\dot{\theta})^2. \tag{6.23}$$

Hence, the Lagrangian of the system takes the form

$$\mathcal{L} = \frac{1}{2}m\left(\dot{r}^2 + r^2\dot{\theta}^2\right) - U(r). \tag{6.24}$$

Note that

$$\frac{\partial \mathcal{L}}{\partial \dot{r}} = m\dot{r}, \qquad \frac{\partial \mathcal{L}}{\partial r} = m r\dot{\theta}^2 - \frac{dU}{dr}, \tag{6.25}$$

$$\frac{\partial \mathcal{L}}{\partial \dot{\theta}} = m r^2 \dot{\theta}, \quad \text{and} \quad \frac{\partial \mathcal{L}}{\partial \theta} = 0. \tag{6.26}$$

Now, Lagrange's equation, Equation (6.22), yields the equations of motion,

$$\frac{d}{dt}\left(\frac{\partial \mathcal{L}}{\partial \dot{r}}\right) - \frac{\partial \mathcal{L}}{\partial r} = 0 \tag{6.27}$$

and

$$\frac{d}{dt}\left(\frac{\partial \mathcal{L}}{\partial \dot{\theta}}\right) - \frac{\partial \mathcal{L}}{\partial \theta} = 0. \tag{6.28}$$

Hence, we obtain

$$\frac{d}{dt}(m\,\dot{r}) - m\,r\,\dot{\theta}^2 + \frac{dU}{dr} = 0 \tag{6.29}$$

and

$$\frac{d}{dt}\left(m\,r^2\,\dot{\theta}\right) = 0, \tag{6.30}$$

or

$$\ddot{r} - r\,\dot{\theta}^2 = -\frac{dV}{dr} \tag{6.31}$$

and

$$r^2\,\dot{\theta} = h, \tag{6.32}$$

where $V = U/m$ and h is a constant. We recognize Equations (6.31) and (6.32) as the equations that we derived in Chapter 3 for motion in a central potential. The advantage of the Lagrangian method of deriving these equations is that we avoid having to express the acceleration in terms of the generalized coordinates r and θ.

6.5 Generalized momenta

Consider the motion of a single particle moving in one dimension. The kinetic energy is

$$K = \frac{1}{2}\,m\,\dot{x}^2, \tag{6.33}$$

where m is the mass of the particle and x its displacement. The particle's linear momentum is $p = m\,\dot{x}$. However, this can also be written

$$p = \frac{\partial K}{\partial \dot{x}} = \frac{\partial \mathcal{L}}{\partial \dot{x}}, \tag{6.34}$$

because $\mathcal{L} = K - U$ and the potential energy U is independent of $\dot{x}$.

Consider a dynamical system described by $\mathcal{F}$ generalized coordinates q_i for $i = 1, \mathcal{F}$. By analogy with the above expression, we can define *generalized momenta* of the form

$$p_i = \frac{\partial \mathcal{L}}{\partial \dot{q}_i} \tag{6.35}$$

for $i = 1, \mathcal{F}$. Here, p_i is sometimes called the momentum *conjugate* to the coordinate q_i. Hence, Lagrange's equation, Equation (6.22), can be written

$$\frac{dp_i}{dt} = \frac{\partial \mathcal{L}}{\partial q_i} \qquad (6.36)$$

for $i = 1, \mathcal{F}$. Note that a generalized momentum does not necessarily have the dimensions of linear momentum.

Suppose that the Lagrangian $\mathcal{L}$ does not depend explicitly on some coordinate q_k. It follows from Equation (6.36) that

$$\frac{dp_k}{dt} = \frac{\partial \mathcal{L}}{\partial q_k} = 0. \qquad (6.37)$$

Hence,

$$p_k = \text{const.} \qquad (6.38)$$

The coordinate q_k is said to be *ignorable* in this case. Thus, we conclude that the generalized momentum associated with an ignorable coordinate is a constant of the motion.

For example, the Lagrangian [Equation (6.24)] for a particle moving in a central potential is independent of the angular coordinate θ. Thus, θ is an ignorable coordinate, and

$$p_\theta = \frac{\partial \mathcal{L}}{\partial \dot{\theta}} = m r^2 \dot{\theta} \qquad (6.39)$$

is a constant of the motion. Of course, p_θ is the angular momentum about the origin. This is conserved because a central force exerts no torque about the origin.

Exercises

6.1 A horizontal rod AB rotates with constant angular velocity ω about its midpoint O. A particle P is attached to it by equal strings AP and BP. If θ is the inclination of the plane APB to the vertical, prove that

$$\frac{d^2\theta}{dt^2} - \omega^2 \sin\theta \cos\theta = -\frac{g}{l} \sin\theta,$$

where $l = OP$. Deduce the condition that the vertical position of OP should be stable.

6.2 A double pendulum consists of two simple pendula, with one pendulum suspended from the bob of the other. Suppose that the two pendula have equal lengths, l, and bobs of equal mass, m, and are confined to move in the same vertical plane. Let θ and ϕ—the angles that the upper and lower pendula make with the downward vertical (respectively)—be the generalized coordinates. Demonstrate that Lagrange's equations of motion for the system are

$$2\ddot{\theta} + \cos(\theta - \phi)\ddot{\phi} + \sin(\theta - \phi)\dot{\phi}^2 + \frac{2g}{l} \sin\theta = 0$$

and

$$\ddot{\phi} + \cos(\theta - \phi)\,\ddot{\theta} - \sin(\theta - \phi)\,\dot{\theta}^2 + \frac{g}{l}\,\sin\phi = 0.$$

6.3 Consider an elastic pendulum consisting of a bob of mass m attached to a light elastic string of stiffness k and unstretched length l. Let x be the extension of the string, and θ the angle that the string makes with the downward vertical. Assume that any motion is confined to a vertical plane. Demonstrate that Lagrange's equations of motion for the system are

$$\ddot{x} - (l + x)\,\dot{\theta}^2 - g\,\cos\theta + \frac{k}{m}\,x = 0$$

and

$$\ddot{\theta} + \frac{2\,\dot{x}\,\dot{\theta}}{l + x} + \frac{g}{l + x}\,\sin\theta = 0.$$

6.4 A disk of mass M and radius R rolls without slipping down a plane inclined at an angle α to the horizontal. The disk has a short weightless axle of negligible radius. From this axle is suspended a simple pendulum of length $l < R$ whose bob is of mass m. Assume that the motion of the pendulum takes place in the plane of the disk. Let s be the displacement of the center of mass of the disk down the slope, and let θ be the angle subtended between the pendulum and the downward vertical. Demonstrate that Lagrange's equations of motion for the system are

$$\left(\frac{3}{2}\,M + m\right)\ddot{s} + m\,l\,\cos(\alpha + \theta)\,\ddot{\theta} - m\,l\,\sin(\alpha + \theta)\,\dot{\theta}^2 - (M + m)\,g\,\sin\alpha = 0$$

and

$$\ddot{\theta} + \cos(\alpha + \theta)\,\frac{\ddot{s}}{l} + \frac{g}{l}\,\sin\theta = 0.$$

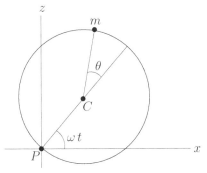

6.5 A vertical circular hoop of radius a is rotated in a vertical plane about a point P on its circumference at the constant angular velocity ω. A bead of mass m slides without friction on the hoop. Let the generalized coordinate be the angle θ shown in the diagram. Here, x is a horizontal Cartesian coordinate, z a vertical Cartesian coordinate, and C the center of the hoop. Demonstrate that the equation of motion of the system is

$$\ddot{\theta} + \omega^2\,\sin\theta + \frac{g}{a}\,\cos(\omega\,t + \theta) = 0.$$

(Modified from Fowles and Cassiday 2005.)

6.6 The kinetic energy of a rotating rigid object with an axis of symmetry can be written

$$K = \frac{1}{2}\left[\mathcal{I}_\perp \dot{\theta}^2 + (\mathcal{I}_\perp \sin^2\theta + \mathcal{I}_\| \cos^2\theta)\dot{\phi}^2 + 2\,\mathcal{I}_\| \cos\theta\,\dot{\phi}\dot{\psi} + \mathcal{I}_\| \dot{\psi}^2\right],$$

where $\mathcal{I}_\|$ is the moment of inertia about the symmetry axis, $\mathcal{I}_\perp$ is the moment of inertia about an axis perpendicular to the symmetry axis, and θ, ϕ, ψ are the three Euler angles. (See Chapter 7.) Suppose that the object is rotating freely. Find the momenta conjugate to the Euler angles. Which of these momenta are conserved? Find Lagrange's equations of motion for the system. Demonstrate that if the system is precessing steadily (which implies that θ, $\dot{\phi}$, and $\dot{\psi}$ are constants), then

$$\dot{\psi} = \left(\frac{\mathcal{I}_\perp - \mathcal{I}_\|}{\mathcal{I}_\|}\right)\cos\theta\,\dot{\phi}.$$

6.7 Demonstrate that the components of acceleration in the spherical coordinate system are

$$a_r = \ddot{r} - r\dot{\theta}^2 - r\sin^2\theta\,\dot{\phi}^2,$$

$$a_\theta = \frac{1}{r}\frac{d}{dt}(r^2\dot{\theta}) - r\sin\theta\cos\theta\,\dot{\phi}^2,$$

$$a_\phi = \frac{1}{r\sin\theta}\frac{d}{dt}(r^2\sin^2\theta\,\dot{\phi}).$$

(From Lamb 1923.)

6.8 A particle is constrained to move on a smooth spherical surface of radius a. Suppose that the particle is projected with velocity v along the horizontal great circle. Demonstrate that the particle subsequently falls a vertical height $a\,e^{-u}$, where

$$\sinh u = \frac{v^2}{4\,g\,a}.$$

Show that if v^2 is large compared with $4\,g\,a$, then this height becomes approximately $2\,g\,a^2/v^2$. (From Lamb 1923.)

6.9 Consider a nonconservative system in which the dissipative forces take the form $f_i = -k_i\dot{x}_i$, where the x_i are Cartesian coordinates, and the k_i are all positive. Demonstrate that the dissipative forces can be incorporated into the Lagrangian formalism provided that Lagrange's equations of motion are modified to read

$$\frac{d}{dt}\left(\frac{\partial \mathcal{L}}{\partial \dot{q}_i}\right) - \frac{\partial \mathcal{L}}{\partial q_i} + \frac{\partial \mathcal{R}}{\partial \dot{q}_i} = 0,$$

where

$$\mathcal{R} = \frac{1}{2}\sum_i k_i\dot{x}_i^2$$

is termed the *Rayleigh dissipation function*.

7   Rigid body rotation

7.1   Introduction

This chapter examines the rotation of rigid bodies (e.g., the planets) in three dimensions. The analysis presented here is largely due to Euler (1707–1783).

7.2   Fundamental equations

We can think of a rigid body as a collection of a large number of small mass elements that all maintain a fixed spatial relationship with respect to one another. Let there be N elements, and let the ith element be of mass m_i, and instantaneous position vector $\mathbf{r}_i$. The equation of motion of the ith element is written

$$m_i \frac{d^2 \mathbf{r}_i}{dt^2} = \sum_{j=1,N}^{j \neq i} \mathbf{f}_{ij} + \mathbf{F}_i. \tag{7.1}$$

Here, $\mathbf{f}_{ij}$ is the internal force exerted on the ith element by the jth element, and $\mathbf{F}_i$ the external force acting on the ith element. The internal forces $\mathbf{f}_{ij}$ represent the stresses that develop within the body to ensure that its various elements maintain a fixed spatial relationship with respect to one another. Of course, $\mathbf{f}_{ij} = -\mathbf{f}_{ji}$, by Newton's third law. The external forces represent forces that originate outside the body.

Repeating the analysis of Section 1.6, we can sum Equation (7.1) over all mass elements to obtain

$$M \frac{d^2 \mathbf{r}_{cm}}{dt^2} = \mathbf{F}. \tag{7.2}$$

Here, $M = \sum_{i=1,N} m_i$ is the total mass, $\mathbf{r}_{cm}$ the position vector of the center of mass [see Equation (1.27)], and $\mathbf{F} = \sum_{i=1,N} \mathbf{F}_i$ the total external force. It can be seen that the center of mass of a rigid body moves under the action of the external forces like a point particle whose mass is identical with that of the body.

Again repeating the analysis of Section 1.6, we can sum $\mathbf{r}_i \times$ Equation (7.1) over all mass elements to obtain

$$\frac{d\mathbf{L}}{dt} = \boldsymbol{\tau}. \tag{7.3}$$

Here, $\mathbf{L} = \sum_{i=1,N} m_i \, \mathbf{r}_i \times d\mathbf{r}_i/dt$ is the total angular momentum of the body (about the origin), and $\boldsymbol{\tau} = \sum_{i=1,N} \mathbf{r}_i \times \mathbf{F}_i$ is the total external torque (about the origin). The

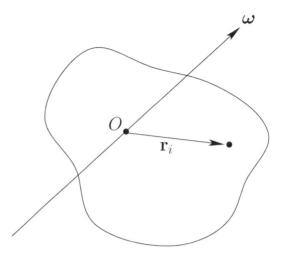

Fig. 7.1 A rigid rotating body.

preceding equation is valid only if the internal forces are *central* in nature. However, this is not a particularly onerous constraint. Equation (7.3) describes how the angular momentum of a rigid body evolves in time under the action of the external torques.

In the following, we shall consider only the *rotational* motion of rigid bodies, as their translational motion is similar to that of point particles [see Equation (7.2)] and, therefore, is fairly straightforward in nature.

7.3 Moment of inertia tensor

Consider a rigid body rotating with fixed angular velocity ω about an axis that passes through the origin. (See Figure 7.1.) Let $\mathbf{r}_i$ be the position vector of the ith mass element, whose mass is m_i. We expect this position vector to *precess* about the axis of rotation (which is parallel to ω) with angular velocity ω. It, therefore, follows from Section A.7 that

$$\frac{d\mathbf{r}_i}{dt} = \omega \times \mathbf{r}_i. \tag{7.4}$$

Thus, Equation (7.4) specifies the velocity, $\mathbf{v}_i = d\mathbf{r}_i/dt$, of each mass element as the body rotates with fixed angular velocity ω about an axis passing through the origin.

The total angular momentum of the body (about the origin) is written

$$\mathbf{L} = \sum_{i=1,N} m_i\, \mathbf{r}_i \times \frac{d\mathbf{r}_i}{dt} = \sum_{i=1,N} m_i\, \mathbf{r}_i \times (\omega \times \mathbf{r}_i) = \sum_{i=1,N} m_i \left[r_i^2\, \omega - (\mathbf{r}_i \cdot \omega)\, \mathbf{r}_i \right], \tag{7.5}$$

where use has been made of Equation (7.4) and some standard vector identities. (See Section A.4.) The preceding formula can be written as a matrix equation of the form

$$\begin{pmatrix} L_x \\ L_y \\ L_z \end{pmatrix} = \begin{pmatrix} \mathcal{I}_{xx} & \mathcal{I}_{xy} & \mathcal{I}_{xz} \\ \mathcal{I}_{yx} & \mathcal{I}_{yy} & \mathcal{I}_{yz} \\ \mathcal{I}_{zx} & \mathcal{I}_{zy} & \mathcal{I}_{zz} \end{pmatrix} \begin{pmatrix} \omega_x \\ \omega_y \\ \omega_z \end{pmatrix}, \tag{7.6}$$

where

$$\mathcal{I}_{xx} = \sum_{i=1,N} (y_i^2 + z_i^2)\, m_i = \int (y^2 + z^2)\, dm, \tag{7.7}$$

$$\mathcal{I}_{yy} = \sum_{i=1,N} (x_i^2 + z_i^2)\, m_i = \int (x^2 + z^2)\, dm, \tag{7.8}$$

$$\mathcal{I}_{zz} = \sum_{i=1,N} (x_i^2 + y_i^2)\, m_i = \int (x^2 + y^2)\, dm, \tag{7.9}$$

$$\mathcal{I}_{xy} = \mathcal{I}_{yx} = -\sum_{i=1,N} x_i\, y_i\, m_i = -\int x\, y\, dm, \tag{7.10}$$

$$\mathcal{I}_{yz} = \mathcal{I}_{zy} = -\sum_{i=1,N} y_i\, z_i\, m_i = -\int y\, z\, dm, \tag{7.11}$$

and

$$\mathcal{I}_{xz} = \mathcal{I}_{zx} = -\sum_{i=1,N} x_i\, z_i\, m_i = -\int x\, z\, dm. \tag{7.12}$$

Here, $\mathcal{I}_{xx}$ is called the *moment of inertia* about the x-axis, $\mathcal{I}_{yy}$ the moment of inertia about the y-axis, $\mathcal{I}_{xy}$ the *xy product of inertia*, $\mathcal{I}_{yz}$ the *yz* product of inertia, and so on. The matrix of the $\mathcal{I}_{ij}$ values is known as the *moment of inertia tensor*. Each component of the moment of inertia tensor can be written as either a sum over separate mass elements or as an integral over infinitesimal mass elements. In the integrals, $dm = \rho\, d^3\mathbf{r}$, where ρ is the mass density, and $d^3\mathbf{r}$ a volume element. Equation (7.6) can be written more succinctly as

$$\mathbf{L} = \boldsymbol{I}\,\boldsymbol{\omega}. \tag{7.13}$$

Here, it is understood that $\mathbf{L}$ and $\boldsymbol{\omega}$ are both *column vectors*, and $\boldsymbol{I}$ is the *matrix* of the $\mathcal{I}_{ij}$ values. Note that $\boldsymbol{I}$ is a *real symmetric* matrix: $\mathcal{I}_{ij}^* = \mathcal{I}_{ij}$ and $\mathcal{I}_{ji} = \mathcal{I}_{ij}$.

In general, the angular momentum vector, $\mathbf{L}$, obtained from Equation (7.13), points in a different direction from the angular velocity vector, $\boldsymbol{\omega}$. In other words, $\mathbf{L}$ is generally not parallel to $\boldsymbol{\omega}$.

Finally, although the preceding results were obtained assuming a fixed angular velocity, they remain valid, at each instant in time, if the angular velocity varies.

7.4 Rotational kinetic energy

The instantaneous rotational kinetic energy of a rotating rigid body is written

$$K = \frac{1}{2} \sum_{i=1,N} m_i \left(\frac{d\mathbf{r}_i}{dt}\right)^2. \tag{7.14}$$

Making use of Equation (7.4) and some vector identities (see Section A.4), we find that
the kinetic energy takes the form

$$K = \frac{1}{2} \sum_{i=1,N} m_i \, (\boldsymbol{\omega} \times \mathbf{r}_i) \cdot (\boldsymbol{\omega} \times \mathbf{r}_i) = \frac{1}{2} \, \boldsymbol{\omega} \cdot \sum_{i=1,N} m_i \, \mathbf{r}_i \times (\boldsymbol{\omega} \times \mathbf{r}_i). \qquad (7.15)$$

Hence, it follows from Equation (7.5) that

$$K = \frac{1}{2} \, \boldsymbol{\omega} \cdot \mathbf{L}. \qquad (7.16)$$

Making use of Equation (7.13), we can also write

$$K = \frac{1}{2} \, \boldsymbol{\omega}^T \, \mathbf{I} \, \boldsymbol{\omega}. \qquad (7.17)$$

Here, $\boldsymbol{\omega}^T$ is the *row vector* of the Cartesian components ω_x, ω_y, ω_z, which is, of course,
the transpose (denoted ^{T}) of the column vector $\boldsymbol{\omega}$. When written in component form,
Equation (7.17) yields

$$K = \frac{1}{2} \left(\mathcal{I}_{xx} \, \omega_x^2 + \mathcal{I}_{yy} \, \omega_y^2 + \mathcal{I}_{zz} \, \omega_z^2 + 2 \, \mathcal{I}_{xy} \, \omega_x \, \omega_y + 2 \, \mathcal{I}_{yz} \, \omega_y \, \omega_z + 2 \, \mathcal{I}_{xz} \, \omega_x \, \omega_z \right). \quad (7.18)$$

7.5 Principal axes of rotation

We have seen that, for a general orientation of the Cartesian coordinate axes, the moment of inertia tensor, $\mathbf{I}$, defined in Section 7.3, takes the form of a *real symmetric* 3×3
matrix. It therefore follows, from the standard matrix theory discussed in Section A.11,
that the moment of inertia tensor possesses three mutually orthogonal eigenvectors,
which are associated with three real eigenvalues. Let the ith eigenvector (which can be
normalized to be a unit vector) be denoted $\hat{\omega}_i$, and the ith eigenvalue λ_i. It then follows
that

$$\mathbf{I} \, \hat{\omega}_i = \lambda_i \, \hat{\omega}_i \qquad (7.19)$$

for $i = 1, 3$.

The directions of the three mutually orthogonal unit vectors $\hat{\omega}_i$ define the three so-
called *principal axes of rotation* of the rigid body under investigation. These axes are
special because when the body rotates about one of them (i.e., when $\boldsymbol{\omega}$ is parallel to
one of them), the angular momentum vector $\mathbf{L}$ becomes *parallel* to the angular velocity
vector $\boldsymbol{\omega}$. This can be seen from a comparison of Equation (7.13) with Equation (7.19).

Suppose that we reorient our Cartesian coordinate axes so they coincide with the
mutually orthogonal principal axes of rotation. In this new reference frame, the eigen-
vectors of $\mathbf{I}$ are the unit vectors, $\mathbf{e}_x$, $\mathbf{e}_y$, and $\mathbf{e}_z$, and the eigenvalues are the moments of
inertia about these axes, $\mathcal{I}_{xx}$, $\mathcal{I}_{yy}$, and $\mathcal{I}_{zz}$, respectively. These latter quantities are referred to as the *principal moments of inertia*. The products of inertia are all zero in the
new reference frame. Hence, in this frame, the moment of inertia tensor takes the form

of a *diagonal* matrix:

$$\mathbf{I} = \begin{pmatrix} \mathcal{I}_{xx} & 0 & 0 \\ 0 & \mathcal{I}_{yy} & 0 \\ 0 & 0 & \mathcal{I}_{zz} \end{pmatrix}. \tag{7.20}$$

Incidentally, it is easy to verify that $\mathbf{e}_x$, $\mathbf{e}_y$, and $\mathbf{e}_z$ are indeed the eigenvectors of this matrix, with the eigenvalues $\mathcal{I}_{xx}$, $\mathcal{I}_{yy}$, and $\mathcal{I}_{zz}$, respectively, and that $\mathbf{L} = \mathbf{I}\,\omega$ is indeed parallel to ω whenever ω is directed along $\mathbf{e}_x$, $\mathbf{e}_y$, or $\mathbf{e}_z$.

When expressed in our new coordinate system, Equation (7.13) yields

$$\mathbf{L} = \left(\mathcal{I}_{xx}\,\omega_x,\ \mathcal{I}_{yy}\,\omega_y,\ \mathcal{I}_{zz}\,\omega_z \right), \tag{7.21}$$

whereas Equation (7.18) reduces to

$$K = \frac{1}{2} \left(\mathcal{I}_{xx}\,\omega_x^2 + \mathcal{I}_{yy}\,\omega_y^2 + \mathcal{I}_{zz}\,\omega_z^2 \right). \tag{7.22}$$

In conclusion, we may obtain many great simplifications by choosing a coordinate system whose axes coincide with the principal axes of rotation of the rigid body under investigation.

7.6 Euler's equations

The fundamental equation of motion of a rotating body [see Equation (7.3)],

$$\tau = \frac{d\mathbf{L}}{dt}, \tag{7.23}$$

is valid only in an *inertial* frame. However, we have seen that $\mathbf{L}$ is most simply expressed in a frame of reference whose axes are aligned along the principal axes of rotation of the body. Such a frame of reference rotates with the body and is therefore noninertial. Thus, it is helpful to define two Cartesian coordinate systems with the same origins. The first, with coordinates x, y, z, is a fixed inertial frame—let us denote this the *fixed frame*. The second, with coordinates x', y', z', co-rotates with the body in such a manner that the x'-, y'-, and z'-axes are always pointing along its principal axes of rotation—we shall refer to this as the *body frame*. Because the body frame co-rotates with the body, its instantaneous angular velocity is the same as that of the body. Hence, it follows from the analysis in Section 5.2 that

$$\frac{d\mathbf{L}}{dt} = \frac{d\mathbf{L}}{dt'} + \omega \times \mathbf{L}. \tag{7.24}$$

Here, d/dt is the time derivative in the fixed frame, and d/dt' the time derivative in the body frame. Combining Equations (7.23) and (7.24), we obtain

$$\tau = \frac{d\mathbf{L}}{dt'} + \omega \times \mathbf{L}. \tag{7.25}$$

In the body frame, let $\tau = (\tau_{x'}, \tau_{y'}, \tau_{z'})$ and $\omega = (\omega_{x'}, \omega_{y'}, \omega_{z'})$. It follows that $\mathbf{L} = (\mathcal{I}_{x'x'}\,\omega_{x'}, \mathcal{I}_{y'y'}\,\omega_{y'}, \mathcal{I}_{z'z'}\,\omega_{z'})$, where $\mathcal{I}_{x'x'}$, $\mathcal{I}_{y'y'}$, and $\mathcal{I}_{z'z'}$ are the principal moments of inertia. Hence, in the body frame, the components of Equation (7.25) yield

$$\tau_{x'} = \mathcal{I}_{x'x'}\,\dot{\omega}_{x'} - (\mathcal{I}_{y'y'} - \mathcal{I}_{z'z'})\,\omega_{y'}\,\omega_{z'}, \tag{7.26}$$

$$\tau_{y'} = \mathcal{I}_{y'y'}\,\dot{\omega}_{y'} - (\mathcal{I}_{z'z'} - \mathcal{I}_{x'x'})\,\omega_{z'}\,\omega_{x'}, \tag{7.27}$$

and

$$\tau_{z'} = \mathcal{I}_{z'z'}\,\dot{\omega}_{z'} - (\mathcal{I}_{x'x'} - \mathcal{I}_{y'y'})\,\omega_{x'}\,\omega_{y'}, \tag{7.28}$$

where $\dot{} = d/dt'$. Here, we have made use of the fact that the moments of inertia of a rigid body are constant in time in the co-rotating body frame. The preceding three equations are known as *Euler's equations*.

Consider a body that is *freely rotating*—that is, in the absence of external torques. Furthermore, let the body be *rotationally symmetric* about the z'-axis. It follows that $\mathcal{I}_{x'x'} = \mathcal{I}_{y'y'} = \mathcal{I}_{\perp}$. Likewise, we can write $\mathcal{I}_{z'z'} = \mathcal{I}_{\|}$. In general, however, $\mathcal{I}_{\perp} \neq \mathcal{I}_{\|}$. Thus, Euler's equations yield

$$\mathcal{I}_{\perp}\frac{d\omega_{x'}}{dt'} + (\mathcal{I}_{\|} - \mathcal{I}_{\perp})\,\omega_{z'}\,\omega_{y'} = 0, \tag{7.29}$$

$$\mathcal{I}_{\perp}\frac{d\omega_{y'}}{dt'} - (\mathcal{I}_{\|} - \mathcal{I}_{\perp})\,\omega_{z'}\,\omega_{x'} = 0, \tag{7.30}$$

and

$$\frac{d\omega_{z'}}{dt'} = 0. \tag{7.31}$$

Clearly, $\omega_{z'}$ is a constant of the motion. Equations (7.29) and (7.30) can be written

$$\frac{d\omega_{x'}}{dt'} + \Omega\,\omega_{y'} = 0 \tag{7.32}$$

and

$$\frac{d\omega_{y'}}{dt'} - \Omega\,\omega_{x'} = 0, \tag{7.33}$$

where $\Omega = (\mathcal{I}_{\|}/\mathcal{I}_{\perp} - 1)\,\omega_{z'}$. As is easily demonstrated, the solution to these equations is

$$\omega_{x'} = \omega_{\perp}\,\cos(\Omega\,t') \tag{7.34}$$

and

$$\omega_{y'} = \omega_{\perp}\,\sin(\Omega\,t'), \tag{7.35}$$

where $\omega_{\perp}$ is a constant. Thus, the projection of the angular velocity vector onto the x'–y' plane has the fixed length $\omega_{\perp}$, and rotates steadily about the z'-axis with angular velocity Ω. It follows that the length of the angular velocity vector, $\omega = (\omega_{x'}^2 + \omega_{y'}^2 + \omega_{z'}^2)^{1/2}$, is a constant of the motion. Clearly, the angular velocity vector subtends some constant angle, α, with the z'-axis, which implies that $\omega_{z'} = \omega\cos\alpha$ and $\omega_{\perp} = \omega\sin\alpha$. Hence, the components of the angular velocity vector are

$$\omega_{x'} = \omega\,\sin\alpha\,\cos(\Omega\,t'), \tag{7.36}$$

$$\omega_{y'} = \omega\,\sin\alpha\,\sin(\Omega\,t'), \tag{7.37}$$

and

$$\omega_{z'} = \omega \cos \alpha, \tag{7.38}$$

where

$$\Omega = \omega \cos \alpha \left(\frac{\mathcal{I}_{\parallel}}{\mathcal{I}_{\perp}} - 1 \right). \tag{7.39}$$

We conclude that, in the body frame, the angular velocity vector *precesses* about the symmetry axis (i.e., the z'-axis) with the angular frequency Ω. The components of the angular momentum vector are

$$L_{x'} = \mathcal{I}_{\perp} \, \omega \, \sin \alpha \, \cos(\Omega t'), \tag{7.40}$$

$$L_{y'} = \mathcal{I}_{\perp} \, \omega \, \sin \alpha \, \sin(\Omega t'), \tag{7.41}$$

and

$$L_{z'} = \mathcal{I}_{\parallel} \, \omega \, \cos \alpha. \tag{7.42}$$

Thus, in the body frame, the angular momentum vector is also of constant length, and precesses about the symmetry axis with the angular frequency Ω. Furthermore, the angular momentum vector subtends a constant angle θ with the symmetry axis, where

$$\tan \theta = \frac{\mathcal{I}_{\perp}}{\mathcal{I}_{\parallel}} \tan \alpha. \tag{7.43}$$

The angular momentum vector, the angular velocity vector, and the symmetry axis all lie in the same plane, that is, $\mathbf{e}_{z'} \cdot \mathbf{L} \times \boldsymbol{\omega} = 0$, as can easily be verified. Moreover, the angular momentum vector lies between the angular velocity vector and the symmetry axis (i.e, $\theta < \alpha$) for a flattened (or oblate) body (i.e., $\mathcal{I}_{\perp} < \mathcal{I}_{\parallel}$), whereas the angular velocity vector lies between the angular momentum vector and the symmetry axis (i.e., $\theta > \alpha$) for an elongated (or prolate) body (i.e., $\mathcal{I}_{\perp} > \mathcal{I}_{\parallel}$). (See Figure 7.2.)

7.7 Euler angles

We have seen how we can solve Euler's equations to determine the properties of a rotating body in the co-rotating body frame. Let us now investigate how we can determine the same properties in the inertial *fixed frame*.

The fixed frame and the body frame share the same origin. Hence, we can transform from one to the other by means of an appropriate rotation of our coordinate axes. In general, if we restrict ourselves to rotations about one of the Cartesian axes, three successive rotations are required to transform the fixed frame into the body frame. There are, in fact, many different ways to combined three successive rotations to achieve this goal. In the following, we shall describe the most widely used method, which is due to Euler.

We start in the fixed frame, which has coordinates x, y, z, and unit vectors $\mathbf{e}_x$, $\mathbf{e}_y$, $\mathbf{e}_z$. Our first rotation is counterclockwise (if we look down the axis) through an angle ϕ

about the z-axis. The new frame has coordinates x'', y'', z'' and unit vectors $\mathbf{e}_{x''}$, $\mathbf{e}_{y''}$, $\mathbf{e}_{z''}$. According to Section A.6, the transformation of coordinates can be represented as follows:

$$\begin{pmatrix} x'' \\ y'' \\ z'' \end{pmatrix} = \begin{pmatrix} \cos\phi & \sin\phi & 0 \\ -\sin\phi & \cos\phi & 0 \\ 0 & 0 & 1 \end{pmatrix} \begin{pmatrix} x \\ y \\ z \end{pmatrix}. \tag{7.44}$$

The angular velocity vector associated with ϕ has the magnitude $\dot{\phi}$ and is directed along $\mathbf{e}_z$ (i.e., along the axis of rotation). Hence, we can write

$$\boldsymbol{\omega}_\phi = \dot{\phi}\,\mathbf{e}_z. \tag{7.45}$$

Clearly, $\dot{\phi}$ is the precession rate about the z-axis, as seen in the fixed frame.

The second rotation is counterclockwise (if we look down the axis) through an angle θ about the x''-axis. The new frame has coordinates x''', y''', z''' and unit vectors $\mathbf{e}_{x'''}$, $\mathbf{e}_{y'''}$, $\mathbf{e}_{z'''}$. By analogy with Equation (7.44), the transformation of coordinates can be represented as follows:

$$\begin{pmatrix} x''' \\ y''' \\ z''' \end{pmatrix} = \begin{pmatrix} 1 & 0 & 0 \\ 0 & \cos\theta & \sin\theta \\ 0 & -\sin\theta & \cos\theta \end{pmatrix} \begin{pmatrix} x'' \\ y'' \\ z'' \end{pmatrix}. \tag{7.46}$$

The angular velocity vector associated with θ has the magnitude $\dot{\theta}$ and is directed along $\mathbf{e}_{x''}$ (i.e., along the axis of rotation). Hence, we can write

$$\boldsymbol{\omega}_\theta = \dot{\theta}\,\mathbf{e}_{x''}. \tag{7.47}$$

The third rotation is counterclockwise (if we look down the axis) through an angle ψ about the z'''-axis. The new frame is the body frame, which has coordinates x', y', z' and unit vectors $\mathbf{e}_{x'}$, $\mathbf{e}_{y'}$, $\mathbf{e}_{z'}$. The transformation of coordinates can be represented as follows:

$$\begin{pmatrix} x' \\ y' \\ z' \end{pmatrix} = \begin{pmatrix} \cos\psi & \sin\psi & 0 \\ -\sin\psi & \cos\psi & 0 \\ 0 & 0 & 1 \end{pmatrix} \begin{pmatrix} x''' \\ y''' \\ z''' \end{pmatrix}. \tag{7.48}$$

The angular velocity vector associated with ψ has the magnitude $\dot{\psi}$ and is directed along $\mathbf{e}_{z'''}$ (i.e., along the axis of rotation). Note that $\mathbf{e}_{z'''} = \mathbf{e}_{z'}$, since the third rotation is about $\mathbf{e}_{z'''}$. Hence, we can write

$$\boldsymbol{\omega}_\psi = \dot{\psi}\,\mathbf{e}_{z'}. \tag{7.49}$$

Clearly, $\dot{\psi}$ is *minus* the precession rate about the z'-axis, as seen in the body frame.

The full transformation between the fixed frame and the body frame is rather complicated. However, the following results can easily be verified:

$$\mathbf{e}_z = \sin\psi\,\sin\theta\,\mathbf{e}_{x'} + \cos\psi\,\sin\theta\,\mathbf{e}_{y'} + \cos\theta\,\mathbf{e}_{z'}, \tag{7.50}$$

$$\mathbf{e}_{x''} = \cos\psi\,\mathbf{e}_{x'} - \sin\psi\,\mathbf{e}_{y'}. \tag{7.51}$$

It follows from Equation (7.50) that $\mathbf{e}_z \cdot \mathbf{e}_{z'} = \cos\theta$. In other words, θ is the angle of inclination between the z- and z'-axes. Finally, because the total angular velocity can be written

$$\omega = \omega_\phi + \omega_\theta + \omega_\psi, \tag{7.52}$$

Equations (7.45), (7.47), and (7.49)–(7.51) yield

$$\omega_{x'} = \sin\psi \sin\theta \, \dot\phi + \cos\psi \, \dot\theta, \tag{7.53}$$

$$\omega_{y'} = \cos\psi \sin\theta \, \dot\phi - \sin\psi \, \dot\theta, \tag{7.54}$$

and

$$\omega_{z'} = \cos\theta \, \dot\phi + \dot\psi. \tag{7.55}$$

The angles ϕ, θ, and ψ are termed *Euler angles*. Each has a clear physical interpretation: ϕ is the angle of precession about the z-axis in the fixed frame, ψ is minus the angle of precession about the z'-axis in the body frame, and θ is the angle of inclination between the z- and z'-axes. Moreover, we can express the components of the angular velocity vector ω in the body frame entirely in terms of the Eulerian angles and their time derivatives [see Equations (7.53)–(7.55)].

Consider a freely rotating body that is rotationally symmetric about one axis (the z'-axis). In the absence of an external torque, the angular momentum vector $\mathbf{L}$ is a constant of the motion [see Equation (7.3)]. Let $\mathbf{L}$ point along the z-axis. In the previous section, we saw that the angular momentum vector subtends a constant angle θ with the axis of symmetry, that is, with the z'-axis. Hence, the time derivative of the Eulerian angle θ is zero. We also saw that the angular momentum vector, the axis of symmetry, and the angular velocity vector are coplanar. Consider an instant in time at which all these vectors lie in the y'–z' plane. This implies that $\omega_{x'} = 0$. According to the previous section, the angular velocity vector subtends a constant angle α with the symmetry axis. It follows that $\omega_{y'} = \omega \sin\alpha$ and $\omega_{z'} = \omega \cos\alpha$. Equation (7.53) gives $\psi = 0$. Hence, Equation (7.54) yields

$$\omega \sin\alpha = \sin\theta \, \dot\phi. \tag{7.56}$$

This can be combined with Equation (7.43) to give

$$\dot\phi = \omega \left[1 + \left(\frac{\mathcal{I}_\parallel^2}{\mathcal{I}_\perp^2} - 1 \right) \cos^2\alpha \right]^{1/2}. \tag{7.57}$$

Finally, Equation (7.55), together with Equations (7.43) and (7.56), yields

$$\dot\psi = \omega \cos\alpha - \cos\theta \, \dot\phi = \omega \cos\alpha \left(1 - \frac{\tan\alpha}{\tan\theta} \right) = \omega \cos\alpha \left(1 - \frac{\mathcal{I}_\parallel}{\mathcal{I}_\perp} \right). \tag{7.58}$$

A comparison of this equation with Equation (7.39) gives

$$\dot\psi = -\Omega. \tag{7.59}$$

Thus, as expected, $\dot\psi$ is minus the precession rate (of the angular momentum and angular velocity vectors) in the body frame. On the other hand, $\dot\phi$ is the precession rate (of the angular velocity vector and the symmetry axis) in the fixed frame. Note that $\dot\phi$ and Ω are quite dissimilar. For instance, Ω is negative for elongated bodies ($\mathcal{I}_\parallel < \mathcal{I}_\perp$) whereas $\dot\phi$ is positive definite. It follows that the precession is always in the same sense as L_z in the

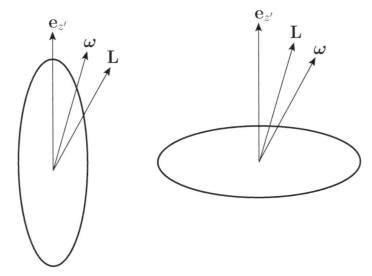

A freely rotating object that is elongated along its axis of symmetry, $\mathbf{e}_{z'}$ (left), and a freely rotating object that is flattened along its axis of symmetry (right). The $\mathbf{L}$ vector is fixed.

fixed frame, whereas the precession in the body frame is in the opposite sense to $L_{z'}$ for elongated bodies. We found, in the previous section, that for a flattened body the angular momentum vector lies between the angular velocity vector and the symmetry axis. This means that, in the fixed frame, the angular velocity vector and the symmetry axis lie on opposite sides of the fixed angular momentum vector, about which they precess. (See Figure 7.2.) On the other hand, for an elongated body we found that the angular velocity vector lies between the angular momentum vector and the symmetry axis. This means that, in the fixed frame, the angular velocity vector and the symmetry axis lie on the *same* side of the fixed angular momentum vector, about which they precess. (See Figure 7.2.) Recall that the angular momentum vector, the angular velocity vector, and the symmetry axis are coplanar.

7.8 Free precession of the Earth

It is known that the Earth's axis of rotation is very slightly inclined to its symmetry axis (which passes through the two geographic poles). The angle α is approximately 0.2 seconds of an arc (which corresponds to a distance of about 6 m on the Earth's surface). It is also known that the ratio of the terrestrial moments of inertia is about $\mathcal{I}_{\parallel}/\mathcal{I}_{\perp} = 1.00327$, as determined from the Earth's oblateness (Yoder 1995). (See Section 7.9.) Hence, from Equation (7.39), the precession rate of the angular velocity vector about the symmetry axis, as viewed in a geostationary reference frame, is

$$\Omega = 0.00327\,\omega, \tag{7.60}$$

giving a precession period of

$$T' = \frac{2\pi}{\Omega} = 305 \text{ days}. \tag{7.61}$$

[Of course, $2\pi/\omega = 1$ (sidereal) day.] The observed period of precession is about 434 days (Yoder 1995). The disagreement between theory and observation is attributed to the fact that the Earth is not perfectly rigid (Bertotti et al. 2003). The Earth's symmetry axis subtends an angle $\theta \simeq \alpha = 0.2''$ [see Equation (7.43)] with its angular momentum vector, but it lies on the opposite side of this vector to the angular velocity vector. This implies that, as viewed from space, the Earth's angular velocity vector is almost parallel to its fixed angular momentum vector, whereas its symmetry axis subtends an angle of $0.2''$ with both vectors and precesses about them. The (theoretical) precession rate of the Earth's symmetry axis, as seen from space, is given by Equation (7.57):

$$\dot{\phi} = 1.00327\,\omega. \tag{7.62}$$

The associated precession period is

$$T = \frac{2\pi}{\dot{\phi}} = 0.997 \text{ days}. \tag{7.63}$$

The free precession of the Earth's symmetry axis in space, which is known as the *Chandler wobble*—because it was discovered by the American astronomer S.C. Chandler (1846–1913) in 1891—is superimposed on a much slower forced precession, with a period of about 26,000 years, caused by the small gravitational torque exerted on the Earth by the Sun and Moon, as a consequence of the Earth's slight oblateness. (See Section 7.10.)

7.9 MacCullagh's formula

According to Equations (2.59) and (2.64), if the Earth is modeled as spheroid of uniform density γ, its ellipticity is given by

$$\epsilon = -\int r^2\,\gamma\,P_2(\cos\theta)\,d^3\mathbf{r}\bigg/ \mathcal{I}_0 = -\frac{1}{2}\int r^2\,\gamma\left(3\cos^2\theta - 1\right)d^3\mathbf{r}\bigg/ \mathcal{I}_0, \tag{7.64}$$

where the integral is over the whole volume of the Earth, and $\mathcal{I}_0 = (2/5)\,M\,R^2$ would be the Earth's moment of inertia were it exactly spherical. The Earth's moment of inertia about its axis of rotation is given by

$$\mathcal{I}_\parallel = \int (x^2 + y^2)\,\gamma\,d^3\mathbf{r} = \int r^2\,\gamma\,(1 - \cos^2\theta)\,d^3\mathbf{r}. \tag{7.65}$$

Here, use has been made of Equations (2.24)–(2.26). Likewise, the Earth's moment of inertia about an axis perpendicular to its axis of rotation (and passing through the Earth's center) is

$$
\mathcal{I}_\perp = \int (y^2 + z^2) \, \gamma \, d^3\mathbf{r} = \int r^2 \, \gamma \, (\sin^2\theta \, \sin^2\phi + \cos^2\theta) \, d^3\mathbf{r}
$$

$$
= \int r^2 \, \gamma \left(\frac{1}{2} \sin^2\theta + \cos^2\theta \right) d^3\mathbf{r} = \frac{1}{2} \int r^2 \, \gamma \, (1 + \cos^2\theta) \, d^3\mathbf{r}, \qquad (7.66)
$$

as the average of $\sin^2\phi$ is $1/2$ for an axisymmetric mass distribution. It follows from the preceding three equations that

$$
\epsilon = \frac{\mathcal{I}_\parallel - \mathcal{I}_\perp}{\mathcal{I}_0} \simeq \frac{\mathcal{I}_\parallel - \mathcal{I}_\perp}{\mathcal{I}_\parallel}. \qquad (7.67)
$$

This formula demonstrates that the Earth's ellipticity is directly related to the difference between its principal moments of inertia. Actually, the formula holds for *any* axially symmetric mass distribution, not just a spheroidal distribution of uniform density (this is discussed later). When Equation (7.67) is combined with Equation (2.65), we get

$$
\Phi(r, \theta) \simeq -\frac{G M}{r} + \frac{G (\mathcal{I}_\parallel - \mathcal{I}_\perp)}{r^3} P_2(\cos\theta), \qquad (7.68)
$$

which is the general expression for the gravitational potential generated outside an axially symmetric mass distribution. The first term on the right-hand side is the *monopole* gravitational potential that would be generated if all the mass in the distribution were concentrated at its center of mass, whereas the second term is the *quadrupole* potential generated by any deviation from spherical symmetry in the distribution.

More generally, consider an asymmetric mass distribution consisting of N mass elements. Suppose that the ith element has mass m_i and position vector $\mathbf{r}_i$, where i runs from 1 to N. Let us define a Cartesian coordinate system x, y, z such that the origin coincides with the center of mass of the distribution. It follows that

$$
\sum_{i=1,N} m_i \, x_i = \sum_{i=1,N} m_i \, y_i = \sum_{i=1,N} m_i \, z_i = 0. \qquad (7.69)
$$

Suppose that the x-, y-, and z-axes coincide with the mass distribution's principal axes of rotation (for rotation about an axis that passes through the origin). It follows from Section 7.5 that

$$
\sum_{i=1,N} m_i \, y_i \, z_i = \sum_{i=1,N} m_i \, x_i \, z_i = \sum_{i=1,N} m_i \, x_i \, y_i = 0, \qquad (7.70)
$$

and that the distribution's principal moments of inertia about the x-, y-, and z-axes take the form

$$
\mathcal{I}_{xx} = \sum_{i=1,N} m_i \, (y_i^2 + z_i^2), \qquad (7.71)
$$

$$
\mathcal{I}_{yy} = \sum_{i=1,N} m_i \, (x_i^2 + z_i^2), \qquad (7.72)
$$

and

$$
\mathcal{I}_{zz} = \sum_{i=1,N} m_i \, (x_i^2 + y_i^2), \qquad (7.73)
$$

respectively.

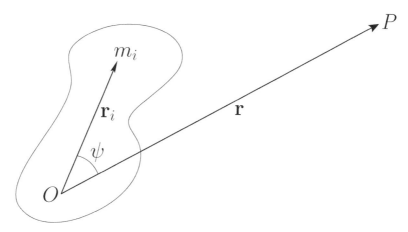

Fig. 7.3 A general mass distribution.

Consider the gravitational potential, Φ, generated by the mass distribution at some external point P whose position vector is $\mathbf{r} \equiv (x, y, z)$. According to Section 2.2,

$$\Phi(\mathbf{r}) = -G \sum_{i=1,N} \frac{m_i}{|\mathbf{r}_i - \mathbf{r}|}, \tag{7.74}$$

which can also be written

$$\Phi(\mathbf{r}) = -\frac{G}{r} \sum_{i=1,N} m_i \left(1 - 2\frac{r_i}{r} \cos\psi + \frac{r_i^2}{r^2}\right)^{-1/2}, \tag{7.75}$$

where ψ is the angle subtended between the vectors $\mathbf{r}$ and $\mathbf{r}_i$. (See Figure 7.3.) Suppose that the distance $OP \equiv r$ is much larger than the characteristic radius of the mass distribution, which implies that $r_i/r \ll 1$ for all i. Expanding up to second order in r_i/r, we obtain

$$\Phi(\mathbf{r}) \simeq -\frac{G}{r} \sum_{i=1,N} m_i \left[1 + \frac{r_i}{r} \cos\psi + \frac{1}{2}\frac{r_i^2}{r^2}\left(3 \cos^2\psi - 1\right)\right]. \tag{7.76}$$

However,

$$\sum_{i=1,N} m_i \frac{r_i}{r} \cos\psi = \sum_{i=1,N} m_i \frac{\mathbf{r}_i \cdot \mathbf{r}}{r^2} = \sum_{i=1,N} m_i \frac{x_i\,x + y_i\,y + z_i\,z}{r^2} = 0, \tag{7.77}$$

where we have made use of Equation (7.69). Hence, we are left with

$$\Phi(\mathbf{r}) \simeq -\frac{G\,M}{r} - \frac{G}{2\,r^3} \sum_{i=1,N} m_i\,r_i^2\,(2 - 3\,\sin^2\psi). \tag{7.78}$$

Here, $M = \sum_{i=1,N}$ is the total mass of the distribution. Now,

$$\sum_{i=1,N} 2\,m_i\,r_i^2 = \sum_{i=1,N} m_i\,(y_i^2 + z_i^2) + \sum_{i=1,N} m_i\,(x_i^2 + z_i^2) + \sum_{i=1,N} m_i\,(x_i^2 + y_i^2)$$

$$= \mathcal{I}_{xx} + \mathcal{I}_{yy} + \mathcal{I}_{zz}, \tag{7.79}$$

where we have made use of Equations (7.70)–(7.73). Furthermore,

$$\mathcal{I} \equiv \sum_{i=1,N} m_i \, r_i^2 \, \sin^2 \psi \tag{7.80}$$

is the distribution's moment of inertia about the axis OP. Thus, we deduce that

$$\Phi(\mathbf{r}) \simeq -\frac{G\,M}{r} - \frac{G\,(\mathcal{I}_{xx} + \mathcal{I}_{yy} + \mathcal{I}_{zz} - 3\,\mathcal{I})}{2\,r^3}. \tag{7.81}$$

This famous result is known as *MacCullagh's formula*, after its discoverer, the Irish mathematician James MacCullagh (1809–1847). Actually,

$$\sum_{i=1,N} m_i \, r_i^2 \, \sin^2 \psi = \sum_{i=1,N} m_i \, \frac{(y_i^2 + z_i^2)\,x^2 + (x_i^2 + z_i^2)\,y^2 + (x_i^2 + y_i^2)\,z^2}{r^2}$$

$$= \frac{\mathcal{I}_{xx}\,x^2 + \mathcal{I}_{yy}\,y^2 + \mathcal{I}_{zz}\,z^2}{r^2}. \tag{7.82}$$

Hence, MacCullagh's formula can also be written in the alternative form

$$\Phi(\mathbf{r}) \simeq -\frac{G\,M}{r} - \frac{G\,(\mathcal{I}_{xx} + \mathcal{I}_{yy} + \mathcal{I}_{zz})}{2\,r^3} + \frac{3\,G\,(\mathcal{I}_{xx}\,x^2 + \mathcal{I}_{yy}\,y^2 + \mathcal{I}_{zz}\,z^2)}{2\,r^5}. \tag{7.83}$$

Finally, for an axisymmetric distribution, such that $\mathcal{I}_{xx} = \mathcal{I}_{yy} = \mathcal{I}_\perp$ and $\mathcal{I}_{zz} = \mathcal{I}_\|$, MacCullagh's formula reduces to

$$\Phi(\mathbf{r}) \simeq -\frac{G\,M}{r} + \frac{G\,(\mathcal{I}_\| - \mathcal{I}_\perp)}{r^3} \, P_2(\cos\theta), \tag{7.84}$$

where $\cos\theta = z/r$. Of course, this expression is the same as Equation (7.68), which justifies our earlier assertion that Equation (7.67) is valid for a general axisymmetric mass distribution. Incidentally, a comparison of the above expression with Equation (2.66) reveals that

$$J_2 = \frac{\mathcal{I}_\| - \mathcal{I}_\perp}{M\,R^2}, \tag{7.85}$$

where R is the mean radius of the distribution, and the dimensionless parameter J_2 characterizes the quadrupole gravitational field external to the distribution.

7.10 Forced precession and nutation of the Earth

Consider the Earth–Sun system. (See Figure 7.4.) From a geocentric viewpoint, the Sun orbits the Earth counterclockwise (if we look from the north), once per year, in an approximately circular orbit of radius $a_s = 1.50 \times 10^{11}$ m (Yoder 1995). In astronomy, the plane of the Sun's apparent orbit relative to the Earth is known as the *ecliptic plane*. Let us define *nonrotating* Cartesian coordinates, centered on the Earth, which are such that the x- and y-axes lie in the ecliptic plane, and the z-axis is normal to this plane (in the sense that the Earth's north pole lies at positive z). It follows that the z-axis is directed toward a point in the sky (located in the constellation Draco) known as the *ecliptic north pole*. (See Figure 7.5.) In the following, we shall treat the x, y, z coordinate system as

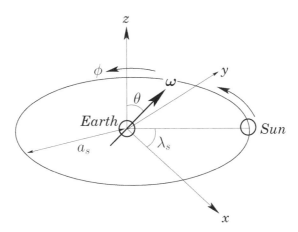

The Earth–Sun system.

inertial. This is a reasonable approximation because the orbital acceleration of the Earth is much smaller than the acceleration due to its diurnal rotation. It is convenient to parameterize the instantaneous position of the Sun in terms of a counterclockwise (if we look from the north) azimuthal angle λ_s that is zero on the positive x-axis. (See Figure 7.4.)

Let $\boldsymbol{\omega}$ be the Earth's angular velocity vector due to its daily rotation. This vector makes an angle θ with the z-axis, where $\theta = 23.44°$ is the mean inclination of the ecliptic to the Earth's equatorial plane (Yoder 1995). Suppose that the projection of $\boldsymbol{\omega}$ onto the ecliptic plane subtends an angle ϕ with the y-axis, where ϕ is measured in a counterclockwise (if we look from the north) sense. (See Figure 7.4.) The orientation of the Earth's axis of rotation (which is, of course, parallel to $\boldsymbol{\omega}$) is thus determined by the two angles θ and ϕ. Note, however, that these two angles are also *Euler angles*, in the sense given in Section 7.7. Let us examine the Earth–Sun system at an instant in time, $t = 0$, when $\phi = 0$: that is, when $\boldsymbol{\omega}$ lies in the y–z plane. At this particular instant, the x-axis points toward the so-called vernal equinox, which is defined as the point in the sky where the Sun's apparent orbit crosses the projection of the Earth's equator (i.e., the plane normal to $\boldsymbol{\omega}$) from south to north. A counterclockwise (if we look from the north) angle in the ecliptic plane that is zero at the vernal equinox is generally known as an *ecliptic longitude*. Thus, λ_s is the Sun's ecliptic longitude.

According to Equation (7.68), the potential energy of the Earth–Sun system is written

$$U = M_s\,\Phi = -\frac{G\,M_s\,M}{a_s} + \frac{G\,M_s\,(\mathcal{I}_\parallel - \mathcal{I}_\perp)}{a_s^3}\,P_2[\cos(\gamma_s)], \qquad (7.86)$$

where M_s is the mass of the Sun, M the mass of the Earth, $\mathcal{I}_\parallel$ the Earth's moment of inertia about its axis of rotation, $\mathcal{I}_\perp$ the Earth's moment of inertia about an axis lying in its equatorial plane, and $P_2(x) = (1/2)\,(3\,x^2 - 1)$. Furthermore, γ_s is the angle subtended between $\boldsymbol{\omega}$ and $\mathbf{r}_s$, where $\mathbf{r}_s$ is the position vector of the Sun relative to the Earth.

It is easily demonstrated that (with $\phi = 0$)

$$\boldsymbol{\omega} = \omega\,(0,\,\sin\theta,\,\cos\theta) \qquad (7.87)$$

and

$$\mathbf{r}_s = a_s \, (\cos \lambda_s, \, \sin \lambda_s, \, 0). \tag{7.88}$$

Hence,

$$\cos \gamma_s = \frac{\boldsymbol{\omega} \cdot \mathbf{r}_s}{|\boldsymbol{\omega}| \, |\mathbf{r}_s|} = \sin \theta \, \sin \lambda_s, \tag{7.89}$$

giving

$$U = -\frac{G \, M_s \, M}{a_s} + \frac{G \, M_s \, (\mathcal{I}_\parallel - \mathcal{I}_\perp)}{2 \, a_s^3} \, (3 \, \sin^2 \theta \, \sin^2 \lambda_s - 1). \tag{7.90}$$

Given that we are primarily interested in the motion of the Earth's axis of rotation on timescales that are much longer than a year, we can average the preceding expression over the Sun's orbit to give

$$U = -\frac{G \, M_s \, M}{a_s} + \frac{G \, M_s \, (\mathcal{I}_\parallel - \mathcal{I}_\perp)}{2 \, a_s^3} \left(\frac{3}{2} \, \sin^2 \theta - 1 \right) \tag{7.91}$$

(because the average of $\sin^2 \lambda_s$ over a year is $1/2$). Thus, we obtain

$$U = U_0 - \epsilon \, \alpha_s \, \cos(2\,\theta), \tag{7.92}$$

where U_0 is a constant, and

$$\alpha_s = \frac{3}{8} \, \mathcal{I}_\parallel \, n_s^2. \tag{7.93}$$

Here,

$$\epsilon = \frac{\mathcal{I}_\parallel - \mathcal{I}_\perp}{\mathcal{I}_\parallel} = 0.00335 \tag{7.94}$$

is the Earth's ellipticity (Yoder 1995), and

$$n_s = \frac{d\lambda_s}{dt} = \left(\frac{G \, M_s}{a_s^3} \right)^{1/2} \tag{7.95}$$

is the Sun's apparent orbital angular velocity.

The rotational kinetic energy of the Earth can be written (see Section 7.4)

$$K = \frac{1}{2} \left(\mathcal{I}_\perp \, \omega_{x'}^2 + \mathcal{I}_\perp \, \omega_{y'}^2 + \mathcal{I}_\parallel \, \omega_{z'}^2 \right), \tag{7.96}$$

which reduces to

$$K = \frac{1}{2} \left(\mathcal{I}_\perp \, \dot{\theta}^2 + \mathcal{I}_\perp \, \sin^2 \theta \, \dot{\phi}^2 + \mathcal{I}_\parallel \, \omega^2 \right) \tag{7.97}$$

with the aid of Equations (7.53)–(7.55). Here,

$$\omega = \cos \theta \, \dot{\phi} + \dot{\psi} \tag{7.98}$$

and ψ is the third Euler angle. Hence, the Earth's Lagrangian takes the form

$$\mathcal{L} = K - U = \frac{1}{2} \left(\mathcal{I}_\perp \, \dot{\theta}^2 + \mathcal{I}_\perp \, \sin^2 \theta \, \dot{\phi}^2 + \mathcal{I}_\parallel \, \omega^2 \right) + \epsilon \, \alpha_s \, \cos(2\,\theta), \tag{7.99}$$

where any constant terms have been neglected. The Lagrangian does not depend explicitly on the angular coordinate ψ. It follows that the conjugate momentum is a constant of the motion (see Section 6.5). In other words,

$$p_\psi = \frac{\partial \mathcal{L}}{\partial \dot{\psi}} = \mathcal{I}_\| \, \omega \tag{7.100}$$

is a constant of the motion, implying that ω is also a constant of the motion. Note that ω is effectively the Earth's angular velocity of rotation about its axis [because $|\omega_{x'}|, |\omega_{y'}| \ll \omega_{z'} = \omega$, which follows because $|\dot{\phi}|, |\dot{\theta}| \ll \dot{\psi}$; see Equations (7.53)–(7.55)]. Another equation of motion that can be derived from the Lagrangian is

$$\frac{d}{dt}\left(\frac{\partial \mathcal{L}}{\partial \dot{\theta}}\right) - \frac{\partial \mathcal{L}}{\partial \theta} = 0, \tag{7.101}$$

which reduces to

$$\mathcal{I}_\perp \, \ddot{\theta} - \frac{\partial \mathcal{L}}{\partial \theta} = 0. \tag{7.102}$$

Consider *steady precession* of the Earth's rotational axis, which is characterized by $\ddot{\theta} = 0$, with both $\dot{\phi}$ and $\dot{\psi}$ constant. It follows, from Equation (7.102), that such motion must satisfy the constraint

$$\frac{\partial \mathcal{L}}{\partial \theta} = 0. \tag{7.103}$$

Thus, we obtain

$$\frac{1}{2}\mathcal{I}_\perp \, \sin(2\,\theta)\,\dot{\phi}^2 - \mathcal{I}_\| \, \sin\theta \, \omega \, \dot{\phi} - 2\,\epsilon\,\alpha_s \, \sin(2\,\theta) = 0, \tag{7.104}$$

where we have made use of Equations (7.98) and (7.99). As we can easily verify after the fact, $|\dot{\phi}| \ll \omega$, so Equation (7.104) reduces to

$$\dot{\phi} \simeq -\frac{4\,\epsilon\,\alpha_s \, \cos\theta}{\mathcal{I}_\| \, \omega} = \Omega_\phi, \tag{7.105}$$

which can be integrated to give

$$\phi \simeq -\Omega_\phi \, t, \tag{7.106}$$

where

$$\Omega_\phi = \frac{3}{2}\,\frac{\epsilon\,n_s^2}{\omega}\,\cos\theta \tag{7.107}$$

and we have made use of Equation (7.93). According to the preceding expressions, the mutual interaction between the Sun and the quadrupole gravitational field generated by the Earth's slight oblateness causes the Earth's axis of rotation to precess steadily about the normal to the ecliptic plane at the rate $-\Omega_\phi$. The fact that $-\Omega_\phi$ is negative implies that the precession is in the *opposite* sense to that of the Earth's diurnal rotation and the Sun's apparent orbit about the Earth. Incidentally, the interaction causes a precession of the Earth's rotational axis, rather than the plane of the Sun's orbit, because the Earth's axial moment of inertia is much less than the Sun's orbital moment of inertia. The precession period in (sidereal) years is given by

$$T_\phi(\text{yr}) = \frac{n_s}{\Omega_\phi} = \frac{2\,T_s(\text{day})}{3\,\epsilon\,\cos\theta}, \tag{7.108}$$

where $T_s(\text{day}) = \omega/n_s = 365.26$ is the length of a sidereal year in days. Thus, given that $\epsilon = 0.00335$ and $\theta = 23.44°$, we obtain

$$T_\phi \simeq 79,200 \text{ years.} \tag{7.109}$$

Unfortunately, the observed precession period of the Earth's axis of rotation about the normal to the ecliptic plane is approximately 25,800 years (Yoder 1995), so something is clearly missing from our model. It turns out that the missing factor is the influence of the *Moon*.

Using analogous arguments to those given previously, the potential energy of the Earth–Moon system can be written

$$U = -\frac{G M_m M}{a_m} + \frac{G M_m (\mathcal{I}_\| - \mathcal{I}_\perp)}{a_m^3} P_2[\cos(\gamma_m)], \tag{7.110}$$

where M_m is the lunar mass, and a_m the radius of the Moon's (approximately circular) orbit. Furthermore, γ_m is the angle subtended between ω and $\mathbf{r}_m$, where

$$\omega = \omega \, (-\sin\theta \, \sin\phi, \, \sin\theta \, \cos\phi, \, \cos\theta) \tag{7.111}$$

is the Earth's angular velocity vector and $\mathbf{r}_m$ is the position vector of the Moon relative to the Earth. Here, for the moment, we have retained the ϕ dependence in our expression for ω (because we shall presently differentiate by t before setting $\phi = 0$). The Moon's orbital plane is actually slightly inclined to the ecliptic plane, the (mean) angle of inclination being $I_m = 5.16°$ (Yoder 1995). Hence, we can write

$$\mathbf{r}_m \simeq a_m \, (\cos\lambda_m, \, \sin\lambda_m, \, I_m \, \sin(\lambda_m - \alpha_n)), \tag{7.112}$$

to first order in I_m, where λ_m is the Moon's ecliptic longitude and α_n is the ecliptic longitude of the lunar *ascending node*, which is defined as the point on the lunar orbit where the Moon crosses the ecliptic plane from south to north. Of course, λ_m increases at the rate n_m, where

$$n_m = \frac{d\lambda_m}{dt} \simeq \left(\frac{G M}{a_m^3}\right)^{1/2} \tag{7.113}$$

is the Moon's mean orbital angular velocity. It turns out that the lunar ascending node precesses steadily, in the opposite direction to the Moon's orbital rotation, in such a manner that it completes a full circuit every 18.6 years (Yoder 1995). This precession is caused by the perturbing influence of the Sun. (See Chapter 10.) It follows that

$$\frac{d\alpha_n}{dt} = -\Omega_n, \tag{7.114}$$

where $2\pi/\Omega_n = 18.6$ years. From Equations (7.111) and (7.112),

$$\cos\gamma_m = \frac{\omega \cdot \mathbf{r}_m}{|\omega| \, |\mathbf{r}_m|} = \sin\theta \, \sin(\lambda_m - \phi) + I_m \, \cos\theta \, \sin(\lambda_m - \alpha_n), \tag{7.115}$$

so Equation (7.110) yields

$$U \simeq -\frac{G M_m M}{a_m} + \frac{G M_m (\mathcal{I}_\| - \mathcal{I}_\perp)}{2 \, a_m^3} \Big[3 \, \sin^2\theta \, \sin^2(\lambda_m - \phi)$$
$$+3 \, I_m \, \sin(2\theta) \, \sin(\lambda_m - \phi) \, \sin(\lambda_m - \alpha_n) - 1\Big] \tag{7.116}$$

to first order in I_m. Given that we are interested in the motion of the Earth's axis of rotation on timescales that are much longer than a month, we can average this expression over the Moon's orbit to give

$$U \simeq U_0' - \epsilon \alpha_m \cos(2\theta) + \epsilon \beta_m \sin(2\theta) \cos(\alpha_n - \phi), \tag{7.117}$$

[because the average of $\sin^2(\lambda_m - \phi)$ over a month is $1/2$, whereas that of $\sin(\lambda_m - \phi) \sin(\lambda_m - \alpha_n)$ is $(1/2) \cos(\alpha_n - \phi)$]. Here, U_0' is a constant,

$$\alpha_m = \frac{3}{8} \mathcal{I}_{\parallel} \mu_m n_m^2, \tag{7.118}$$

$$\beta_m = \frac{3}{4} \mathcal{I}_{\parallel} I_m \mu_m n_m^2, \tag{7.119}$$

and

$$\mu_m = \frac{M_m}{M} = 0.0123 \tag{7.120}$$

is the ratio of the lunar to the terrestrial mass (Yoder 1995). Gravity is a superposable force, so the total potential energy of the Earth–Moon–Sun system is the sum of Equations (7.92) and (7.117). In other words,

$$U = U_0'' - \epsilon \alpha \cos(2\theta) + \epsilon \beta_m \sin(2\theta) \cos(\alpha_n - \phi), \tag{7.121}$$

where U_0'' is a constant and

$$\alpha = \alpha_s + \alpha_m. \tag{7.122}$$

Finally, making use of Equation (7.97), the Lagrangian of the Earth is written

$$\mathcal{L} = \frac{1}{2} \left(\mathcal{I}_{\perp} \dot{\theta}^2 + \mathcal{I}_{\perp} \sin^2\theta \, \dot{\phi}^2 + \mathcal{I}_{\parallel} \omega^2 \right) + \epsilon \alpha \cos(2\theta) - \epsilon \beta_m \sin(2\theta) \cos(\alpha_n - \phi), \tag{7.123}$$

where any constant terms have been neglected. Recall that ω is given by Equation (7.98) and is a constant of the motion.

Two equations of motion that can immediately be derived from the preceding Lagrangian are

$$\frac{d}{dt}\left(\frac{\partial \mathcal{L}}{\partial \dot{\theta}} \right) - \frac{\partial \mathcal{L}}{\partial \theta} = 0 \tag{7.124}$$

and

$$\frac{d}{dt}\left(\frac{\partial \mathcal{L}}{\partial \dot{\phi}} \right) - \frac{\partial \mathcal{L}}{\partial \phi} = 0. \tag{7.125}$$

(The third equation, involving ψ, merely confirms that ω is a constant of the motion.) These above two equations yield

$$0 = \mathcal{I}_{\perp} \ddot{\theta} - \frac{1}{2} \mathcal{I}_{\perp} \sin(2\theta) \, \dot{\phi}^2 + \mathcal{I}_{\parallel} \sin\theta \, \omega \dot{\phi} + 2\epsilon\alpha \sin(2\theta)$$

$$+ 2\epsilon\beta_m \cos(2\theta) \cos(\alpha_n - \phi), \tag{7.126}$$

and

$$0 = \frac{d}{dt}\left(\mathcal{I}_{\perp} \sin^2\theta \, \dot{\phi} + \mathcal{I}_{\parallel} \cos\theta \, \omega \right) + \epsilon\beta_m \sin(2\theta) \sin(\alpha_n - \phi), \tag{7.127}$$

respectively. Let

$$\theta(t) = \theta_0 + \epsilon\,\theta_1(t), \tag{7.128}$$

and

$$\phi(t) = \epsilon\,\phi_1(t), \tag{7.129}$$

where $\theta_0 = 23.44°$ is the mean inclination of the ecliptic to the Earth's equatorial plane. To first order in ϵ, Equations (7.126) and (7.127) reduce to

$$0 \simeq \mathcal{I}_\perp\,\ddot{\theta}_1 + \mathcal{I}_\parallel \sin\theta_0\,\omega\,\dot{\phi}_1 + 2\,\alpha\,\sin(2\,\theta_0) + 2\beta_m\,\cos(2\,\theta_0)\cos(\Omega_n\,t) \tag{7.130}$$

and

$$0 \simeq \mathcal{I}_\perp\,\sin^2\theta_0\,\ddot{\phi}_1 - \mathcal{I}_\parallel \sin\theta_0\,\omega\,\dot{\theta}_1 - \beta_m\,\sin(2\,\theta_0)\sin(\Omega_n\,t), \tag{7.131}$$

respectively, where use has been made of Equation (7.114). However, as can easily be verified after the fact, $d/dt \ll \omega$, so we obtain

$$\dot{\phi}_1 \simeq -\frac{4\,\alpha\,\cos\theta_0}{\mathcal{I}_\parallel\,\omega} - \frac{2\beta_m\,\cos(2\,\theta_0)}{\mathcal{I}_\parallel\,\omega\,\sin\theta_0}\,\cos(\Omega_n\,t) \tag{7.132}$$

and

$$\dot{\theta}_1 \simeq -\frac{2\beta_m\,\cos\theta_0}{\mathcal{I}_\parallel\,\omega}\,\sin(\Omega_n\,t). \tag{7.133}$$

These equations can be integrated and then combined with Equations (7.128) and (7.129) to give

$$\phi(t) = -\Omega_\phi\,t - \delta\phi\,\sin(\Omega_n\,t) \tag{7.134}$$

and

$$\theta(t) = \theta_0 + \delta\theta\,\cos(\Omega_n\,t), \tag{7.135}$$

where

$$\Omega_\phi = \frac{3}{2}\,\frac{\epsilon\,(n_s^2 + \mu_m\,n_m^2)}{\omega}\,\cos\theta_0, \tag{7.136}$$

$$\delta\phi = \frac{3}{2}\,\frac{\epsilon\,I_m\,\mu_m\,n_m^2}{\omega\,\Omega_n}\,\frac{\cos(2\,\theta_0)}{\sin\theta_0}, \tag{7.137}$$

and

$$\delta\theta = \frac{3}{2}\,\frac{\epsilon\,I_m\,\mu_m\,n_m^2}{\omega\,\Omega_n}\,\cos\theta_0. \tag{7.138}$$

Incidentally, in these expressions, we have assumed that the lunar ascending node coincides with the vernal equinox at time $t = 0$ (i.e., $\alpha_n = 0$ at $t = 0$), in accordance with our previous assumption that $\phi = 0$ at $t = 0$.

According to Equation (7.134), the combined gravitational interaction of the Sun and the Moon with the gravitational quadrupole field generated by the Earth's slight oblateness causes the Earth's axis of rotation to precess steadily about the normal to the ecliptic plane at the rate $-\Omega_\phi$. As before, the negative sign indicates that the precession

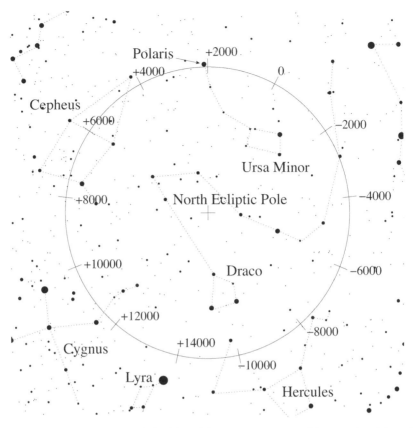

Fig. 7.5 Path of the north celestial pole against the backdrop of the stars as consequence of the precession of the equinoxes (calculated assuming constant precessional speed and obliquity). Numbers indicate years relative to the start of the common era. Stellar positions and magnitudes from Hoffleit and Warren 1991.

is in the opposite direction to the (apparent) orbital motion of the Sun and the Moon. The period of this so-called *luni-solar precession* in (sidereal) years is given by

$$T_\phi(\text{yr}) = \frac{n_s}{\Omega_\phi} = \frac{2\,T_s(\text{day})}{3\,\epsilon\,\{1 + \mu_m/[T_m(\text{yr})]^2\}\,\cos\theta_0}, \tag{7.139}$$

where $T_m(\text{yr}) = n_s/n_m = 0.081$ is the Moon's (synodic) orbital period in years. Given that $\epsilon = 0.00335$, $\theta_0 = 23.44°$, $T_s(\text{day}) = 365.26$, and $\mu_m = 0.0123$, we obtain

$$T_\phi \simeq 27,600 \text{ years.} \tag{7.140}$$

This prediction is fairly close to the observed precession period of 25,800 years (Yoder 1995). The main reason that our estimate is slightly inaccurate is because we have neglected to take into account the small eccentricities of the Earth's orbit around the Sun and the Moon's orbit around the Earth.

The point in the sky toward which the Earth's axis of rotation points is known as the *north celestial pole*. Currently, this point lies within about a degree of the fairly bright star Polaris, which is consequently sometimes known as the *north star* or *pole star*. (See Figure 7.5.) It follows that Polaris appears to be almost stationary in the sky, always

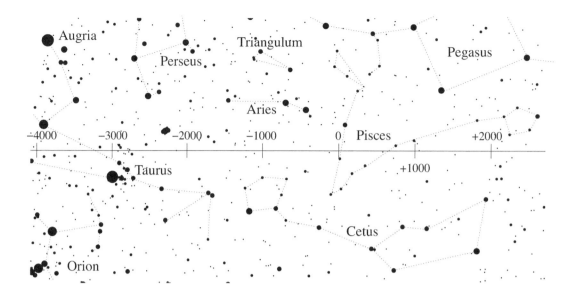

Augria
Perseus
Triangulum
Pegasus
Aries
−4000   −3000   −2000   −1000   0   Pisces   +2000
Taurus
+1000
Cetus
Orion

Fig. 7.6 Path of the vernal equinox against the backdrop of the stars as a consequence of the precession of the equinoxes (calculated assuming constant precessional speed and obliquity). Numbers indicate years relative to the start of the common era. Stellar positions and magnitudes from Hoffleit and Warren 1991.

lying due north, and can thus be used for navigational purposes. Indeed, mariners have relied on the north star for many hundreds of years to determine direction at sea. Unfortunately, because of the precession of the Earth's axis of rotation, the north celestial pole is not a fixed point in the sky but instead traces out a circle, of angular radius $23.44°$, about the north ecliptic pole, with a period of 25,800 years. (See Figure 7.5.) Hence, a few thousand years from now, the north celestial pole will no longer coincide with Polaris, and there will be no convenient way of telling direction from the stars.

The projection of the ecliptic plane onto the sky is called the *ecliptic circle* and coincides with the apparent path of the Sun against the backdrop of the stars. The projection of the Earth's equator onto the sky is known as the *celestial equator*. As has been previously mentioned, the ecliptic is inclined at $23.44°$ to the celestial equator. The two points in the sky at which the ecliptic crosses the celestial equator are called the *equinoxes*, as night and day are equally long when the Sun lies at these points. Thus, the Sun reaches the vernal equinox on about March 20, and this traditionally marks the beginning of spring. Likewise, the Sun reaches the autumnal equinox on about September 22, and this traditionally marks the beginning of autumn. However, the precession of the Earth's axis of rotation causes the celestial equator (which is always normal to this axis) to precess in the sky; it thus also causes the equinoxes to precess along the ecliptic. This effect is known as the *precession of the equinoxes*. The precession is in the opposite direction to the Sun's apparent motion around the ecliptic and is of magnitude $1.4°$ per century. Amazingly, this minuscule effect was discovered by the ancient Greeks (with the help of ancient Babylonian observations; Heath 1991). In about 2000 BCE, when the science of astronomy originated in ancient Egypt and Babylonia, the vernal equinox lay in the constellation Aries. (See Figure 7.6.) Indeed, the vernal equinox is still sometimes called the *first point of Aries* in astronomical texts. About 90 BCE, the vernal equinox moved into the constellation Pisces, where it still remains. The equinox will move into the constellation Aquarius (marking the beginning of the much-heralded "Age of Aquarius") in about 2600 CE. Incidentally, the position of the vernal equinox in the sky is of great significance in astronomy, as it is used as the zero of celestial longitude (much as the Greenwich meridian is used as the zero of terrestrial longitude).

Equations (7.134) and (7.135) indicate that the small inclination of the lunar orbit to the ecliptic plane, combined with the precession of the lunar ascending node, causes the Earth's axis of rotation to wobble sightly. This wobble is known as *nutation* (from the Latin *nutare*, to nod) and is superimposed on the aforementioned precession. In the absence of precession, nutation would cause the north celestial pole to periodically trace out a small ellipse on the sky, the sense of rotation being counterclockwise. The nutation period is 18.6 years—the same as the precession period of the lunar ascending node. The nutation amplitudes in the polar and azimuthal angles θ and ϕ are

$$\delta\theta = \frac{3}{2} \frac{\epsilon\, I_m\, \mu_m\, T_n(\mathrm{yr})}{T_s(\mathrm{day})\,[T_m(\mathrm{yr})]^2} \cos\theta_0 \tag{7.141}$$

and

$$\delta\phi = \frac{3}{2} \frac{\epsilon\, I_m\, \mu_m\, T_n(\mathrm{yr})}{T_s(\mathrm{day})\,[T_m(\mathrm{yr})]^2} \frac{\cos(2\,\theta_0)}{\sin\theta_0}, \tag{7.142}$$

respectively, where $T_n(\mathrm{yr}) = n_s/\Omega_n = 18.6$. Given that $\epsilon = 0.00335$, $\theta_0 = 23.44°$, $I_m = 5.16°$, $T_s(\mathrm{day}) = 365.26$, $T_m(\mathrm{yr}) = 0.081$, and $\mu_m = 0.0123$, we obtain

$$\delta\theta = 8.2'' \tag{7.143}$$

and

$$\delta\phi = 15.3''. \tag{7.144}$$

The observed nutation amplitudes are $9.2''$ and $17.2''$, respectively (Meeus 2005). Hence, our estimates are quite close to the mark. Any inaccuracy is mainly due to the fact that we have neglected to take into account the small eccentricities of the Earth's orbit around the Sun and the Moon's orbit around the Earth. The nutation of the Earth was discovered in 1728 by the English astronomer James Bradley (1693–1748); it was explained theoretically about 20 years later by d'Alembert and Euler. Nutation is important because the corresponding gyration of the Earth's rotation axis appears to be transferred to celestial objects when they are viewed using terrestrial telescopes. This effect causes the celestial longitudes and latitudes of heavenly objects to oscillate sinusoidally by up to $20''$ (i.e., about the maximum apparent angular size of Saturn) with a period of 18.6 years. It is necessary to correct for this oscillation (as well as for the precession of the equinoxes) to accurately guide terrestrial telescopes to particular celestial objects.

7.11 Spin-orbit coupling

Let us investigate the spinning motion (i.e., the rotational motion about an axis passing through the center of mass) of an *aspherical* moon in a Keplerian elliptical orbit about a spherically symmetric planet of mass m_p. It is convenient to analyze this motion in a frame of reference whose origin always coincides with the moon's center of mass, O. Let us define a Cartesian coordinate system x, y, z whose axes are aligned with the moon's principal axes of rotation, and let $\mathcal{I}_{xx}$, $\mathcal{I}_{yy}$, $\mathcal{I}_{zz}$ be the corresponding principal moments of inertia. Suppose that $\mathcal{I}_{zz} > \mathcal{I}_{yy} > \mathcal{I}_{xx}$, which implies that the moon's radius attains its greatest and least values at those points where the x- and z-axes pierce its surface, respectively (assuming that the moon's shape is roughly ellipsoidal). Let the planet, P, be located at position vector $\mathbf{r} \equiv (x, y, z)$. We can treat the planet as a point mass, as it is spherically symmetric. Incidentally, we are assuming that the moon's deviations from spherical symmetry are of a permanent nature, being maintained by internal tensile strength rather than being induced by tidal or rotational effects.

According to MacCullagh's formula, the gravitational potential produced at P by the gravitational field of the moon is (see Section 7.9)

$$\Phi(\mathbf{r}) \simeq -\frac{G M}{r} - \frac{G (\mathcal{I}_{xx} + \mathcal{I}_{yy} + \mathcal{I}_{zz})}{2 r^3} + \frac{3 G (\mathcal{I}_{xx} x^2 + \mathcal{I}_{yy} y^2 + \mathcal{I}_{zz} z^2)}{2 r^5}. \tag{7.145}$$

Thus, the gravitational force, $\mathbf{f}$, exerted on the planet by the moon has the components

$$f_x = -m_p \frac{\partial \Phi}{\partial x}, \tag{7.146}$$

$$f_y = -m_p \frac{\partial \Phi}{\partial y}, \tag{7.147}$$

and

$$f_z = -m_p \frac{\partial \Phi}{\partial z}. \tag{7.148}$$

Furthermore, the components of the torque, $\boldsymbol{\tau}$, acting on the planet about point O are

$$\tau_x = y f_z - z f_y = -\frac{3 G m_p (\mathcal{I}_{zz} - \mathcal{I}_{yy}) z y}{r^5}, \tag{7.149}$$

$$\tau_y = z f_x - x f_z = -\frac{3 G m_p (\mathcal{I}_{xx} - \mathcal{I}_{zz}) x z}{r^5}, \tag{7.150}$$

and

$$\tau_z = x f_y - y f_x = -\frac{3 G m_p (\mathcal{I}_{yy} - \mathcal{I}_{xx}) y x}{r^5}. \tag{7.151}$$

Of course, an equal and opposite torque, $-\boldsymbol{\tau}$, acts on the moon.

Euler's equations for the moon's spinning motion take the form (see Section 7.6)

$$\mathcal{I}_{xx} \dot{\omega}_x - (\mathcal{I}_{yy} - \mathcal{I}_{zz}) \omega_z \omega_y = -\tau_x, \tag{7.152}$$

$$\mathcal{I}_{yy} \dot{\omega}_y - (\mathcal{I}_{zz} - \mathcal{I}_{xx}) \omega_x \omega_z = -\tau_y, \tag{7.153}$$

and

$$\mathcal{I}_{zz} \dot{\omega}_z - (\mathcal{I}_{xx} - \mathcal{I}_{yy}) \omega_y \omega_x = -\tau_z, \tag{7.154}$$

where $\boldsymbol{\omega} \equiv (\omega_x, \omega_y, \omega_z)$ is the associated angular velocity vector. Suppose that the moon is actually spinning about the z-axis (i.e., the principal axis of rotation with the largest

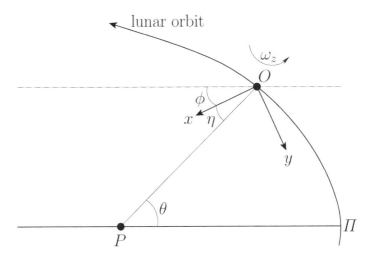

Fig. 7.7 Geometry of spin-orbit coupling.

associated moment of inertia), and that this axis is directed *normal* to the moon's orbital plane (which is assumed to be fixed). It follows that

$$\boldsymbol{\omega} = (0,\ 0,\ \omega_z) \qquad (7.155)$$

and

$$\mathbf{r} = r\,(\cos\eta,\ \sin\eta,\ 0), \qquad (7.156)$$

where η is the relative position angle of the planet in the x–y plane. (See Figure 7.7.) We can write $\omega_z = \dot{\phi}$, where ϕ is the angle subtended between the x-axis (say) and some fixed (with respect to distant stars) direction in the x–y plane. Let this direction be parallel to the major axis $P\Pi$ of the moon's orbit, where Π is the *pericenter* (i.e., the point of closest approach of the moon to the planet). In this case, it is clear from Figure 7.7 that

$$\eta = \theta - \phi, \qquad (7.157)$$

where θ is the moon's orbital true anomaly. (See Chapter 3.) Hence, Equations (7.151), (7.154), (7.155), and (7.156) yield

$$\ddot{\phi} + \frac{3}{2}\,n^2\left(\frac{\mathcal{I}_{yy} - \mathcal{I}_{xx}}{\mathcal{I}_{zz}}\right)\left(\frac{a}{r}\right)^3 \sin[2\,(\phi - \theta)] = 0, \qquad (7.158)$$

where use has been made of the standard Keplerian result $G\,m_p = n^2\,a^3$ (assuming that the mass of the moon is much less than that of the planet). (See Chapter 3.) Here, a and n are the moon's orbital major radius and mean angular velocity, respectively.

Assuming that the eccentricity, e, of the moon's orbit is low—so that $0 \le e \ll 1$—it follows from Equations (3.85) and (3.86), as well as the trigonometric inequalities listed in Section A.3, that

$$\left(\frac{a}{r}\right)^3 = 1 + 3\,e\,\cos\mathcal{M} + \mathcal{O}(e^2), \tag{7.159}$$

$$\theta = \mathcal{M} + 2\,e\,\sin\mathcal{M} + \mathcal{O}(e^2), \tag{7.160}$$

$$\cos 2\theta = \cos 2\mathcal{M} + 2\,e\,(\cos 3\mathcal{M} - \cos\mathcal{M}) + \mathcal{O}(e^2), \tag{7.161}$$

and

$$\sin 2\theta = \sin 2\mathcal{M} + 2\,e\,(\sin 3\mathcal{M} - \sin\mathcal{M}) + \mathcal{O}(e^2), \tag{7.162}$$

where $\mathcal{M}$ is the moon's mean anomaly. Note that $d\mathcal{M}/dt = n$. Hence, Equation (7.158) gives

$$\ddot{\phi} \simeq -\frac{3}{2}n^2\left(\frac{\mathcal{I}_{yy} - \mathcal{I}_{xx}}{\mathcal{I}_{zz}}\right)\left[-\frac{e}{2}\,\sin(2\,\phi - \mathcal{M}) + \sin(2\,\phi - 2\,\mathcal{M})\right. $$
$$\left. + \frac{7\,e}{2}\,\sin(2\,\phi - 3\,\mathcal{M})\right], \tag{7.163}$$

where any $\mathcal{O}(e^2)$ terms have been neglected. Here, use has again been made of the trigonometric identities in Section A.3.

Suppose that the moon passes through its pericenter at time $t = 0$, so that

$$\mathcal{M} = n\,t. \tag{7.164}$$

In this case, the previous equation becomes

$$\frac{d^2\phi}{dt^2} \simeq -\frac{n_0^2}{2}\left[-\frac{e}{2}\,\sin(2\,\phi - n\,t) + \sin(2\,\phi - 2\,n\,t) + \frac{7\,e}{2}\,\sin(2\,\phi - 3\,n\,t)\right], \tag{7.165}$$

where the so-called *asphericity parameter*,

$$\alpha = \left[3\left(\frac{\mathcal{I}_{yy} - \mathcal{I}_{xx}}{\mathcal{I}_{zz}}\right)\right]^{1/2}, \tag{7.166}$$

is a measure of the moon's departure from spherical symmetry, and

$$n_0 = \alpha\,n. \tag{7.167}$$

Equation (7.165) is highly nonlinear in nature. Consequently, it does not possess a general analytic solution. Fortunately, Equation (7.165) is relatively straightforward to solve numerically. In fact, the solution can be represented as a trajectory in ϕ, $n^{-1}\,d\phi/dt$, $n\,t$ space. Because Equation (7.165) is deterministic, a trajectory that corresponds to a unique set of initial conditions cannot intersect a second trajectory that corresponds to a different set of initial conditions. Unfortunately, it is difficult to visualize a trajectory in three dimensions. However, we can alleviate this problem by plotting only those points where the trajectory pierces a set of equally spaced planes normal to the $n\,t$ axis. These planes are located at $n\,t = i\,2\pi$, where i is an integer. This procedure is equivalent to projecting the trajectory onto the ϕ, $n^{-1}\,d\phi/dt$ plane each time the moon passes through its pericenter. The resulting plot is known as a *surface of section*. Figure 7.8 shows the surface of section for a set of trajectories corresponding to many different initial conditions. All of the trajectories are calculated from Equation (7.165) using $\alpha = 0.15$

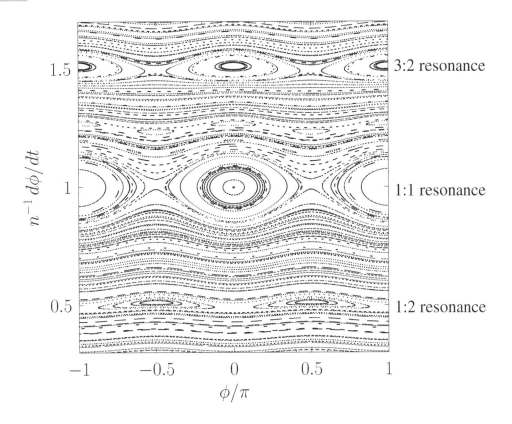

Fig. 7.8 Surface of section plot for solutions of Equation (7.165) with $\alpha = 0.15$ and $e = 0.05$. The major spin-orbit resonances are labeled.

and $e = 0.05$. The relatively small value adopted for the eccentricity, e, is consistent with our earlier assumption that the moon's orbit is nearly circular. On the other hand, the relatively small value adopted for the asphericity parameter, α, implies that the moon is almost spherical. It can be seen, from Figure 7.8, that a trajectory corresponding to a given set of initial conditions generates a series of closely spaced points that trace out a closed curve running roughly parallel to the ϕ-axis. Actually, there are two distinct types of curve. The majority of curves extend over all values of ϕ and represent trajectories for which there is no particular correlation between the moon's spin and orbital motions. However, a relatively small number of curves extend over only a limited range of ϕ values. These curves represent trajectories for which a *resonant interaction* between the moon's spin and orbital motions produces a strong correlation between these two types of motion. The exact resonances correspond to the centers of the eye-shaped structures that can be seen in Figure 7.8. The three principal spin-orbit resonances evident in the figure are the 1:2, 1:1, and 3:2 resonances. Here, a j_o:j_s resonance, where j_o and j_s are positive integers, is such that j_o times the moon's spin period is equal to j_s times its orbital period. At such a resonance, the moon's principal axes of rotation point in the *same* direction every j_s pericenter passages.

Consider the $j_o:j_s$ spin-orbit resonance. It is helpful to define

$$\gamma = \phi - p\,\mathcal{M} = \phi - p\,n\,t, \tag{7.168}$$

where $p = j_o/j_s$. Here, γ is minus the angle subtended between the moon's x-axis and the major axis of its orbit every j_s passages through the pericenter. Note that $\gamma = \phi$ at such passages. In the vicinity of the resonance, we expect γ to be a relatively slowly varying function of time. When expressed in terms of γ, Equation (7.165) yields

$$\frac{d^2\gamma}{dt^2} \simeq -\frac{n_0^2}{2}\left(\left\{-\frac{e}{2}\cos[(2\,p-1)\,n\,t] + \cos[2\,(p-1)\,n\,t] + \frac{7\,e}{2}\cos[(2\,p-3)\,n\,t]\right\}\sin 2\gamma\right.$$

$$\left. + \left\{-\frac{e}{2}\sin[(2\,p-1)\,n\,t] + \sin[2\,(p-1)\,n\,t] + \frac{7\,e}{2}\sin[(2\,p-3)\,n\,t]\right\}\cos 2\gamma\right). \tag{7.169}$$

Let us now average the right-hand side of the above equation over j_s orbital periods, treating the relatively slowly varying quantity γ as a constant. For the 1:1 resonance, for which $p = 1$, we are left with

$$\frac{d^2\gamma}{dt^2} \simeq -\frac{1}{2}\,n_0^2\,\sin 2\gamma. \tag{7.170}$$

This follows because $\langle\cos[2\,(p-1)\,n\,t]\rangle = \langle 1\rangle = 1$, when $p = 1$, whereas all the other averages over rapidly varying terms are zero: e.g., $\langle\cos[(2\,p-1)\,n\,t]\rangle = \langle\cos(n\,t)\rangle = 0$. For the 1:2 resonance, for which $p = 1/2$, we are left with

$$\frac{d^2\gamma}{dt^2} \simeq \frac{e}{4}\,n_0^2\,\sin 2\gamma. \tag{7.171}$$

Finally, for the 3:2 resonance, for which $p = 3/2$, we are left with

$$\frac{d^2\gamma}{dt^2} \simeq -\frac{7\,e}{2}\,n_0^2\,\sin 2\gamma. \tag{7.172}$$

Consider the 1:1 resonance. Multiplying Equation (7.170) by $d\gamma/dt$, and integrating, we obtain

$$\frac{1}{2}\left(\frac{d\gamma}{dt}\right)^2 - \frac{1}{4}\,n_0^2\,\cos 2\gamma \simeq \frac{1}{4}\,n_0^2\,\mathcal{E}, \tag{7.173}$$

where the constant $\mathcal{E}$ is related to the moon's spin energy per unit mass. Now, $n^{-1}\,d\gamma/dt = n^{-1}\,d\phi/dt - 1$. Furthermore, $\phi = \gamma$ at the times of pericenter passage. Hence, at such times,

$$\mathcal{E} \simeq \frac{2}{\alpha^2}\left(n^{-1}\,\frac{d\phi}{dt} - 1\right)^2 - \cos 2\phi. \tag{7.174}$$

Similar arguments reveal that

$$\mathcal{E} \simeq \frac{4}{e\,\alpha^2}\left(n^{-1}\,\frac{d\phi}{dt} - \frac{1}{2}\right)^2 + \cos 2\phi \tag{7.175}$$

for the 1:2 resonance and

$$\mathcal{E} \simeq \frac{4}{7\,e\,\alpha^2}\left(n^{-1}\,\frac{d\phi}{dt} - \frac{3}{2}\right)^2 - \cos 2\phi \tag{7.176}$$

for the 3:2 resonance.

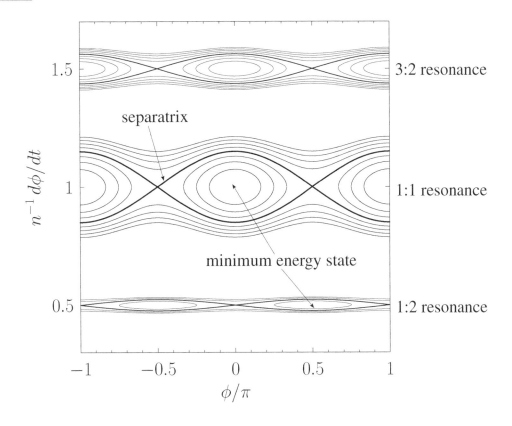

Fig. 7.9 Contours of $\mathcal{E}$, for $-1 \leq \mathcal{E} \leq 2$, plotted in $\phi, n^{-1} d\phi/dt$ space for the 1:2, 1:1, and 3:2 spin-orbit resonances. The contours are calculated with $\alpha = 0.15$ and $e = 0.05$.

Figure 7.9 shows contours of $\mathcal{E}$ plotted in ϕ, $n^{-1} d\phi/dt$ space for the 1:2, 1:1, and 3:2 spin-orbit resonances. These contours are calculated from Equations (7.174)–(7.176) by using $\alpha = 0.15$ and $e = 0.05$. It can be seen that the contours shown in Figure 7.9 are very similar to the surface of section curves displayed in Figure 7.8—at least, in the vicinity of the resonances. This suggests that the analytic expressions in Equations (7.174)–(7.176) can be used to efficiently map closed surface of section curves in the vicinity of the 1:1, 1:2, and 3:2 spin-orbit resonances. Moreover, it is clear from these expressions that solutions to Equation (7.165) are effectively trapped on such curves. This implies that a solution initially close to (say) a 1:1 resonance will remain close to this resonance indefinitely.

For a given spin-orbit resonance, there exists a *separatrix*, corresponding to the $\mathcal{E} = 1$ contour, dividing contours that span the whole range of ϕ values from those that only span a restricted range of ϕ values. (See Figure 7.9.) The former contours are characterized by $\mathcal{E} > 1$, whereas the latter are characterized by $\mathcal{E} < 1$. As the energy integral, $\mathcal{E}$, is reduced below the critical value $\mathcal{E} = 1$, the range of allowed values of ϕ becomes narrower and narrower. Eventually, when $\mathcal{E}$ attains its minimum possible value (i.e., $\mathcal{E} = -1$), ϕ is constrained to take a fixed value. This situation corresponds to an exact spin-orbit resonance. For the case of the 1:1 resonance, the minimum energy state corresponds to $\phi = 0$, $\pm\pi$ and $n^{-1}\,d\phi/dt = 1$, which implies that, at the exact resonance, the moon's x-axis points directly toward (or away from) the planet at the time of pericenter passage. Because we previously assumed that $\mathcal{I}_{yy} > \mathcal{I}_{xx}$, which means that the moon is more elongated in the x-direction than in the y-direction, it follows that the long axis (in the x–y plane) is directed toward the planet each time the moon passes through its pericenter. In this respect, the 3:2 resonance is similar to the 2:1 resonance. However, for a moon with a low-eccentricity orbit locked in a 1:1 spin-orbit resonance, the x-axis always points in the general vicinity of the planet, even when the moon is far from its pericenter. The same is not true for a moon trapped in a 3:2 resonance. For the case of the 1:2 resonance, the minimum energy state corresponds to $\phi = \pm\pi/2$ and $n^{-1}\,d\phi/dt = 1/2$, which implies that, at the exact resonance, the moon's y-axis points directly toward (or away from) the planet at the time of pericenter passage. In other words, the short axis (in the x–y plane) is directed toward the planet each time the moon passes through its pericenter.

It can be seen from Equations (7.174)–(7.176) that the *resonance widths* (i.e., maximum extent, in the $n^{-1}\,d\phi/dt$ direction, of the eyelike structure enclosed by the separatrix) of the 1:1, 1:2, and 3:2 spin-orbit resonances are $2\,\alpha$, $\sqrt{2}\,e\,\alpha$, and $\sqrt{14}\,e\,\alpha$, respectively. As long as these widths are significantly less than the interresonance spacing (which is $1/2$), the three resonances remain relatively widely separated and are thus distinct from one another. A rough criterion for the overlap of the 1:1 and 3:2 resonances is

$$\alpha > \frac{1}{2 + \sqrt{14\,e}}. \tag{7.177}$$

The best example of a celestial body trapped in a 3:2 spin-orbit resonance is the planet Mercury, whose spin period is 58.65 days, and whose orbital period is 87.97 = 1.5 × 58.65 days (Yoder 1995). Note that Mercury's axial tilt (with respect to the normal to its orbital plane) is only 2′ (Margot et al. 2007). In other words, Mercury is effectively rotating about an axis that is directed normal to its orbital plane, in accordance with our earlier assumption. It is thought that Mercury was originally spinning faster than at present, but that its spin rate was gradually reduced by the tidal de-spinning effect of the Sun (see Section 5.7), until it fell into a 3:2 spin-orbit resonance. As we have seen, once established, such a resonance is maintained by the locking torque exerted by the Sun on Mercury because of the latter body's small permanent asphericity. However, this is possible only because, close to the resonance, the locking torque exceeds the de-spinning torque.

The best example of a celestial body trapped in a 1:1 spin-orbit resonance is the Moon, whose spin and orbital periods are both 27.32 days. The Moon's axial tilt (with respect to the normal to its orbital plane) is 1.59°. In other words, the Moon is rotating about an axis that is (almost) normal to its orbital plane, in accordance with our previous assumption. Like Mercury, it is thought that the Moon was originally spinning faster than at present, but that its spin rate was gradually reduced by tidal de-spinning until it fell into a 1:1 spin-orbit resonance. This resonance is maintained by the locking torque exerted by the Earth on the Moon because of the latter body's small permanent asphericity, rather than by tidal effects, as (when the eccentricity of the lunar orbit is taken into account) tidal effects alone would actually cause the moon's spin rate to exceed its mean orbital rotation rate by about 3 percent (Murray and Dermott 1999).

Consider a moon whose spin state is close to an exact 1:1 spin-orbit resonance. According to the full (i.e., nonaveraged) equation of motion, Equation (7.169),

$$\frac{d^2\gamma}{dt^2} + n_0^2 \, \gamma \simeq 2 \, e \, n_0^2 \, \sin(n \, t), \tag{7.178}$$

where we have assumed that $|\gamma| \ll 1$ (because the moon is close to the exact resonance) and have also neglected terms of order $e \, \gamma$ with respect to unity. The preceding equation has the standard solution

$$\gamma = -\gamma_0 \, \sin(n_0 \, t - \varphi_0) - 2 \, e \, \frac{n_0^2}{n^2 - n_0^2} \, \sin \mathcal{M}, \tag{7.179}$$

where γ_0 and φ_0 are arbitrary. This expression is more conveniently written

$$\gamma = -\gamma_0 \, \sin(n_0 \, t - \varphi_0) - 2 \, e \, \frac{\alpha^2}{1 - \alpha^2} \, \sin \mathcal{M}. \tag{7.180}$$

From Equations (7.157), (7.160), and (7.168), we have

$$\eta \simeq 2 \, e \, \sin \mathcal{M} - \gamma, \tag{7.181}$$

which implies that

$$\eta \simeq 2 \, e \, \sin \mathcal{M} + \gamma_0 \, \sin(n_0 \, t - \varphi_0) + 2 \, e \, \frac{\alpha^2}{1 - \alpha^2} \, \sin \mathcal{M}. \tag{7.182}$$

Here, η is the angle subtended between the moon's x-axis and the line joining the center of the moon to the planet. (See Figure 7.7.) According to the preceding equation, this angle *librates* (i.e., oscillates). The first term on the right-hand side of the preceding expression describes so-called *optical libration* (in longitude). This is merely a perspective effect due to the eccentricity of the moon's orbit; it does not imply any irregularity in the moon's axial spin rate. The final two terms describe so-called *physical libration* (in longitude) and are associated with real irregularities in the moon's spin rate. To be more exact, the first of these terms describes *free libration* (in longitude), whereas the second describes *forced libration* (in longitude). Optical libration causes an oscillation in η whose period matches the moon's orbital period, whose amplitude (in radians) is $2 \, e$, and whose phase is such that $\eta = 0$ as the moon passes through its pericenter. Forced libration causes a similar oscillation of much smaller amplitude (assuming that $\alpha \ll 1$). Free libration, on the other hand, causes an oscillation in η whose period is α^{-1} times the moon's orbital period, and whose amplitude and phase are arbitrary.

Consider the Moon, whose spin state is close to a 1:1 spin-orbit resonance. According to data from the Lunar Prospector probe (Konopliv et al. 1998),

$$\frac{\mathcal{I}_{yy} - \mathcal{I}_{xx}}{\mathcal{I}_{zz}} = 2.279 \times 10^{-4}, \tag{7.183}$$

which implies that

$$\alpha = 2.615 \times 10^{-2}. \tag{7.184}$$

Hence, given that the Moon's orbital period is 27.322 days (i.e., a sidereal month), the predicted free libration period is 2.86 years. Because of the comparatively rapid precession of the Moon's perigee (which completes a full circuit about the Earth every 8.85 years—see Chapter 10), the expected period of both optical and forced libration (in longitude) is 27.555 days (i.e., an anomalistic month). Moreover, the predicted amplitudes of these librations are 6.5° [when evaluated up to $\mathcal{O}(e^2)$], and 15.8″, respectively. Optical libration (in longitude) has been observed for hundreds of years and does indeed have the characteristics described earlier. Furthermore, despite having an amplitude that is a thousand times less than that of optical libration, the forced libration (in longitude) of the Moon (due to the eccentricity of the lunar orbit) is measurable by means of laser ranging. The observed period and amplitude are 27.555 days and 16.8″, respectively (Williams and Dickey 2003), and are in good agreement with the above predictions. Finally, a free libration (in longitude) mode of the Moon with a period of 2.87 years and an amplitude of 1.87″ has been observed via laser ranging (Jin and Li 1996). The period of this libration is, thus, in good agreement with our analysis. Note that, since the Moon's orbit has significant non-Keplerian elements, due to the perturbing action of the Sun, the Moon's forced libration (in longitude) also has important non-Keplerian elements. (See Exercise 10.7.) Furthermore, the Moon also possesses free and forced modes of libration in latitude. (See Section 7.12.)

The forced libration of the Moon is a tiny effect because of the Moon's relatively small departures from sphericity. There exist other moons in the solar system, however, that are locked in a 1:1 spin-orbit resonance (like the Moon) and whose departures from sphericity are substantial. For such moons, forced libration can attain quite large amplitudes. A prime example is the Martian moon Phobos. The shape of this moon, which is highly irregular (see Figure 2.2), has been measured to high precision by the Mars Express probe, allowing the computation of the relative magnitudes of its principal moments of inertia (on the assumption that the moon is homogeneous). According to this calculation (Wilner et al. 2010),

$$\frac{\mathcal{I}_{yy} - \mathcal{I}_{xx}}{\mathcal{I}_{zz}} = 0.129, \tag{7.185}$$

which implies that

$$\alpha = 0.623. \tag{7.186}$$

Because the observed eccentricity of Phobos' orbit is $e = 0.0151$ (Yoder 1995), the predicted amplitude of its forced physical libration is 1.1°. The measured amplitude is 1.2° (Wilner et al. 2010). Note that the α and e values for Phobos satisfy the resonance overlap criterion in Equation (7.177). Figure 7.10 shows a surface of section plot for Phobos. It can be seen that resonance overlap leads to the destruction of many of the closed curves that are a feature of Figure 7.8. Nevertheless, some closed curves remain intact, especially in the vicinity of the 1:1 spin-orbit resonance, that is, around $\phi = 0$, $\pm\pi$, and $n^{-1} d\phi/dt = 1$. Consequently, it is possible for Phobos to remain close to a 1:1 spin-orbit resonant state for an indefinite period of time.

The most extreme example of spin-orbit coupling in the solar system occurs in Hyperion, which is a small moon of Saturn. Hyperion has a highly irregular shape, with an

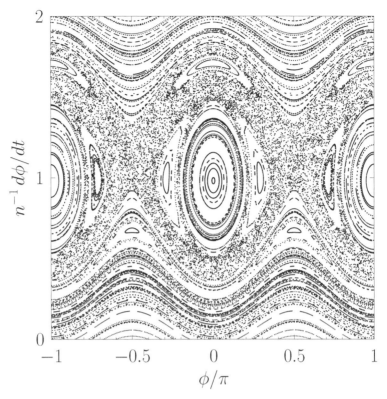

Surface of section plot for various solutions of Equation (7.165) with the Phobos-like parameters $\alpha = 0.623$ and $e = 0.0151$.

asphericity parameter of

$$\alpha \simeq 0.89 \qquad (7.187)$$

(calculated on the assumption that Hyperion is homogeneous), and is in a fairly eccentric orbit of eccentricity

$$e = 0.123 \qquad (7.188)$$

(Thomas et al. 1995). Hence, Hyperion easily satisfies the resonance overlap criterion in Equation (7.177). Figure 7.11 shows a surface of section plot for Hyperion. It can be seen that resonance overlap leads to the complete destruction of all the closed curves associated with the 1:1 spin-orbit resonance. This would seem to imply that Hyperion cannot remain trapped in a 1:1 resonance for any appreciable length of time. Figure 7.12 shows the time evolution of a solution of Equation (7.165), with Hyperion-like values of α and e, that starts off in an exact 1:1 spin-orbit resonance. If the solution were to stay close to the resonant state, the angle γ—and, hence, $\sin\gamma$—would remain close to zero. It can be seen, from the figure, that this is not the case. In fact, $\sin\gamma$ quickly becomes of order unity, indicating a strong deviation from the resonant state. Moreover, $\sin\gamma$—and, hence, γ itself—subsequently varies in a markedly irregular manner. The time variation

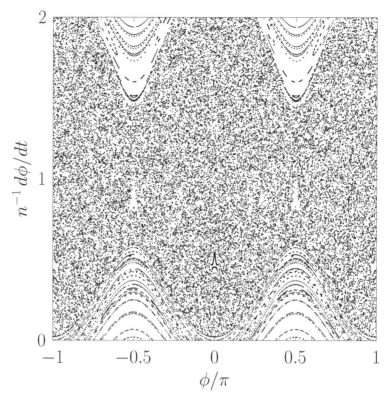

Surface of section plot for various solutions of Equation (7.165) with the Hyperion-like parameters $\alpha = 0.89$ and $e = 0.123$.

of γ is in fact *chaotic*: that is, it is quasi-random, never repeats itself, and exhibits extreme sensitivity to initial conditions (Strogatz 2001). This suggests that Hyperion is spinning in a chaotic manner. Data from the Voyager 2 probe seem to confirm that this is indeed the case (Black et al. 1995). (Note, however, that Hyperion's spinning motion involves large chaotic variations in its axial tilt that are not taken into account in our analysis.)

7.12 Cassini's laws

Consider Figure 7.13. Here, M, E, and N represent the center of the Moon, the center of the Earth, and the north ecliptic pole, respectively. Moreover, MZ is the instantaneous normal to the Moon's equatorial plane, and MP the instantaneous normal to the Moon's orbital plane. Let $\mathbf{e}$, ζ, ν, and $\mathbf{p}$ be unit vectors parallel to ME, MZ, MN, and MP, respectively. The *fixed* angle, $I = 5.16°$, subtended between the directions of $\mathbf{p}$ and ν, represents the fixed inclination of the Moon's orbital plane to the ecliptic plane. Furthermore, as is well known, because of the perturbing action of the Sun, the normal to the Moon's orbital plane, $\mathbf{p}$, precesses about the normal to the ecliptic plane, ν, in

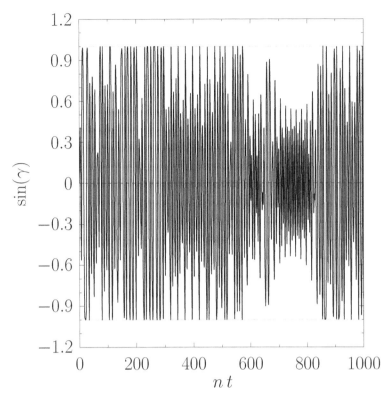

Fig. 7.12 Solution of Equation (7.165) with the Hyperion-like parameters $\alpha = 0.89$ and $e = 0.123$. The initial conditions are $\phi = 0$ and $n^{-1} d\phi/dt = 1$. Here, $\gamma = \phi - nt$.

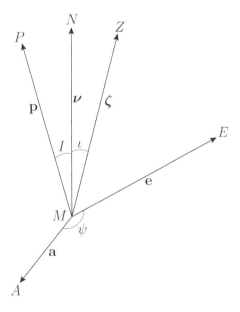

Fig. 7.13 Geometry of Cassini's laws.

the opposite sense to the Moon's orbital motion, such that it completes a full circuit every 18.6 years. (See Chapter 10.) Recall that precession in the opposite sense to orbital motion is usually termed *regression*.

In 1693, the astronomer Gian Domenico Cassini (1625–1712) formulated a set of empirical laws that succinctly describe the Moon's axial rotation. According to these laws:

1. The Moon spins at a *uniform* rate that matches its mean orbital rotation rate.
2. The normal to the Moon's equatorial plane subtends a *fixed* angle, $\iota = 1.59°$, with the normal to the ecliptic plane.
3. The normal to the Moon's equatorial plane, the normal to the Moon's orbital plane, and the normal to the ecliptic plane are *coplanar* vectors that are oriented such that the latter vector lies between the other two.

Law 1 effectively states that the Moon is locked in a 1:1 spin-orbit resonance. (See Section 7.11.) Let the x-, y-, and z-axes be the Moon's principal axes of rotation, and let $\mathcal{I}_{xx}$, $\mathcal{I}_{yy}$, and $\mathcal{I}_{zz}$ be the corresponding principal moments of inertia. Furthermore, let us label the principal axes such that the Moon's equatorial plane corresponds to the x–y plane, the normal to the equatorial plane corresponds to the z-axis, and $\mathcal{I}_{yy} > \mathcal{I}_{xx}$. In this case, as we saw in the previous section, a 1:1 spin-orbit resonant state is such that the Moon's x-axis always points approximately in the direction of the Earth: that is, **e** is almost parallel to the x-axis.

Law 2 states that the angle, ι, subtended between ζ and ν is fixed. Moreover, because the angles I and ι are both small (when expressed in radians), we deduce that the vectors ν and **p** are almost parallel to ζ.

Law 3 states that the vectors ζ, ν, and **p** all lie in the same plane, with ζ and **p** on opposite sides of ν. In other words, as the normal to the Moon's orbital plane, **p**, regresses about the normal to the ecliptic plane, ν, the normal to the Moon's equatorial plane, ζ, regresses at the same rate, such that ζ is always directly opposite **p** with respect to ν.

Cassini's first law was accounted for in the previous section. The ultimate aim of this section is to account for Cassini's second and third laws. Our approach is largely based on that of Danby (1992). To simplify the analysis, we shall assume that the Moon orbits around the Earth, at the uniform angular velocity, n, in a circular orbit of major radius a. When expressed in terms of the x, y, z coordinate system, $\zeta = (0, 0, 1)$. Furthermore, because the unit vectors ν and **p** are almost parallel to ζ, we can write

$$\nu \simeq (\nu_x, \nu_y, 1) \qquad\qquad (7.189)$$

and

$$\mathbf{p} \simeq (p_x, p_y, 1), \qquad\qquad (7.190)$$

where $|\nu_x|, |\nu_y|, |p_x|, |p_y| \ll 1$. Similarly, because the unit vector **e** is almost parallel to the x-axis, we have

$$\mathbf{e} \simeq (1, e_y, e_z), \qquad\qquad (7.191)$$

where $|e_y|$, $|e_z| \ll 1$. The position vector, $\mathbf{r} = (x, y, z)$, of the center of the Earth with respect to the center of the Moon is written

$$\mathbf{r} = a\,\mathbf{e}. \tag{7.192}$$

Finally, given Cassini's first law, and assuming that the Moon's spin axis is almost parallel to the z-axis, we find that the Moon's spin angular velocity takes the form

$$\omega = n\,\mathbf{w}. \tag{7.193}$$

Here, $\mathbf{w}$ is a unit vector such that

$$\mathbf{w} \simeq (w_x,\, w_y,\, 1), \tag{7.194}$$

where $|w_x|$, $|w_y| \ll 1$.

According to Equations (7.149), (7.150), (7.152), and (7.153),

$$\dot{\omega}_x + \sigma\,\omega_y\,\omega_z = 3\,n^2\,\sigma\,\frac{y\,z}{a^2} \tag{7.195}$$

and

$$\dot{\omega}_y - \tau\,\omega_x\,\omega_z = -3\,n^2\,\tau\,\frac{x\,z}{a^2}, \tag{7.196}$$

because $r = a$ and $G\,m_p = n^2\,a^3$. Here,

$$\sigma = \frac{\mathcal{I}_{zz} - \mathcal{I}_{yy}}{\mathcal{I}_{xx}} \tag{7.197}$$

and

$$\tau = \frac{\mathcal{I}_{zz} - \mathcal{I}_{xx}}{\mathcal{I}_{yy}}. \tag{7.198}$$

Note that $|\sigma|$, $|\tau| \ll 1$ because the Moon is almost spherically symmetric. To second order in small quantities, Equations (7.195) and (7.196) yield

$$n^{-1}\,\dot{w}_x + \sigma\,w_y \simeq 0 \tag{7.199}$$

and

$$n^{-1}\,\dot{w}_y - \tau\,w_x \simeq -3\,\tau\,e_z, \tag{7.200}$$

where use has been made of Equations (7.191)–(7.194).

The unit vector $\mathbf{v}$ is *stationary* in an inertial frame whose coordinate axes are fixed with respect to distant stars. Hence, in the x, y, z body frame, which rotates with respect to the aforementioned fixed frame at the angular velocity ω, we have (see Section 5.2)

$$\frac{d\mathbf{v}}{dt} + \omega \times \mathbf{v} = \mathbf{0}. \tag{7.201}$$

It follows, from Equations (7.189), (7.193), and (7.194), that

$$w_x \simeq n^{-1}\,\dot{v}_y + v_x \tag{7.202}$$

and

$$w_y \simeq -n^{-1}\,\dot{v}_x + v_y. \tag{7.203}$$

These expressions can be combined with Equations (7.199) and (7.200) to give

$$n^{-2}\ddot{v}_x - (1-\tau)n^{-1}\dot{v}_y + \tau v_x \simeq 3\tau e_z \qquad (7.204)$$

and

$$n^{-2}\ddot{v}_y + (1-\sigma)n^{-1}\dot{v}_x + \sigma v_y \simeq 0. \qquad (7.205)$$

It now remains to express e_z in terms of v_x and v_y.

By definition, $\mathbf{p}$ is normal to $\mathbf{e}$, since the vector $\mathbf{e}$ lies in the plane of the Moon's orbit. Hence, according to Equations (7.190) and (7.191),

$$\mathbf{p} \cdot \mathbf{e} \simeq p_x + e_z = 0, \qquad (7.206)$$

which implies that

$$e_z \simeq -p_x. \qquad (7.207)$$

Let A be the ascending node of the Earth's apparent orbit about the Moon (which implies that A is the descending node of the Moon's actual orbit about the Earth), and let $\mathbf{a}$ be a unit vector parallel to MA. (See Figure 7.13 and Section 3.12.) By definition, $\mathbf{a}$ is normal to both $\mathbf{p}$ and $\mathbf{v}$. In fact, we can write

$$\mathbf{a} = \frac{\mathbf{v} \times \mathbf{p}}{\sin I}, \qquad (7.208)$$

where I is the angle subtended between the vectors $\mathbf{p}$ and $\mathbf{v}$. It follows from Equations (7.189) and (7.190), and the fact that I is small, that

$$\mathbf{a} \simeq I^{-1}\left(v_y - p_y,\ p_x - v_x,\ v_x p_y - v_y p_x\right). \qquad (7.209)$$

Now,

$$\mathbf{a} \cdot \mathbf{e} = \cos\psi \qquad (7.210)$$

and

$$\mathbf{a} \times \mathbf{e} \cdot \mathbf{p} = \sin\psi, \qquad (7.211)$$

where ψ is the angle subtended between $\mathbf{a}$ and $\mathbf{e}$. (See Figure 7.13.) Thus, Equations (7.190), (7.191), and (7.209) yield

$$v_x = p_x + I\sin\psi, \qquad (7.212)$$
$$v_y = p_y + I\cos\psi, \qquad (7.213)$$

and

$$\mathbf{a} \simeq \left(\cos\psi,\ -\sin\psi,\ v_y\sin\psi - v_x\cos\psi\right). \qquad (7.214)$$

In fact, ψ is the longitude of the Earth relative to the ascending node of its apparent orbit around the Moon. It follows that

$$\psi = (n+g)t, \qquad (7.215)$$

where g is the uniform regression rate of the Earth's apparent ascending node (which is the same as the regression rate of the true ascending node of the Moon's orbit around the Earth). Here, for the sake of simplicity, we have assumed that the Earth passes through its apparent ascending node at time $t = 0$. Hence, Equations (7.204), (7.205), (7.207), (7.212), and (7.215) can be combined to give

$$n^{-2} \ddot{v}_x - (1 - \tau) n^{-1} \dot{v}_y + 4 \tau v_x \simeq 3 \tau I \sin[(n + g) t] \tag{7.216}$$

and

$$n^{-2} \ddot{v}_y + (1 - \sigma) n^{-1} \dot{v}_x + \sigma v_y \simeq 0. \tag{7.217}$$

The previous two equations govern the Moon's physical libration in *latitude*. As is the case for libration in longitude, there are both free and forced modes. The free modes satisfy

$$n^{-2} \ddot{v}_x - (1 - \tau) n^{-1} \dot{v}_y + 4 \tau v_x \simeq 0 \tag{7.218}$$

and

$$n^{-2} \ddot{v}_y + (1 - \sigma) n^{-1} \dot{v}_x + \sigma v_y \simeq 0. \tag{7.219}$$

Let us search for solutions of the form

$$v_x(t) = \hat{v}_x \sin(s n t - \phi) \tag{7.220}$$

and

$$v_y(t) = \hat{v}_y \cos(s n t - \phi), \tag{7.221}$$

where $\hat{v}_x$, $\hat{v}_y$, ϕ are constants. It follows that

$$\frac{\hat{v}_y}{\hat{v}_x} = \frac{s^2 - 4 \tau}{(1 - \tau) s} = \frac{(1 - \sigma) s}{s^2 - \sigma}. \tag{7.222}$$

Given that $|\sigma|$ and $|\tau|$ are both small compared to unity, two independent free libration modes can be derived from the preceding expression. The first mode is such that $s \simeq 1 + 3 \tau / 2$ and $\hat{v}_y / \hat{v}_x \simeq 1 - 3 \tau / 2$, whereas the second is such that $s \simeq 2 \sqrt{\sigma \tau}$ and $\hat{v}_y / \hat{v}_x \simeq -2 \sqrt{\tau / \sigma}$. In the Moon's body frame, these modes cause the normal to the ecliptic plane, $\mathbf{v} = (v_x, v_y, 1)$, to precess about the normal to the Moon's equatorial plane, $\boldsymbol{\zeta} = (0, 0, 1)$, in such a manner that

$$v_x \simeq A_1 \sin(\omega_1 t - \phi_1) + A_2 \sin(\omega_2 t - \phi_2) \tag{7.223}$$

and

$$v_y \simeq K_1 A_1 \cos(\omega_1 t - \phi_1) + K_2 A_2 \cos(\omega_2 t - \phi_2), \tag{7.224}$$

where $\omega_1 \simeq n (1 + 3 \tau / 2)$, $\omega_2 \simeq 2 n \sqrt{\sigma \tau}$, $K_1 \simeq 1 - 3 \tau / 2$, $K_2 \simeq -2 \sqrt{\tau / \sigma}$, and the constants A_1, A_2, ϕ_1, ϕ_2 are arbitrary. The observed values of n, σ, and τ are $13.1764°$ per day, 4.0362×10^{-4}, and 6.3149×10^{-4}, respectively (Konopliv et al. 1998). Thus, it follows that $\omega_1 = 13.1889°$ day, $\omega_2 = 1.3304 \times 10^{-2°}$ day, $K_1 = 0.9991$, and $K_2 = -2.5017$. In the body frame, the first mode causes $\mathbf{v}$ to regress about $\boldsymbol{\zeta}$ with a period of 27.2957 days, whereas the second mode causes $\mathbf{v}$ to precess about $\boldsymbol{\zeta}$ with a period of 74.1 years. Both these modes of libration have been detected by means of lunar laser ranging. The measured amplitude of the first mode is $A_1 = 0.37''$, whereas that of the second mode is $A_2 = 3.25''$ (Jin and Li 1996). Incidentally, the second mode is very similar in nature to the Chandler wobble of the Earth. (See Section 7.8.) Note that if σ and τ were of opposite sign—that is, if $\mathcal{I}_{zz}$ were intermediate between $\mathcal{I}_{xx}$ and $\mathcal{I}_{yy}$—the second mode of libration would grow exponentially in time, rather than oscillate at a constant amplitude: in other words, the Moon's spin state would be unstable. In fact, the Moon's principal axes of rotation are oriented such that $\mathcal{I}_{zz} > \mathcal{I}_{yy} > \mathcal{I}_{xx}$, which ensures that the Moon spins in a stable manner.

Let us now search for forced solutions of Equations (7.216) and (7.217) of the form

$$v_x = \hat{v}_x \sin[(n + g)\,t] \tag{7.225}$$

and

$$v_y = \hat{v}_y \cos[(n + g)\,t], \tag{7.226}$$

where $\hat{v}_x$, $\hat{v}_y$ are constants. It follows that

$$\left[4\,\tau - (1 + g/n)^2\right]\hat{v}_x + (1 - \tau)(1 + g/n)\,\hat{v}_y \simeq 3\,\tau\,I \tag{7.227}$$

and

$$(1 - \sigma)(1 + g/n)\,\hat{v}_x + \left[\sigma - (1 + g/n)^2\right]\hat{v}_y \simeq 0. \tag{7.228}$$

Hence, recalling that σ, τ, and g/n are all small compared to unity, we obtain the following mode of forced libration:

$$\frac{\hat{v}_y}{\hat{v}_x} \simeq 1 - \frac{g}{n} \tag{7.229}$$

and

$$\hat{v}_x \simeq \frac{3\,\tau\,I}{3\,\tau - g/n}. \tag{7.230}$$

In the Moon's body frame, this mode causes the vectors $\mathbf{v}$ and $\mathbf{p}$ to regress about $\boldsymbol{\zeta}$ (i.e., the z-axis) in such a manner that

$$\mathbf{v} \simeq (-\iota \sin\psi, \ -\iota \cos\psi, \ 1) \tag{7.231}$$

and

$$\mathbf{p} \simeq (-(I + \iota) \sin\psi, \ -(I + \iota) \cos\psi, \ 1), \tag{7.232}$$

where

$$\iota = \frac{3\,\tau\,I}{2\,g/n - 3\,\tau}, \tag{7.233}$$

and use has been made of Equations (7.212), (7.213), and (7.215). Because the observed values of I, τ, and g/n are $I = 5.16°$, $\tau = 6.3149 \times 10^{-4}$, and $g/n = 4.0185 \times 10^{-3}$ (Konopliv et al. 1998; Yoder 1995), we deduce that

$$\iota = 1.59°. \tag{7.234}$$

According to Equation (7.231), $\mathbf{v}$ regresses around $\boldsymbol{\zeta}$, with a period of 27.2123 days (i.e., a draconic month), in such a manner that $\mathbf{v}$ subtends a *fixed* angle of $\iota = 1.59°$ with respect to $\boldsymbol{\zeta}$. This accounts for Cassini's second law. According to Equation (7.232), $\mathbf{p}$ simultaneously regress around $\boldsymbol{\zeta}$, with the same period, in such a manner that $\boldsymbol{\zeta} \cdot \mathbf{v} \times \mathbf{p} = 0$.

In other words, the three vectors $\boldsymbol{\zeta}$, $\boldsymbol{\nu}$, and $\mathbf{p}$ always lie in the same plane. Moreover, it is clear that $\boldsymbol{\nu}$ is intermediate between $\boldsymbol{\zeta}$ and $\mathbf{p}$. This accounts for Cassini's third law. The angle $I' = I + \iota$, subtended between $\boldsymbol{\zeta}$ and $\mathbf{p}$, which is also the angle of inclination between the Moon's equatorial and orbital planes, takes the fixed value

$$I' = \frac{2\,(g/n)\,I}{2\,g/n - 3\,\tau} = 6.75°. \tag{7.235}$$

This angle would be zero in the absence of any regression of the Moon's orbital ascending node (i.e., if g/n were zero). In other words, the nonzero angle of inclination between the Moon's equatorial and orbital planes is a direct consequence of this regression, which is ultimately due to the perturbing action of the Sun. Because the regression of the Moon's orbital ascending node is also responsible for the forced nutation of the Earth's axis of rotation (see Section 7.10), it follows that this nutation is closely related to the forced inclination between the Moon's equatorial and orbital planes.

Exercises

7.1 Let $\mathbf{e}_{z'}$, $\mathbf{L}$, and $\boldsymbol{\omega}$ be the symmetry axis, the angular momentum vector, and the angular velocity vector, respectively, of a rotating body with an axis of symmetry. Demonstrate that these three vectors are coplanar.

7.2 Verify Equations (7.50) and (7.51).

7.3 A rigid body having an axis of symmetry rotates freely about a fixed point under no torques. If α is the angle between the symmetry axis and the instantaneous axis of rotation, show that the angle between the axis of rotation and the invariable line (the $\mathbf{L}$ vector) is

$$\tan^{-1}\left[\frac{(\mathcal{I}_\| - \mathcal{I}_\perp)\,\tan\alpha}{\mathcal{I}_\| + \mathcal{I}_\perp\,\tan^2\alpha}\right],$$

where $\mathcal{I}_\|$ (the moment of inertia about the symmetry axis) is greater than $\mathcal{I}_\perp$ (the moment of inertia about an axis normal to the symmetry axis). (From Fowles and Cassiday 2005.)

7.4 Because the greatest value of $\mathcal{I}_\|/\mathcal{I}_\perp$ is 2 (symmetrical lamina), show from the previous result that the angle between the angular velocity and angular momentum vectors cannot exceed $\tan^{-1}\left(1/\sqrt{8}\right) \simeq 19.5°$. Find the corresponding value of α. (Modified from Fowles and Cassiday 2005.)

7.5 A thin uniform rod of length l and mass m is constrained to rotate with constant angular velocity ω about an axis passing through the center of the rod, and making an angle α with the rod. Show that the angular momentum about the center of the rod is perpendicular to the rod and is of magnitude $(m\,l^2\,\omega/12)\sin\alpha$. Show that the torque is perpendicular to both the rod and the angular momentum vector and is of magnitude $(m\,l^2\omega^2/12)\sin\alpha\cos\alpha$. (From Fowles and Cassiday 2005.)

7.6 A thin uniform disk of radius a and mass m is constrained to rotate with constant angular velocity ω about an axis passing through its center, and making an angle

α with the normal to the disk. Find the angular momentum about the center of the disk, as well as the torque acting on the disk.

7.7 A freely rotating rigid body has principal moments of inertia such that $\mathcal{I}_{zz} > \mathcal{I}_{yy} > \mathcal{I}_{xx}$.

    a. Demonstrate that the rotational energy of the body attains its maximum and minimum values (at fixed ω) when the body rotates about the z- and the x-axes, respectively.

    b. Demonstrate, from Euler's equations, that the rotational state is stable to small perturbations when the body rotates about either the x-axis or the z-axis, but is unstable when it rotates about the y-axis.

7.8 The length of the *mean sidereal year*, which is defined as the average time required for the Sun to (appear to) complete a full orbit around the Earth, relative to the fixed stars, is 365.25636 days (Yoder 1995). The *mean tropical year* is defined as the average time interval between successive vernal equinoxes. Demonstrate that, as a consequence of the precession of the equinoxes, whose period is 25,772 years (Yoder 1995), the length of the mean tropical year is 20.4 minutes shorter than that of the mean sidereal year (i.e., 365.24219 days).

7.9 Consider an artificial satellite in a circular orbit of radius a about the Earth. Suppose that the normal to the plane of the orbit subtends an angle I with the Earth's axis of rotation. By approximating the orbiting satellite as a uniform ring, demonstrate that the Earth's oblateness causes the plane of the satellite's orbit to precess about the Earth's rotational axis at the approximate rate

$$\frac{3}{2} J_2 \, n \left(\frac{R}{a}\right)^2 \cos I.$$

Here, n is the satellite's orbital angular velocity, R is the Earth's mean radius, $J_2 = (\mathcal{I}_\parallel - \mathcal{I}_\perp)/(M R^2)$, M is the Earth's mass, and $\mathcal{I}_\parallel$ and $\mathcal{I}_\perp$ are the Earth's parallel and perpendicular (to the rotation axis) moments of inertia. Is the precession in the same sense as the orbital motion, or the opposite sense?

7.10 A *Sun-synchronous* satellite is one that always passes over a given point on the Earth at the same local solar time. This is achieved by fixing the precession rate of the satellite's orbital plane such that it matches the rate at which the Sun appears to move against the background of the stars. What orbital altitude above the surface of the Earth would such a satellite need to have in order to fly over all latitudes between 50° N and 50° S? Is the direction of the satellite orbit in the same sense as the Earth's rotation (prograde), or the opposite sense (retrograde)? Note that $J_2 = 1.083 \times 10^{-3}$ for the Earth (Yoder 1995).

7.11 Consider an aspherical moon in a low-eccentricity Keplerian orbit about a spherical planet. Suppose that the moon is locked in a 1:1 spin-orbit resonance. Demonstrate that, to lowest order in the eccentricity, the optical libration of the moon can be accounted for by saying that the moon's long axis (in the orbital plane) always points toward the empty focus of the orbit.

Three-body problem

8.1 Introduction

We saw earlier, in Section 1.9, that an isolated dynamical system consisting of two freely moving point masses exerting forces on one another—which is usually referred to as a *two-body problem*—can always be converted into an equivalent one-body problem. In particular, this implies that we can *exactly solve* a dynamical system containing two gravitationally interacting point masses, as the equivalent one-body problem is exactly soluble. (See Sections 1.9 and 3.16.) What about a system containing *three* gravitationally interacting point masses? Despite hundreds of years of research, no useful general solution of this famous problem—which is usually called the *three-body problem*—has ever been found. It is, however, possible to make some progress by severely restricting the problem's scope.

8.2 Circular restricted three-body problem

Consider an isolated dynamical system consisting of three gravitationally interacting point masses, m_1, m_2, and m_3. Suppose, however, that the third mass, m_3, is so much smaller than the other two that it has a negligible effect on their motion. Suppose, further, that the first two masses, m_1 and m_2, execute circular orbits about their common center of mass. In the following, we shall examine this simplified problem, which is usually referred to as the *circular restricted three-body problem*. The problem under investigation has obvious applications to the solar system. For instance, the first two masses might represent the Sun and a planet (recall that a given planet and the Sun do indeed execute almost circular orbits about their common center of mass), whereas the third mass might represent an asteroid or a comet (asteroids and comets do indeed have much smaller masses than the Sun or any of the planets).

Let us define a Cartesian coordinate system ξ, η, ζ in an inertial reference frame whose origin coincides with the center of mass, C, of the two orbiting masses, m_1 and m_2. Furthermore, let the orbital plane of these masses coincide with the ξ–η plane, and let them both lie on the ξ-axis at time $t = 0$. (See Figure 8.1.) Suppose that a is the constant distance between the two orbiting masses, r_1 the constant distance between mass m_1 and the origin, and r_2 the constant distance between mass m_2 and the origin.

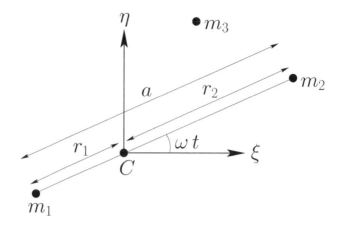

Circular restricted three-body problem.

Moreover, let ω be the constant orbital angular velocity. It follows, from Section 3.16, that

$$\omega^2 = \frac{GM}{a^3} \tag{8.1}$$

and

$$\frac{r_1}{r_2} = \frac{m_2}{m_1}, \tag{8.2}$$

where $M = m_1 + m_2$.

It is convenient to choose our unit of length such that $a = 1$, and our unit of mass such that $GM = 1$. It follows, from Equation (8.1), that $\omega = 1$. However, we shall continue to retain ω in our equations, for the sake of clarity. Let $\mu_1 = G m_1$ and $\mu_2 = G m_2 = 1 - \mu_1$. It is easily demonstrated that $r_1 = \mu_2$ and $r_2 = 1 - r_1 = \mu_1$. Hence, the two orbiting masses, m_1 and m_2, have position vectors

$$\mathbf{r}_1 = (\xi_1, \eta_1, 0) = (-\mu_2 \cos \omega t, -\mu_2 \sin \omega t, 0) \tag{8.3}$$

and

$$\mathbf{r}_2 = (\xi_2, \eta_2, 0) = (\mu_1 \cos \omega t, \mu_1 \sin \omega t, 0), \tag{8.4}$$

respectively. (See Figure 8.1.) Let the third mass have position vector $\mathbf{r} = (\xi, \eta, \zeta)$. The Cartesian components of the equation of motion of this mass are thus

$$\ddot{\xi} = -\mu_1 \frac{(\xi - \xi_1)}{\rho_1^3} - \mu_2 \frac{(\xi - \xi_2)}{\rho_2^3}, \tag{8.5}$$

$$\ddot{\eta} = -\mu_1 \frac{(\eta - \eta_1)}{\rho_1^3} - \mu_2 \frac{(\eta - \eta_2)}{\rho_2^3}, \tag{8.6}$$

$$\ddot{\zeta} = -\mu_1 \frac{\zeta}{\rho_1^3} - \mu_2 \frac{\zeta}{\rho_2^3}, \tag{8.7}$$

where

$$\rho_1^2 = (\xi - \xi_1)^2 + (\eta - \eta_1)^2 + \zeta^2 \tag{8.8}$$

and

$$\rho_2^2 = (\xi - \xi_2)^2 + (\eta - \eta_2)^2 + \zeta^2. \tag{8.9}$$

8.3 Jacobi integral

Consider the function

$$C = 2\left(\frac{\mu_1}{\rho_1} + \frac{\mu_2}{\rho_2}\right) + 2\,\omega\,(\xi\,\dot\eta - \eta\,\dot\xi) - \dot\xi^2 - \dot\eta^2 - \dot\zeta^2. \tag{8.10}$$

The time derivative of this function is written

$$\dot C = -\frac{2\mu_1\,\dot\rho_1}{\rho_1^2} - \frac{2\mu_2\,\dot\rho_2}{\rho_2^2} + 2\,\omega\,(\xi\,\ddot\eta - \eta\,\ddot\xi) - 2\,\dot\xi\,\ddot\xi - 2\,\dot\eta\,\ddot\eta - 2\,\dot\zeta\,\ddot\zeta. \tag{8.11}$$

Moreover, it follows, from Equations (8.3)–(8.4) and (8.8)–(8.9), that

$$\rho_1\,\dot\rho_1 = -(\xi_1\,\dot\xi + \eta_1\,\dot\eta) + \omega\,(\xi\,\eta_1 - \eta\,\xi_1) + \xi\,\dot\xi + \eta\,\dot\eta + \zeta\,\dot\zeta \tag{8.12}$$

and

$$\rho_2\,\dot\rho_2 = -(\xi_2\,\dot\xi + \eta_2\,\dot\eta) + \omega\,(\xi\,\eta_2 - \eta\,\xi_2) + \xi\,\dot\xi + \eta\,\dot\eta + \zeta\,\dot\zeta. \tag{8.13}$$

Combining Equations (8.5)–(8.7) with the preceding three expressions, after considerable algebra (see Exercise 8.1) we obtain

$$\frac{dC}{dt} = 0. \tag{8.14}$$

In other words, the function C—which is usually referred to as the *Jacobi integral*—is a *constant of the motion.*

We can rearrange Equation (8.10) to give

$$\mathcal{E} \equiv \frac{1}{2}\,(\dot\xi^2 + \dot\eta^2 + \dot\zeta^2) - \left(\frac{\mu_1}{\rho_1} + \frac{\mu_2}{\rho_2}\right) = \boldsymbol{\omega}\cdot\mathbf{h} - \frac{C}{2}, \tag{8.15}$$

where $\mathcal{E}$ is the energy (per unit mass) of mass m_3, $\mathbf{h} = \mathbf{r}\times\dot{\mathbf{r}}$ the angular momentum (per unit mass) of mass m_3, and $\boldsymbol{\omega} = (0, 0, \omega)$ the orbital angular velocity of the other two masses. However, $\mathbf{h}$ is *not* a constant of the motion. Hence, $\mathcal{E}$ is not a constant of the motion either. In fact, the Jacobi integral is the *only* constant of the motion in the circular restricted three-body problem. Incidentally, the energy of mass m_3 is not a conserved quantity because the other two masses in the system are *moving.*

8.4 Tisserand criterion

Consider a dynamical system consisting of three gravitationally interacting point masses, m_1, m_2, and m_3. Let mass m_1 represent the Sun, mass m_2 the planet Jupiter, and mass m_3 a comet. Because the mass of a comet is very much less than that of the

Sun or Jupiter, and the Sun and Jupiter are in (almost) circular orbits about their common center of mass, the dynamical system in question satisfies all the necessary criteria to be considered an example of a restricted three-body problem.

The mass of the Sun is much greater than that of Jupiter. It follows that the gravitational effect of Jupiter on the cometary orbit is *negligible* unless the comet makes a very close approach to Jupiter. Hence, as described in Chapter 3, before and after such an approach, the comet executes a Keplerian elliptical orbit about the Sun with fixed orbital parameters: fixed major radius, eccentricity, and inclination to the ecliptic plane. However, in general, the orbital parameters before and after the close approach will *not* be the same as one another. Let us investigate further.

Because $m_1 \gg m_2$, we have $\mu_1 = G m_1 \simeq G(m_1 + m_2) = 1$, and $\rho_1 \simeq r$. Hence, according to Equations (3.34) and (3.44), the (approximately) conserved energy (per unit mass) of the comet before and after its close approach to Jupiter is

$$\mathcal{E} \equiv \frac{1}{2}\left(\dot{\xi}^2 + \dot{\eta}^2 + \dot{\zeta}^2\right) - \frac{1}{r} = -\frac{1}{2a}. \tag{8.16}$$

The comet's orbital energy is determined entirely by its major radius, a. (Incidentally, we are working in units such that the major radius of Jupiter's orbit is unity.) Furthermore, the (approximately) conserved angular momentum (per unit mass) of the comet before and after its approach to Jupiter is written $\mathbf{h}$, where $\mathbf{h}$ is directed *normal* to the comet's orbital plane, and, from Equations (3.31) and (A.107),

$$h^2 = a(1 - e^2). \tag{8.17}$$

Here, e is the comet's orbital eccentricity. It follows that

$$\boldsymbol{\omega} \cdot \mathbf{h} = \omega h \cos I = \sqrt{a(1 - e^2)} \cos I, \tag{8.18}$$

because $\omega = 1$ in our adopted system of units. Here, I is the angle of inclination of the normal to the comet's orbital plane to that of Jupiter's orbital plane.

Let a, e, and I be the major radius, eccentricity, and inclination angle of the cometary orbit before the close encounter with Jupiter, and let a', e', and I' be the corresponding parameters after the encounter. It follows from Equations (8.15), (8.16), and (8.18), and the fact that C is conserved during the encounter whereas $\mathcal{E}$ and h are not, that

$$\frac{1}{2a} + \sqrt{a(1 - e^2)} \cos I = \frac{1}{2a'} + \sqrt{a'(1 - e'^2)} \cos I'. \tag{8.19}$$

This result is known as the *Tisserand criterion* after its discoverer, the French astronomer Felix Tisserand (1845–1896); it restricts the possible changes in the orbital parameters of a comet due to a close encounter with Jupiter (or any other massive planet).

The Tisserand criterion is extremely useful. For instance, whenever a new comet is discovered, astronomers immediately calculate its *Tisserand parameter*,

$$T_J = \frac{1}{a} + 2\sqrt{a(1 - e^2)} \cos I. \tag{8.20}$$

If this parameter has the same value as that of a previously observed comet, it is quite likely that the new comet is, in fact, the same comet, but that its orbital parameters have changed since it was last observed, as a result of a close encounter with Jupiter. Incidentally, the subscript J in the preceding formula is to remind us that we are dealing

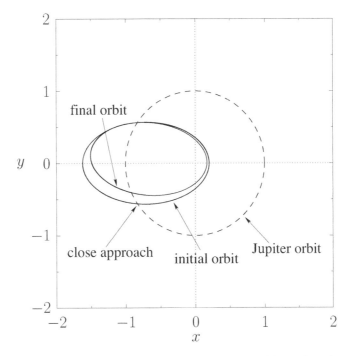

Fig. 8.2
Changing orbit of a hypothetical comet with a close approach to Jupiter. The solid curve shows the cometary orbit, and the dashed curve shows the jovian orbit. x and y are Cartesian coordinates in Jupiter's orbital plane. The origin is the center of mass of the Sun–Jupiter system. The system of units is such that the Jovian major radius is unity, and $\mu_2 = 9.533 \times 10^{-4}$. Both the comet and Jupiter orbit in a counterclockwise sense.

with the Tisserand parameter for close encounters with Jupiter. (The parameter is thus evaluated in a system of units in which the major radius of Jupiter's orbit is unity.) Obviously, it is also possible to calculate Tisserand parameters for close encounters with other massive planets.

Figure 8.2 shows the changing orbit of a hypothetical comet that has a close approach to Jupiter. The initial orbit is such that $a = 0.916$ (in units in which the major radius of the Jovian orbit is unity), $e = 0.781$, and $I = 0$, whereas the final orbit is such that $a = 0.841$, $e = 0.800$, and $I = 0$. Figure 8.3 shows the comet's major radius, a, eccentricity, e, and Tisserand parameter, $\mathcal{T}_J$, as functions of time before, during, and after the encounter with Jupiter. It can be seen that the major radius and the eccentricity are both modified by the encounter (which occurs when $t \simeq 12$), whereas the Tisserand parameter remains constant in time. This remains true even when the small eccentricity of the Jovian orbit is taken into account in the calculation.

The Tisserand parameter is often employed to distinguish between comets and aster- oids in the solar system. This idea is illustrated in Figure 8.4, which shows the Jovian Tisserand parameter, $\mathcal{T}_J$, plotted against the major radius, a, of the principal asteroids and comets in the solar system. The Tisserand parameter of Jupiter (which is almost exactly 3) is also shown. It can be seen that, roughly speaking, asteroids have higher Tisserand parameters than Jupiter, whereas comets have lower Tisserand parameters. The only major exception to this rule is the so-called *Trojan asteroids* (see Section 8.8),

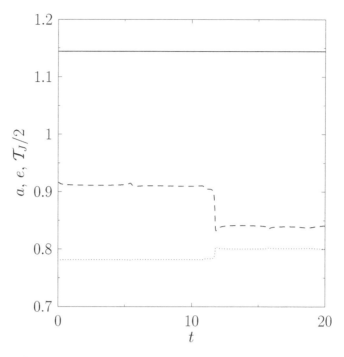

Fig. 8.3 Major radius, a (dashed curve) and eccentricity, e (dotted curve) of the hypothetical comet shown in Figure 8.2, plotted as functions of time. The adopted system of units is such that the Jovian major radius and orbital period are both unity. The solid curve shows the comet's Tisserand parameter, $\mathcal{T}_J$, divided by 2.

which all have very similar major radii to Jupiter (because, by definition, they must have the same orbital period as Jupiter), and consequently have lower Tisserand parameters (because they generally have higher eccentricities and inclinations than Jupiter). The lower Tisserand parameters of comets with respect to Jupiter, and of Jupiter with respect to regular asteroids, is indicative of the fact that comets generally originated *beyond* the Jovian orbit, whereas regular asteroids generally originated *within* the Jovian orbit.

The Tisserand criterion is also applicable to so-called *gravity assists*, in which a spacecraft gains energy as a result of a close encounter with a moving planet. Such assists are often employed in missions to the outer planets to reduce the amount of fuel that the spacecraft must carry in order to reach its destination. In fact, it is clear, from Equations (8.16) and (8.19), that a spacecraft can make use of a close encounter with a moving planet to increase (or decrease) its orbital major radius a, and, hence, to increase (or decrease) its total orbital energy.

8.5 Co-rotating frame

Let us transform to a noninertial frame of reference rotating with angular velocity ω about an axis normal to the orbital plane of masses m_1 and m_2, and passing through their center of mass. It follows that masses m_1 and m_2 appear stationary in this new

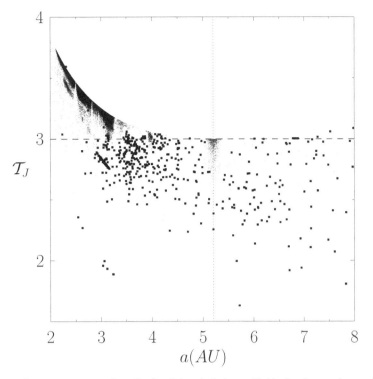

Fig. 8.4 Jovian Tisserand parameter versus major radius for all the principal asteroids (dots) and comets (squares) in the solar system. The dashed line indicates the Tisserand parameter of Jupiter. The dotted line shows the Jovian major radius. Raw data from JPL Small-Body Database.

reference frame. Let us define a Cartesian coordinate system x, y, z in the rotating frame of reference that is such that masses m_1 and m_2 always lie on the x-axis, and the z-axis is parallel to the previously defined ζ-axis. It follows that masses m_1 and m_2 have the fixed position vectors $\mathbf{r}_1 = (-\mu_2, 0, 0)$ and $\mathbf{r}_2 = (\mu_1, 0, 0)$ in our new coordinate system. Finally, let the position vector of mass m_3 be $\mathbf{r} = (x, y, z)$. (See Figure 8.5.)

According to Section 5.2, the equation of motion of mass m_3 in the rotating reference frame takes the form

$$\ddot{\mathbf{r}} + 2\,\boldsymbol{\omega} \times \dot{\mathbf{r}} = -\mu_1\,\frac{(\mathbf{r}-\mathbf{r}_1)}{\rho_1^3} - \mu_2\,\frac{(\mathbf{r}-\mathbf{r}_2)}{\rho_2^3} - \boldsymbol{\omega} \times (\boldsymbol{\omega} \times \mathbf{r}), \qquad (8.21)$$

where $\boldsymbol{\omega} = (0, 0, \omega)$, and

$$\rho_1^2 = (x+\mu_2)^2 + y^2 + z^2, \qquad (8.22)$$
$$\rho_2^2 = (x-\mu_1)^2 + y^2 + z^2. \qquad (8.23)$$

Here, the second term on the left-hand side of Equation (8.21) is the *Coriolis* acceleration, whereas the final term on the right-hand side is the *centrifugal* acceleration.

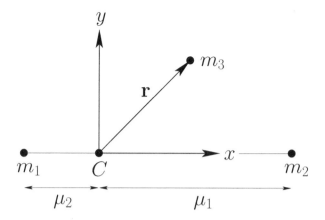

Co-rotating frame.

The components of Equation (8.21) reduce to

$$\ddot{x} - 2\,\omega\,\dot{y} \;=\; -\frac{\mu_1\,(x+\mu_2)}{\rho_1^3} - \frac{\mu_2\,(x-\mu_1)}{\rho_2^3} + \omega^2\,x, \qquad (8.24)$$

$$\ddot{y} + 2\,\omega\,\dot{x} \;=\; -\frac{\mu_1\,y}{\rho_1^3} - \frac{\mu_2\,y}{\rho_2^3} + \omega^2\,y, \qquad (8.25)$$

and

$$\ddot{z} = -\frac{\mu_1\,z}{\rho_1^3} - \frac{\mu_2\,z}{\rho_2^3}, \qquad (8.26)$$

which yield

$$\ddot{x} - 2\,\omega\,\dot{y} \;=\; -\frac{\partial U}{\partial x}, \qquad (8.27)$$

$$\ddot{y} + 2\,\omega\,\dot{x} \;=\; -\frac{\partial U}{\partial y}, \qquad (8.28)$$

and

$$\ddot{z} = -\frac{\partial U}{\partial z}, \qquad (8.29)$$

where

$$U(x, y, z) = -\frac{\mu_1}{\rho_1} - \frac{\mu_2}{\rho_2} - \frac{\omega^2}{2}\,(x^2 + y^2) \qquad (8.30)$$

is the sum of the gravitational and centrifugal potentials.

It follows from Equations (8.27)–(8.29) that

$$\ddot{x}\,\dot{x} - 2\,\omega\,\dot{x}\,\dot{y} \;=\; -\dot{x}\,\frac{\partial U}{\partial x}, \qquad (8.31)$$

$$\ddot{y}\,\dot{y} + 2\,\omega\,\dot{x}\,\dot{y} \;=\; -\dot{y}\,\frac{\partial U}{\partial y}, \qquad (8.32)$$

and

$$\ddot{z}\,\dot{z} = -\dot{z}\,\frac{\partial U}{\partial z}. \tag{8.33}$$

Summing the preceding three equations, we obtain

$$\frac{d}{dt}\left[\frac{1}{2}\left(\dot{x}^2 + \dot{y}^2 + \dot{z}^2\right) + U\right] = 0. \tag{8.34}$$

In other words,

$$C = -2\,U - v^2 \tag{8.35}$$

is a *constant of the motion*, where $v^2 = \dot{x}^2 + \dot{y}^2 + \dot{z}^2$. In fact, C is the *Jacobi integral* introduced in Section 8.3 [it is easily demonstrated that Equations (8.10) and (8.35) are identical—see Exercise 8.4]. Note, finally, that the mass m_3 is restricted to regions in which

$$-2\,U \geq C, \tag{8.36}$$

because v^2 is a positive definite quantity.

8.6 Lagrange points

Let us search for possible *equilibrium points* of the mass m_3 in the rotating reference frame. Such points are termed *Lagrange points*. Hence, in the rotating frame, the mass m_3 would remain at rest if placed at one of the Lagrange points. It is thus clear that these points are fixed in the rotating frame. Conversely, in the inertial reference frame, the Lagrange points rotate about the center of mass with angular velocity ω, and the mass m_3 would consequently also rotate about the center of mass with angular velocity ω if placed at one of these points (with the appropriate velocity). In the following, we shall assume, without loss of generality, that $m_1 \geq m_2$.

The Lagrange points satisfy $\dot{\mathbf{r}} = \ddot{\mathbf{r}} = \mathbf{0}$ in the rotating frame. It thus follows, from Equations (8.27)–(8.29), that the Lagrange points are the solutions of

$$\frac{\partial U}{\partial x} = \frac{\partial U}{\partial y} = \frac{\partial U}{\partial z} = 0. \tag{8.37}$$

It is easily seen that

$$\frac{\partial U}{\partial z} = \left(\frac{\mu_1}{\rho_1^3} + \frac{\mu_2}{\rho_2^3}\right)z. \tag{8.38}$$

Because the term in curved brackets is positive definite, we conclude that the only solution to the above equation is $z = 0$. Hence, all the Lagrange points lie in the x–y plane. If $z = 0$, it is readily demonstrated that

$$\mu_1\,\rho_1^2 + \mu_2\,\rho_2^2 = x^2 + y^2 + \mu_1\,\mu_2, \tag{8.39}$$

where use has been made of the fact that $\mu_1 + \mu_2 = 1$. Hence, Equation (8.30) can also be written

$$U = -\mu_1 \left(\frac{1}{\rho_1} + \frac{\rho_1^2}{2} \right) - \mu_2 \left(\frac{1}{\rho_2} + \frac{\rho_2^2}{2} \right) + \frac{\mu_1 \mu_2}{2}. \tag{8.40}$$

The Lagrange points thus satisfy

$$\frac{\partial U}{\partial x} = \frac{\partial U}{\partial \rho_1} \frac{\partial \rho_1}{\partial x} + \frac{\partial U}{\partial \rho_2} \frac{\partial \rho_2}{\partial x} = 0 \tag{8.41}$$

and

$$\frac{\partial U}{\partial y} = \frac{\partial U}{\partial \rho_1} \frac{\partial \rho_1}{\partial y} + \frac{\partial U}{\partial \rho_2} \frac{\partial \rho_2}{\partial y} = 0, \tag{8.42}$$

which reduce to

$$\mu_1 \left(\frac{1 - \rho_1^3}{\rho_1^2} \right) \left(\frac{x + \mu_2}{\rho_1} \right) + \mu_2 \left(\frac{1 - \rho_2^3}{\rho_2^2} \right) \left(\frac{x - \mu_1}{\rho_2} \right) = 0 \tag{8.43}$$

and

$$\mu_1 \left(\frac{1 - \rho_1^3}{\rho_1^2} \right) \left(\frac{y}{\rho_1} \right) + \mu_2 \left(\frac{1 - \rho_2^3}{\rho_2^2} \right) \left(\frac{y}{\rho_2} \right) = 0. \tag{8.44}$$

One obvious solution of Equation (8.44) is $y = 0$, corresponding to a Lagrange point that lies on the x-axis. It turns out that there are three such points. L_1 lies between masses m_1 and m_2, L_2 lies to the right of mass m_2, and L_3 lies to the left of mass m_1. (See Figure 8.6.) At the L_1 point, we have $x = -\mu_2 + \rho_1 = \mu_1 - \rho_2$ and $\rho_1 = 1 - \rho_2$. Hence, from Equation (8.43),

$$\frac{\mu_2}{3\mu_1} = \frac{\rho_2^3 (1 - \rho_2 + \rho_2^2/3)}{(1 + \rho_2 + \rho_2^2)(1 - \rho_2)^3}. \tag{8.45}$$

Assuming that $\rho_2 \ll 1$, we can find an approximate solution of Equation (8.45) by expanding in powers of ρ_2:

$$\alpha = \rho_2 + \frac{\rho_2^2}{3} + \frac{\rho_2^3}{3} + \frac{53 \rho_2^4}{81} + \mathcal{O}(\rho_2^5). \tag{8.46}$$

This equation can be inverted to give

$$\rho_2 = \alpha - \frac{\alpha^2}{3} - \frac{\alpha^3}{9} - \frac{23 \alpha^4}{81} + \mathcal{O}(\alpha^5), \tag{8.47}$$

where

$$\alpha = \left(\frac{\mu_2}{3\mu_1} \right)^{1/3} \tag{8.48}$$

is assumed to be a small parameter.

At the L_2 point, we have $x = -\mu_2 + \rho_1 = \mu_1 + \rho_2$ and $\rho_1 = 1 + \rho_2$. Hence, from Equation (8.43),

$$\frac{\mu_2}{3\mu_1} = \frac{\rho_2^3 (1 + \rho_2 + \rho_2^2/3)}{(1 + \rho_2)^2 (1 - \rho_2^3)}. \tag{8.49}$$

Again, expanding in powers of ρ_2, we obtain

$$\alpha = \rho_2 - \frac{\rho_2^2}{3} + \frac{\rho_2^3}{3} + \frac{\rho_2^4}{81} + \mathcal{O}(\rho_2^5) \tag{8.50}$$

and

$$\rho_2 = \alpha + \frac{\alpha^2}{3} - \frac{\alpha^3}{9} - \frac{31\,\alpha^4}{81} + \mathcal{O}(\alpha^5). \tag{8.51}$$

Finally, at the L_3 point, we have $x = -\mu_2 - \rho_1 = \mu_1 - \rho_2$ and $\rho_2 = 1 + \rho_1$. Hence, from Equation (8.43),

$$\frac{\mu_2}{\mu_1} = \frac{(1 - \rho_1^3)(1 + \rho_1)^2}{\rho_1^3\,(\rho_1^2 + 3\rho_1 + 3)}. \tag{8.52}$$

Let $\rho_1 = 1 - \beta$. Expanding in powers of β, we obtain

$$\frac{\mu_2}{\mu_1} = \frac{12\beta}{7} + \frac{144\beta^2}{49} + \frac{1567\beta^3}{343} + \mathcal{O}(\beta^4) \tag{8.53}$$

and

$$\beta = \frac{7}{12}\left(\frac{\mu_2}{\mu_1}\right) - \frac{7}{12}\left(\frac{\mu_2}{\mu_1}\right)^2 + \frac{13223}{20736}\left(\frac{\mu_2}{\mu_1}\right)^3 + \mathcal{O}\left(\frac{\mu_2}{\mu_1}\right)^4, \tag{8.54}$$

where μ_2/μ_1 is assumed to be a small parameter.

Let us now search for Lagrange points that *do not* lie on the x-axis. One obvious solution of Equations (8.41) and (8.42) is

$$\frac{\partial U}{\partial \rho_1} = \frac{\partial U}{\partial \rho_2} = 0, \tag{8.55}$$

giving, from Equation (8.40),

$$\rho_1 = \rho_2 = 1 \tag{8.56}$$

or

$$(x + \mu_2)^2 + y^2 = (x - 1 + \mu_2)^2 + y^2 = 1, \tag{8.57}$$

because $\mu_1 = 1 - \mu_2$. The two solutions of this equation are

$$x = \frac{1}{2} - \mu_2 \tag{8.58}$$

and

$$y = \pm\frac{\sqrt{3}}{2}, \tag{8.59}$$

and they specify the positions of the Lagrange points designated L_4 and L_5. Note that the point L_4 and the masses m_1 and m_2 lie at the apexes of an *equilateral triangle*. The same is true for the point L_5. We have now found all of the possible Lagrange points.

Figure 8.6 shows the positions of the two masses, m_1 and m_2, and the five Lagrange points, L_1 to L_5, calculated for the case where $\mu_2 = 0.1$.

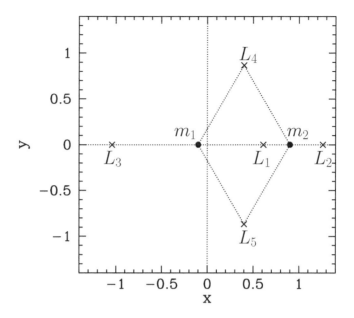

Fig. 8.6 Masses m_1 and m_2, and the five Lagrange points, L_1 to L_5, calculated for $\mu_2 = 0.1$.

8.7 Zero-velocity surfaces

Consider the surface

$$V(x, y, z) = C, \tag{8.60}$$

where

$$V(x, y, z) = -2\,U = \frac{2\,\mu_1}{\rho_1} + \frac{2\,\mu_2}{\rho_2} + x^2 + y^2. \tag{8.61}$$

Note that $V \geq 0$. It follows, from Equation (8.35), that if the mass m_3 has the Jacobi integral C and lies on the surface specified in Equation (8.60), then it must have zero velocity. Hence, such a surface is termed a *zero-velocity surface*. The zero-velocity surfaces are important because they form the boundary of regions from which the mass m_3 is dynamically excluded: that is, regions where $V < C$. Generally speaking, the regions from which m_3 is excluded grow in area as C increases, and vice versa.

Let C_i be the value of V at the L_i Lagrange point, for $i = 1, 5$. When $\mu_2 \ll 1$, it is easily demonstrated that

$$C_1 \simeq 3 + 3^{4/3}\,\mu_2^{2/3} - 10\,\mu_2/3, \tag{8.62}$$

$$C_2 \simeq 3 + 3^{4/3}\,\mu_2^{2/3} - 14\,\mu_2/3, \tag{8.63}$$

$$C_3 \simeq 3 + \mu_2, \tag{8.64}$$

$$C_4 \simeq 3 - \mu_2, \tag{8.65}$$

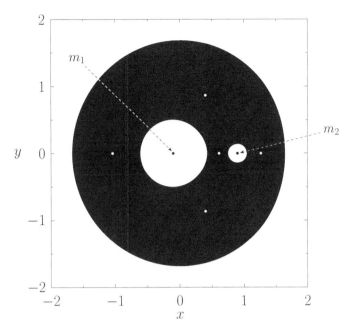

Zero-velocity surface $V = C$, where $C > C_1$, calculated for $\mu_2 = 0.1$. Mass m_3 is excluded from the black region.

and

$$C_5 \simeq 3 - \mu_2. \tag{8.66}$$

Note that $C_1 > C_2 > C_3 > C_4 = C_5$.

Figures 8.7 through 8.11 show the intersection of the zero-velocity surface $V = C$ with the x–y plane for various different values of C, and illustrate how the region from which m_3 is dynamically excluded—which we shall term the *excluded region*—evolves as the value of C is varied. Of course, any point not in the excluded region is in the so-called *allowed region*. For $C > C_1$, the allowed region consists of two separate oval regions centered on m_1 and m_2, respectively, plus an outer region that lies beyond a large circle centered on the origin. All three allowed regions are separated from one another by an excluded region. (See Figure 8.7.) When $C = C_1$, the two inner allowed regions merge at the L_1 point. (See Figure 8.8.) When $C = C_2$, the inner and outer allowed regions merge at the L_2 point, forming a horseshoe-like excluded region. (See Figure 8.9.) When $C = C_3$, the excluded region splits in two at the L_3 point. (See Figure 8.10.) For $C_4 < C < C_3$, the two excluded regions are localized about the L_4 and L_5 points. (See Figure 8.11.) Finally, for $C < C_4$, there is no excluded region.

Figure 8.12 shows the zero-velocity surfaces and Lagrange points calculated for the case $\mu_2 = 0.01$. It can be seen that, at very small values of μ_2, the L_1 and L_2 Lagrange points are almost equidistant from mass m_2. Furthermore, mass m_2 and the L_3, L_4, and L_5 Lagrange points all lie approximately on a *unit circle*, centered on mass m_1. It follows that, when μ_2 is small, the Lagrange points L_3, L_4 and L_5 all share the orbit of mass m_2

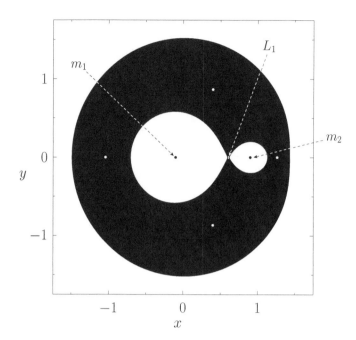

Fig. 8.8 Zero-velocity surface $V = C$, where $C = C_1$, calculated for $\mu_2 = 0.1$. Mass m_3 is excluded from the black region.

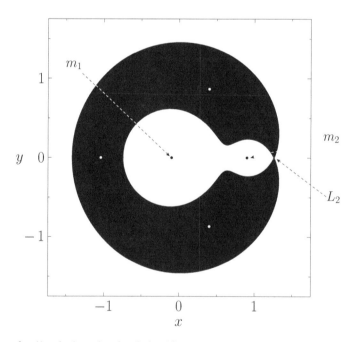

Fig. 8.9 Zero-velocity surface $V = C$, where $C = C_2$, calculated for $\mu_2 = 0.1$. Mass m_3 is excluded from the black region.

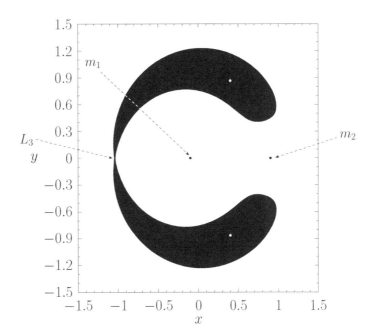

Fig. 8.10   Zero-velocity surface $V = C$, where $C = C_3$, calculated for $\mu_2 = 0.1$. Mass m_3 is excluded from the black region.

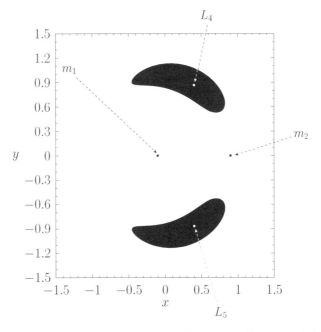

Fig. 8.11   Zero-velocity surface $V = C$, where $C_4 < C < C_3$, calculated for $\mu_2 = 0.1$. Mass m_3 is excluded from the black regions.

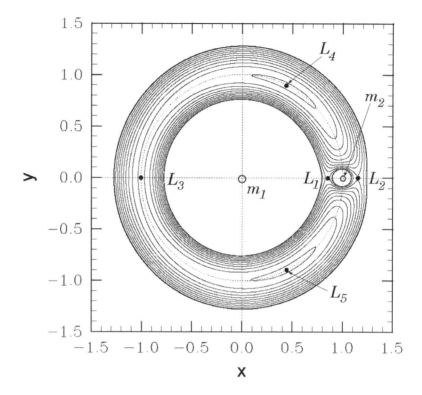

Fig. 8.12 Zero-velocity surfaces and Lagrange points calculated for $\mu_2 = 0.01$.

about m_1 (in the inertial frame) with C_3 being directly opposite m_2, L_4 (by convention) $60°$ ahead of m_2, and L_5 $60°$ behind.

8.8 Stability of Lagrange points

We have seen that the five Lagrange points, L_1 to L_5, are the equilibrium points of mass m_3 in the co-rotating frame. Let us now determine whether these equilibrium points are stable to small displacements.

The equations of motion of mass m_3 in the co-rotating frame are specified in Equations (8.27)–(8.29). The motion in the x–y plane is complicated by presence of the Coriolis acceleration. However, the motion parallel to the z-axis simply corresponds to motion in the potential U. Hence, the condition for the stability of the Lagrange points (which all lie at $z = 0$) to small displacements parallel to the z-axis is simply (see Section 1.7)

$$\left(\frac{\partial^2 U}{\partial z^2}\right)_{z=0} = \frac{\mu_1}{\rho_1^3} + \frac{\mu_2}{\rho_2^3} > 0. \tag{8.67}$$

This condition is satisfied everywhere in the x–y plane. Hence, the Lagrange points are all stable to small displacements parallel to the z-axis. It thus remains to investigate their stability to small displacements lying within the x–y plane.

Suppose that a Lagrange point is situated in the x–y plane at coordinates $(x_0, y_0, 0)$. Let us consider small amplitude x–y motion in the vicinity of this point by writing

$$x = x_0 + \delta x, \tag{8.68}$$

$$y = y_0 + \delta y, \tag{8.69}$$

and

$$z = 0, \tag{8.70}$$

where δx and δy are infinitesimal. Expanding $U(x, y, 0)$ about the Lagrange point as a Taylor series, and retaining terms up to second order in small quantities, we obtain

$$U \simeq U_0 + U_x\,\delta x + U_y\,\delta y + \frac{1}{2}\,U_{xx}\,(\delta x)^2 + U_{xy}\,\delta x\,\delta y + \frac{1}{2}\,U_{yy}\,(\delta y)^2, \tag{8.71}$$

where $U_0 = U(x_0, y_0, 0)$, $U_x = \partial U(x_0, y_0, 0)/\partial x$, $U_{xx} = \partial^2 U(x_0, y_0, 0)/\partial x^2$, and so on. However, by definition, $U_x = U_y = 0$ at a Lagrange point, so the expansion simplifies to

$$U \simeq U_0 + \frac{1}{2}\,U_{xx}\,(\delta x)^2 + U_{xy}\,\delta x\,\delta y + \frac{1}{2}\,U_{yy}\,(\delta y)^2. \tag{8.72}$$

Finally, substituting Equations (8.68)–(8.70) and (8.72) into the equations of x–y motion, (8.27) and (8.28), and only retaining terms up to first order in small quantities, we get

$$\delta \ddot{x} - 2\,\delta \dot{y} \simeq -U_{xx}\,\delta x - U_{xy}\,\delta y \tag{8.73}$$

and

$$\delta \ddot{y} + 2\,\delta \dot{x} \simeq -U_{xy}\,\delta x - U_{yy}\,\delta y, \tag{8.74}$$

as $\omega = 1$.

Let us search for a solution of the preceding pair of equations of the form $\delta x(t) = \delta x_0 \exp(\gamma\, t)$ and $\delta y(t) = \delta y_0 \exp(\gamma\, t)$. We obtain

$$\begin{pmatrix} \gamma^2 + U_{xx} & -2\gamma + U_{xy} \\ 2\gamma + U_{xy} & \gamma^2 + U_{yy} \end{pmatrix} \begin{pmatrix} \delta x_0 \\ \delta y_0 \end{pmatrix} = \begin{pmatrix} 0 \\ 0 \end{pmatrix}. \tag{8.75}$$

This equation only has a nontrivial solution if the determinant of the matrix is zero. Hence, we get

$$\gamma^4 + (4 + U_{xx} + U_{yy})\,\gamma^2 + (U_{xx}\,U_{yy} - U_{xy}^2) = 0. \tag{8.76}$$

It is convenient to define

$$A = \frac{\mu_1}{\rho_1^3} + \frac{\mu_2}{\rho_2^3}, \tag{8.77}$$

$$B = 3 \left[\frac{\mu_1}{\rho_1^5} + \frac{\mu_2}{\rho_2^5} \right] y^2, \tag{8.78}$$

$$C = 3 \left[\frac{\mu_1 (x + \mu_2)}{\rho_1^5} + \frac{\mu_2 (x - \mu_1)}{\rho_2^5} \right] y, \tag{8.79}$$

and

$$D = 3 \left[\frac{\mu_1 (x + \mu_2)^2}{\rho_1^5} + \frac{\mu_2 (x - \mu_1)^2}{\rho_2^5} \right], \tag{8.80}$$

where all terms are evaluated at the point $(x_0, y_0, 0)$. It thus follows that

$$U_{xx} = A - D - 1, \tag{8.81}$$

$$U_{yy} = A - B - 1, \tag{8.82}$$

and

$$U_{xy} = -C. \tag{8.83}$$

Consider the co-linear Lagrange points, L_1, L_2, and L_3. These all lie on the x-axis, and are thus characterized by $y = 0$, $\rho_1^2 = (x + \mu_2)^2$, and $\rho_2^2 = (x - \mu_1)^2$. It follows, from the preceding equations, that $B = C = 0$ and $D = 3A$. Hence, $U_{xx} = -1 - 2A$, $U_{yy} = A - 1$, and $U_{xy} = 0$. Equation (8.76) thus yields

$$\Gamma^2 + (2 - A)\Gamma + (1 - A)(1 + 2A) = 0, \tag{8.84}$$

where $\Gamma = \gamma^2$. For a Lagrange point to be stable to small displacements, all four of the roots, γ, of Equation (8.76) must be *purely imaginary*. This, in turn, implies that the two roots of the preceding equation,

$$\Gamma = \frac{A - 2 \pm \sqrt{A(9A - 8)}}{2}, \tag{8.85}$$

must both be *real* and *negative*. Thus, the stability criterion is

$$\frac{8}{9} \leq A \leq 1. \tag{8.86}$$

Figure 8.13 shows A calculated at the three co-linear Lagrange points as a function of μ_2, for all allowed values of this parameter (i.e., $0 < \mu_2 \leq 0.5$). It can be seen that A is always greater than unity for all three points. Hence, we conclude that the co-linear Lagrange points, L_1, L_2, and L_3, are intrinsically unstable equilibrium points in the co-rotating frame.

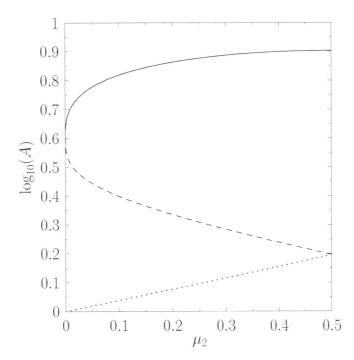

Fig. 8.13   The solid, dashed, and dotted curves show A as a function of μ_2 at L_1, L_2, and L_3 Lagrange points, respectively.

Let us now consider the triangular Lagrange points, L_4 and L_5. These points are characterized by $\rho_1 = \rho_2 = 1$. It follows that $A = 1$, $B = 9/4$, $C = \pm\sqrt{27/16}\,(1 - 2\mu_2)$, and $D = 3/4$. Hence, $U_{xx} = -3/4$, $U_{yy} = -9/4$, and $U_{xy} = \mp\sqrt{27/16}\,(1 - 2\mu_2)$, where the upper and lower signs correspond to L_4 and L_5, respectively. Equation (8.76) thus yields

$$\Gamma^2 + \Gamma + \frac{27}{4}\,\mu_2\,(1 - \mu_2) = 0 \tag{8.87}$$

for both points, where $\Gamma = \gamma^2$. As before, the stability criterion is that the two roots of the preceding equation must both be real and negative. This is the case provided that $1 > 27\,\mu_2\,(1 - \mu_2)$, which yields the stability criterion

$$\mu_2 < \frac{1}{2}\left(1 - \sqrt{\frac{23}{27}}\right) = 0.0385. \tag{8.88}$$

In unnormalized units, this criterion becomes

$$\frac{m_2}{m_1 + m_2} < 0.0385. \tag{8.89}$$

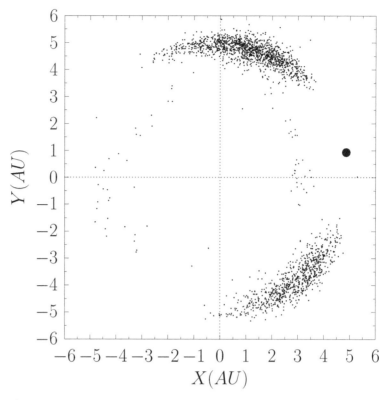

Fig. 8.14 Positions of the Trojan asteroids (small circles) and Jupiter (large circle) projected onto ecliptic plane (viewed from the north) at MJD 55600. The *X*-axis is directed toward the vernal equinox. Raw data from JPL Small-Body Database.

We thus conclude that the L_4 and L_5 Lagrange points are stable equilibrium points, in the co-rotating frame, provided that mass m_2 is less than about 4 percent of mass m_1. If this is the case, then mass m_3 can orbit around these points indefinitely. In the inertial frame, the mass will share the orbit of mass m_2 about mass m_1, but it will stay approximately 60° *ahead* of mass m_2 if it is orbiting the L_4 point, or 60° *behind* if it is orbiting the L_5 point. (See Figure 8.12.) This type of behavior has been observed in the solar system. For instance, there is a subclass of asteroids, known as the *Trojan asteroids*, that are trapped in the vicinity of the L_4 and L_5 points of the Sun–Jupiter system [which easily satisfies the stability criterion in Equation (8.89)], and consequently share Jupiter's orbit around the Sun, staying approximately 60° ahead of and 60° behind Jupiter, respectively. These asteroids are shown in Figures 8.14 and 8.15. The Sun–Jupiter system is not the only dynamical system in the solar system that possesses Trojan asteroids trapped in the vicinity of its L_4 and L_5 points. In fact, the Sun–Neptune system has eight known Trojan asteroids, the Sun–Mars system has four, and the Sun–Earth system has one (designated 2010 TK7) trapped at the L_4 point. The L_4 and L_5 points of the Sun–Earth system are also observed to trap clouds of interplanetary dust.

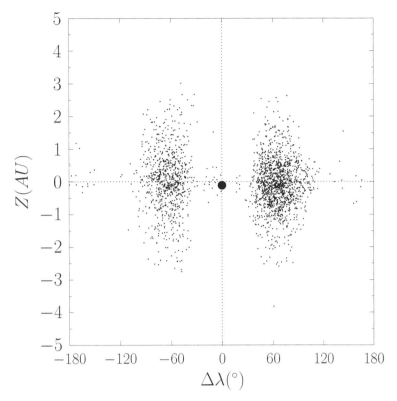

Positions of the Trojan asteroids (small circles) and Jupiter (large circle) at MJD 55600. Z is normal distance from the ecliptic plane. $\Delta\lambda$ is the difference in ecliptic longitude between the asteroids and Jupiter. Raw data from JPL Small-Body Database.

Exercises

8.1 Demonstrate directly from Equations (8.5)–(8.7) and (8.11)–(8.13) that the Jacobi integral C, which is defined in Equation (8.10), is a constant of the motion in the circular restricted three-body problem.

8.2 A comet approaching the Sun in a parabolic orbit of perihelion distance r_p and inclination I (with respect to Jupiter's orbital plane) is disturbed by a close encounter with Jupiter such that its orbit is converted into an ellipse of major radius a', eccentricity e', and inclination I'. Demonstrate that

$$\sqrt{2\,r_p}\,\cos I \simeq \frac{1}{2\,a'} + \sqrt{(1 - e'^{\,2})\,a'}\,\cos I',$$

where all lengths are normalized to the major radius of Jupiter.

8.3 A comet approaching the Sun in a hyperbolic orbit of perihelion distance r_p and inclination I (with respect to Jupiter's orbital plane), whose asymptotes subtend an acute angle ϕ with respect to one another, is disturbed by a close encounter with Jupiter such that its orbit is converted into an ellipse of major radius a', eccentricity e', and inclination I'. Demonstrate that

$$\frac{1}{2\,r_p}\,(1 - e) + \sqrt{(1 + e)\,r_p}\,\cos I \simeq \frac{1}{2\,a'} + \sqrt{.(1 - e'^2)\,a'}\,\cos I',$$

where $e = \sec(\phi/2)$, and all lengths are normalized to the major radius of Jupiter.

8.4 Let $(\xi,\ \eta,\ \zeta)$ be the coordinates of mass m_3 in the inertial frame, and let $(x,\ y,\ z)$ be the corresponding coordinates in the co-rotating frame. It follows that

$$\mathbf{x} = \mathbf{A}\,\boldsymbol{\xi},$$

where $\mathbf{x}$ is the column vector of the co-rotating coordinates, $\boldsymbol{\xi}$ is the column vector of the inertial coordinates, and

$$\mathbf{A} = \begin{pmatrix} \cos\omega t & \sin\omega t & 0 \\ -\sin\omega t & \cos\omega t & 0 \\ 0 & 0 & 1 \end{pmatrix}.$$

Demonstrate that $\mathbf{A}^T\,\mathbf{A} = \mathbf{1}$, where ^{T} denotes a transpose, and

$$\mathbf{1} = \begin{pmatrix} 1 & 0 & 0 \\ 0 & 1 & 0 \\ 0 & 0 & 1 \end{pmatrix}.$$

Hence, deduce that $\mathbf{x}^T\,\mathbf{x} = \boldsymbol{\xi}^T\,\boldsymbol{\xi}$, or $x^2 + y^2 = \xi^2 + \eta^2$.
Show that

$$\dot{\mathbf{x}} = \mathbf{A}\,\dot{\boldsymbol{\xi}} - \omega\,\mathbf{B}\,\boldsymbol{\xi},$$

where $\dot{\mathbf{x}}$ is the column vector of the time derivatives of the co-rotating coordinates, $\dot{\boldsymbol{\xi}}$ is the column vector of the time derivatives of the inertial coordinates, and

$$\mathbf{B} = \begin{pmatrix} \sin\omega t & -\cos\omega t & 0 \\ \cos\omega t & \sin\omega t & 0 \\ 0 & 0 & 0 \end{pmatrix}.$$

Demonstrate that

$$\dot{\mathbf{x}}^T\,\dot{\mathbf{x}} = \dot{\boldsymbol{\xi}}^T\,\mathbf{1}\,\dot{\boldsymbol{\xi}} - \omega\,\boldsymbol{\xi}^T\,\mathbf{C}\,\dot{\boldsymbol{\xi}} - \omega\,\dot{\boldsymbol{\xi}}^T\,\mathbf{C}^T\,\boldsymbol{\xi} + \omega^2\,\boldsymbol{\xi}^T\,\mathbf{1}'\,\boldsymbol{\xi},$$

where

$$\mathbf{1}' = \begin{pmatrix} 1 & 0 & 0 \\ 0 & 1 & 0 \\ 0 & 0 & 0 \end{pmatrix}$$

and

$$\mathbf{C} = \begin{pmatrix} 0 & 1 & 0 \\ -1 & 0 & 0 \\ 0 & 0 & 0 \end{pmatrix}.$$

Hence, deduce that

$$\dot{x}^2 + \dot{y}^2 + \dot{z}^2 = \dot{\xi}^2 + \dot{\eta}^2 + \dot{\zeta}^2 - 2\omega(\xi\dot{\eta} - \eta\dot{\xi}) + \omega^2(\xi^2 + \eta^2).$$

Finally, show that the Jacobi constant in the co-rotating frame,

$$C = 2\left(\frac{\mu_1}{\rho_1} + \frac{\mu_2}{\rho_2}\right) + \omega^2(x^2 + y^2) - \dot{x}^2 - \dot{y}^2 - \dot{z}^2,$$

transforms to

$$C = 2\left(\frac{\mu_1}{\rho_1} + \frac{\mu_2}{\rho_2}\right) + 2\omega(\xi\dot{\eta} - \eta\dot{\xi}) - \dot{\xi}^2 - \dot{\eta}^2 - \dot{\zeta}^2$$

in the inertial frame.

8.5   Derive the first three terms (on the right-hand side) of Equation (8.46) from Equation (8.45), and the first three terms of Equation (8.47) from Equation (8.46).

8.6   Derive the first three terms of Equation (8.50) from Equation (8.49), and the first three terms of Equation (8.51) from Equation (8.50).

8.7   Derive the first two terms of Equation (8.53) from Equation (8.52), and the first two terms of Equation (8.54) from Equation (8.53).

8.8   Derive Equations (8.62)–(8.66).

8.9   Employing the standard system of units for the circular restricted three-body problem, the equation defining the location of a zero-velocity curve in the x–y plane of the co-rotating frame is

$$C = x^2 + y^2 + 2\left(\frac{\mu_1}{\rho_1} + \frac{\mu_2}{\rho_2}\right),$$

where C is the value of the Jacobi constant, and $\rho_1 = [(x + \mu_2)^2 + y^2]^{1/2}$ and $\rho_2 = [(x - \mu_1)^2 + y^2]^{1/2}$ are the distances to the primary and secondary masses, respectively. The critical zero-velocity curve that passes through the L_3 point, when $C \simeq 3 + \mu_2$, has two branches. Defining polar coordinates such that $x = r\cos\theta$ and $y = r\sin\theta$, show that when $\mu_2 \ll 1$ the branches intersect the unit circle $r = 1$ at $\theta = \pi$ and $\theta = \pm 23.9°$. (Modified from Murray and Dermott 1999.)

8.10   In the circular restricted three-body problem (employing the standard system of units) the condition for the three co-linear Lagrange points to be linearly unstable is $A > 1$, where $A = \mu_1/\rho_1^3 + \mu_2/\rho_2^3$. Here, ρ_1 and ρ_2 are the distances to the masses μ_1 and μ_2, respectively. Let $\alpha = (\mu_2/3\mu_1)^{1/3}$ and $\beta = (7/12)(\mu_2/\mu_1)$. Consider the limit $\mu_2 \to 0$. Show that close to L_1, where $\rho_2 \simeq \alpha - \alpha^2/3$ and $\rho_1 = 1 - \rho_2$, the parameter A takes the value $A \simeq 4 + 6\alpha + \mathcal{O}(\alpha^2)$. Likewise, show that close to L_2, where $\rho_2 \simeq \alpha + \alpha^2/3$ and $\rho_1 = 1 + \rho_2$, the parameter A takes the value $4 - 6\alpha + \mathcal{O}(\alpha^2)$. Finally, show that close to L_3, where $\rho_1 \simeq 1 - \beta$ and $\rho_2 = 1 + \rho_1$, the parameter A takes the value $1 + (3/2)\beta + \mathcal{O}(\beta^2)$. Hence, deduce that the three co-linear Lagrange points are all linearly unstable. Demonstrate that, in the case of the L_3 point, the growth-rate of the fastest growing instability is $\gamma \simeq \left[\sqrt{(21/8)\mu_2} + \mathcal{O}(\mu_2)\right]\omega$. (Modified from Murray and Dermott 1999.)

8.11   Consider the circular restricted three-body problem. Demonstrate that if $[x(t), y(t), z(t)]$ is a valid trajectory for m_3 in the co-rotating frame, then $[x(t), y(t), -z(t)]$ and $[x(-t), -y(-t), z(-t)]$ are also valid trajectories. Show that if $[x(t), y(t), z(t)]$

is a valid trajectory when $\mu_2 = \zeta$ (where $0 \le \zeta \le 1$), then $[-x(t), -y(t), z(t)]$ is a valid trajectory when $\mu_2 = 1 - \zeta$.

8.12 Consider the circular restricted three-body problem (adopting the standard system of units). Suppose that $\mu_2 \ll 0.0385$, so that the L_4 and L_5 points are stable equilibrium points (in the co-rotating frame) for the tertiary mass. Consider motion (in the co-rotating frame) of the tertiary mass in the vicinity of L_4 that is confined to the x–y plane. Let

$$x = \frac{1}{2} - \mu_2 + \delta x,$$

$$y = \frac{\sqrt{3}}{2} + \delta y,$$

where $|\delta x|, |\delta y| \ll 1$. It is helpful to rotate the Cartesian axes through $30°$, so that

$$\begin{pmatrix} \delta x \\ \delta y \end{pmatrix} = \begin{pmatrix} \sqrt{3}/2 & 1/2 \\ -1/2 & \sqrt{3}/2 \end{pmatrix} \begin{pmatrix} \delta x' \\ \delta y' \end{pmatrix}.$$

Thus, $\delta x'$ parameterizes displacements from L_4 that are *tangential* to the unit circle on which the mass m_2 and the L_3, L_4, and L_5 points lie, whereas $\delta y'$ parameterizes *radial* displacements. Writing $\delta x'(t) = \delta x_0' \exp(\gamma t)$ and $\delta y'(t) = \delta y_0' \exp(\gamma t)$, where $\delta x_0'$, $\delta y_0'$, γ are constants, demonstrate that

$$\begin{pmatrix} \sqrt{3}\,\gamma^2/2 + \gamma - 3\sqrt{3}\,\mu_2/4 & \gamma^2/2 - \sqrt{3}\,\gamma - 3/2 + 9\,\mu_2/4 \\ -\gamma^2/2 + \sqrt{3}\,\gamma + 9\,\mu_2/4 & \sqrt{3}\,\gamma^2/2 + \gamma - 3\sqrt{3}/2 + 3\sqrt{3}\,\mu_2/4 \end{pmatrix} \begin{pmatrix} \delta x_0' \\ \delta y_0' \end{pmatrix} = \begin{pmatrix} 0 \\ 0 \end{pmatrix},$$

and, hence, that

$$\gamma^4 + \gamma^2 + \frac{27}{4}\,\mu_2\,(1 - \mu_2) = 0.$$

Show that the general solution to the preceding dispersion relation is a linear combination of two normal modes of oscillation, and that the higher-frequency mode takes the form

$$\delta x'(t) \simeq -2\,e\,\sin(\omega_+ t - \phi_+),$$

$$\delta y'(t) \simeq -e\,\cos(\omega_+ t - \phi_+),$$

where

$$\omega_+ \simeq \left(1 - \frac{27}{8}\,\mu_2\right)\omega,$$

and e, ϕ_+ are arbitrary constants. Demonstrate that, in the original inertial reference frame, the addition of the preceding normal mode to the unperturbed orbit of the tertiary mass (in the limit $0 < e \ll 1$) converts a circular orbit into a Keplerian ellipse of eccentricity e. In addition, show that the perihelion point of the new orbit precesses (in the direction of the orbital motion) at the rate

$$\dot{\varpi} = \frac{27}{8}\,\mu_2\,\omega.$$

Demonstrate that (in the co-rotating reference frame) the second normal mode takes the form

$$\delta x'(t) \simeq d \sin(\omega_- t - \phi_-),$$

$$\delta y'(t) \simeq -\sqrt{3\,\mu_2}\, d \sin(\omega_- t - \phi_-),$$

where

$$\omega_- \simeq \frac{3}{2}\sqrt{3\,\mu_2}\,\omega,$$

and d, ϕ_- are arbitrary constants. This type of motion, which entails relatively small-amplitude radial oscillations, combined with much larger-amplitude tangential oscillations, is known as *libration*.

Finally, consider a Trojan asteroid trapped in the vicinity of the L_4 point of the Sun–Jupiter system. Demonstrate that the libration period of the asteroid (in the co-rotating frame) is approximately 148 years, whereas its perihelion precession period (in the inertial frame) is approximately 3,690 years. Show that, in the co-rotating frame, the libration orbit is an ellipse that is elongated in the direction of the tangent to the Jovian orbit in the ratio 18.7:1.

9 Secular perturbation theory

9.1 Introduction

The two-body orbit theory described in Chapter 3 neglects the direct gravitational inter-actions between the planets, while retaining those between each individual planet and the Sun. This is an excellent first approximation, as the former interactions are much weaker than the latter, as a consequence of the small masses of the planets relative to the Sun. (See Table 3.1.) Nevertheless, interplanetary gravitational interactions do have a profound influence on planetary orbits when integrated over long periods of time. In this chapter, a branch of celestial mechanics known as *orbital perturbation theory* is used to examine the *secular* (i.e., long-term) influence of interplanetary gravitational perturbations on planetary orbits. Orbital perturbation theory is also used to investigate the secular influence of planetary perturbations on the orbits of asteroids, as well as the secular influence of the Earth's oblateness on the orbits of artificial satellites.

9.2 Evolution equations for a two-planet solar system

For the moment, let us consider a simplified solar system that consists of the Sun and two planets. (See Figure 9.1.) Let the Sun be of mass M and position vector $\mathbf{R}_s$. Like-wise, let the two planets have masses m and m' and position vectors $\mathbf{R}$ and $\mathbf{R}'$, respec-tively. Here, we are assuming that m, $m' \ll M$. Finally, let $\mathbf{r} = \mathbf{R} - \mathbf{R}_s$ and $\mathbf{r}' = \mathbf{R}' - \mathbf{R}_s$ be the position vectors of each planet relative to the Sun. Without loss of generality, we can assume that $r' > r$.

In an inertial reference frame, the equations of motion of the various elements of our simplified solar system are

$$M \ddot{\mathbf{R}}_s = G M m \frac{\mathbf{r}}{r^3} + G M m' \frac{\mathbf{r}'}{r'^3}, \tag{9.1}$$

$$m \ddot{\mathbf{R}} = G m m' \frac{\mathbf{r}' - \mathbf{r}}{|\mathbf{r}' - \mathbf{r}|^3} - G m M \frac{\mathbf{r}}{r^3}, \tag{9.2}$$

and

$$m' \ddot{\mathbf{R}}' = G m' m \frac{\mathbf{r} - \mathbf{r}'}{|\mathbf{r} - \mathbf{r}'|^3} - G m' M \frac{\mathbf{r}'}{r'^3}. \tag{9.3}$$

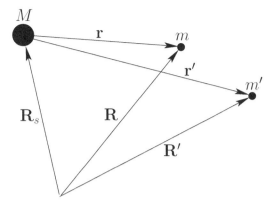

Fig. 9.1 A simplified model of the solar system.

It thus follows that

$$\ddot{\mathbf{r}} + \mu \frac{\mathbf{r}}{r^3} = G\,m'\left(\frac{\mathbf{r}' - \mathbf{r}}{|\mathbf{r}' - \mathbf{r}|^3} - \frac{\mathbf{r}'}{r'^3}\right) \tag{9.4}$$

and

$$\ddot{\mathbf{r}}' + \mu' \frac{\mathbf{r}'}{r'^3} = G\,m\left(\frac{\mathbf{r} - \mathbf{r}'}{|\mathbf{r} - \mathbf{r}'|^3} - \frac{\mathbf{r}}{r^3}\right), \tag{9.5}$$

where $\mu = G\,(M + m)$ and $\mu' = G\,(M + m')$. The right-hand sides of these equations specify the interplanetary interaction forces that were neglected in our previous analysis. These right-hand sides can be conveniently expressed as the gradients of potentials:

$$\ddot{\mathbf{r}} + \mu \frac{\mathbf{r}}{r^3} = \nabla \mathcal{R} \tag{9.6}$$

and

$$\ddot{\mathbf{r}}' + \mu' \frac{\mathbf{r}'}{r'^3} = \nabla' \mathcal{R}', \tag{9.7}$$

where

$$\mathcal{R}(\mathbf{r}, \mathbf{r}') = \tilde{\mu}'\left(\frac{1}{|\mathbf{r} - \mathbf{r}'|} - \frac{\mathbf{r} \cdot \mathbf{r}'}{r'^3}\right) \tag{9.8}$$

and

$$\mathcal{R}'(\mathbf{r}, \mathbf{r}') = \tilde{\mu}\left(\frac{1}{|\mathbf{r} - \mathbf{r}'|} - \frac{\mathbf{r} \cdot \mathbf{r}'}{r^3}\right), \tag{9.9}$$

with $\tilde{\mu} = G\,m$ and $\tilde{\mu}' = G\,m'$. Here, $\mathcal{R}(\mathbf{r}, \mathbf{r}')$ and $\mathcal{R}'(\mathbf{r}, \mathbf{r}')$ are termed *disturbing functions*. Moreover, ∇ and ∇' are the gradient operators involving the unprimed and primed coordinates, respectively.

In the absence of the second planet, the orbit of the first planet is fully described by its six standard orbital elements (which are constants of its motion): the *major radius*, a; the *mean longitude at epoch*, $\bar{\lambda}_0$; the *eccentricity*, e; the *inclination* (to the ecliptic plane), I; the *longitude of the perihelion*, ϖ; and *the longitude of the ascending node*, Ω. (See Section 3.12.) As described in Appendix B, the perturbing influence of the second planet causes these elements to slowly evolve in time. Such time-varying orbital elements are generally known as *osculating elements*.[1] Actually, when describing the aforementioned evolution, it is more convenient to work in terms of an alternative set of osculating elements, namely $a(t) = a^{(0)}[1 + \epsilon' \, a^{(1)}(t)]$, $\bar{\lambda}(t) = \bar{\lambda}_0 + n^{(0)} \, t + \bar{\lambda}^{(1)}(t)$, $h = e \sin \varpi$, $k = e \cos \varpi$, $p = \sin I \sin \Omega$, and $q = \sin I \cos \Omega$. Here, $\epsilon' = \bar{\mu}'/\mu = m'/(M + m) \ll 1$, $n(t) = n^{(0)}[1 - (3/2) \, \epsilon' \, a^{(1)}(t)]$, where $n^{(0)} = (\mu/[a^{(0)}]^3)^{1/2}$ is the unperturbed mean orbital angular velocity. In the following, for ease of notation, $a^{(0)}$ and $n^{(0)}$ are written simply as a and n, respectively. Furthermore, $\bar{\lambda}$ will be used as shorthand for $\bar{\lambda}_0 + n^{(0)} \, t$. The evolution equations for the first planet's osculating orbital elements take the form (see Section C.2)

$$\frac{d\epsilon' a^{(1)}}{dt} = \epsilon' \, n \left[2\,\alpha \, \frac{\partial(\mathcal{S}_0 + \mathcal{S}_1)}{\partial \bar{\lambda}} \right], \tag{9.10}$$

$$\frac{d\bar{\lambda}^{(1)}}{dt} = \epsilon' \, n \left[-\frac{3}{2} \, a^{(1)} - 2\,\alpha^2 \, \frac{\partial(\mathcal{S}_0 + \mathcal{S}_1)}{\partial \alpha} + \alpha \left(h \, \frac{\partial \mathcal{S}_1}{\partial h} + k \, \frac{\partial \mathcal{S}_1}{\partial k} \right) \right], \tag{9.11}$$

$$\frac{dh}{dt} = \epsilon' \, n \left[-\alpha \, h \, \frac{\partial \mathcal{S}_0}{\partial \bar{\lambda}} + \alpha \, \frac{\partial(\mathcal{S}_1 + \mathcal{S}_2)}{\partial k} \right], \tag{9.12}$$

$$\frac{dk}{dt} = \epsilon' \, n \left[-\alpha \, k \, \frac{\partial \mathcal{S}_0}{\partial \bar{\lambda}} - \alpha \, \frac{\partial(\mathcal{S}_1 + \mathcal{S}_2)}{\partial h} \right], \tag{9.13}$$

$$\frac{dp}{dt} = \epsilon' \, n \left[-\frac{\alpha}{2} \, p \, \frac{\partial \mathcal{S}_0}{\partial \bar{\lambda}} + \alpha \, \frac{\partial \mathcal{S}_2}{\partial q} \right], \tag{9.14}$$

and

$$\frac{dq}{dt} = \epsilon' \, n \left[-\frac{\alpha}{2} \, q \, \frac{\partial \mathcal{S}_0}{\partial \bar{\lambda}} - \alpha \, \frac{\partial \mathcal{S}_2}{\partial p} \right], \tag{9.15}$$

where (see Section C.3)

$$\mathcal{S}_0 = \frac{1}{2} \sum_{j=-\infty,\infty} b_{1/2}^{(j)} \, \cos[j \, (\bar{\lambda} - \bar{\lambda}')] - \alpha \, \cos(\bar{\lambda} - \bar{\lambda}'), \tag{9.16}$$

$$\begin{aligned}
\mathcal{S}_1 = \frac{1}{2} \sum_{j=-\infty,\infty} & \left\{ k \, (-2\,j - \alpha \, D) \, b_{1/2}^{(j)} \, \cos[(1-j)\,\bar{\lambda} + j\,\bar{\lambda}'] \right. \\
& + h \, (-2\,j - \alpha \, D) \, b_{1/2}^{(j)} \, \sin[(1-j)\,\bar{\lambda} + j\,\bar{\lambda}'] \\
& + k' \, (-1 + 2\,j + \alpha \, D) \, b_{1/2}^{(j-1)} \, \cos[(1-j)\,\bar{\lambda} + j\,\bar{\lambda}'] \\
& \left. + h' \, (-1 + 2\,j + \alpha \, D) \, b_{1/2}^{(j-1)} \, \sin[(1-j)\,\bar{\lambda} + j\,\bar{\lambda}'] \right\} \\
& + \frac{\alpha}{2} \left\{ -k \, \cos(2\,\bar{\lambda} - \bar{\lambda}') - h \, \sin(2\,\bar{\lambda} - \bar{\lambda}') + 3\,k \, \cos \bar{\lambda}' + 3\,h \, \sin \bar{\lambda}' \right. \\
& \left. - 4\,k' \, \cos(\bar{\lambda} - 2\,\bar{\lambda}') + 4\,h' \, \sin(\bar{\lambda} - 2\,\bar{\lambda}') \right\},
\end{aligned} \tag{9.17}$$

---

[1] In mathematical terminology, two curves are said to osculate when they touch one another so as to have a common tangent at the point of contact. From the Latin *osculatus*, "kissed."

and

$$
\begin{aligned}
S_2 = {} & \frac{1}{8}\,(h^2 + k^2 + h'^2 + k'^2)\,(2\,\alpha\,D + \alpha^2\,D^2)\,b^{(0)}_{1/2} - \frac{1}{8}\,(p^2 + q^2 + p'^2 + q'^2)\,\alpha\,b^{(1)}_{3/2} \\
& + \frac{1}{4}\,(k\,k' + h\,h')\,(2 - 2\,\alpha\,D - \alpha^2\,D^2)\,b^{(1)}_{1/2} + \frac{1}{4}\,(p\,p' + q\,q')\,\alpha\,b^{(1)}_{3/2}.
\end{aligned}
\tag{9.18}
$$

Here, $\alpha = a/a'$, $D \equiv d/d\alpha$, and

$$
b^{(j)}_s(\alpha) = \frac{1}{\pi} \int_0^{2\pi} \frac{\cos(j\,\psi)\,d\psi}{[1 - 2\,\alpha\,\cos\psi + \alpha^2]^s},
\tag{9.19}
$$

where a', $\bar{\lambda}'$, h', k', p', q' are the osculating orbital elements of the second planet. The $b^{(j)}_s$ factors are known as *Laplace coefficients* (Brouwer and Clemence 1961). In deriving these expressions from Equations (9.6) and (9.8), we have expanded to first order in the ratio of the planetary masses to the solar mass; we have then evaluated the secular terms in the disturbing functions (i.e., the terms that are independent of $\bar{\lambda}$ and $\bar{\lambda}'$) to second order in the orbital eccentricities and inclinations. The nonsecular terms in the disturbing functions are evaluated to first order in the eccentricities and inclinations. (See Appendix C.) This expansion procedure is reasonable because the planets all have very small masses compared with that of the Sun, and they also have relatively small orbital eccentricities and inclinations.

There is an analogous set of equations, which can be derived from Equations (9.7) and (9.9), that describe the time evolution of the osculating orbital elements of the second planet due to the perturbing influence of the first. These take the form (see Section C.2)

$$
\frac{d\epsilon'\,a^{(1)'}}{dt} = \epsilon\,n'\left[2\,\alpha^{-1}\,\frac{\partial(S'_0 + S'_1)}{\partial\bar{\lambda}'}\right],
\tag{9.20}
$$

$$
\frac{d\bar{\lambda}^{(1)'}}{dt} = \epsilon\,n'\left[-\frac{3}{2}\,a^{(1)'} + 2\,\frac{\partial(S'_0 + S'_1)}{\partial\alpha} + \alpha^{-1}\left(h'\,\frac{\partial S'_1}{\partial h'} + k'\,\frac{\partial S'_1}{\partial k'}\right)\right],
\tag{9.21}
$$

$$
\frac{dh'}{dt} = \epsilon\,n'\left[-\alpha^{-1}\,h'\,\frac{\partial S'_0}{\partial\bar{\lambda}'} + \alpha^{-1}\,\frac{\partial(S'_1 + S'_2)}{\partial k'}\right],
\tag{9.22}
$$

$$
\frac{dk'}{dt} = \epsilon\,n'\left[-\alpha^{-1}\,k'\,\frac{\partial S'_0}{\partial\bar{\lambda}'} - \alpha^{-1}\,\frac{\partial(S'_1 + S'_2)}{\partial h'}\right],
\tag{9.23}
$$

$$
\frac{dp'}{dt} = \epsilon\,n'\left[-\frac{\alpha^{-1}}{2}\,p'\,\frac{\partial S'_0}{\partial\bar{\lambda}'} + \alpha^{-1}\,\frac{\partial S'_2}{\partial q'}\right],
\tag{9.24}
$$

and

$$
\frac{dq'}{dt} = \epsilon\,n'\left[-\frac{\alpha^{-1}}{2}\,q'\,\frac{\partial S'_0}{\partial\bar{\lambda}'} - \alpha^{-1}\,\frac{\partial S'_2}{\partial p'}\right],
\tag{9.25}
$$

where (see Section C.3)

$$\mathcal{S}_0' = \frac{\alpha}{2} \sum_{j=-\infty,\infty} b_{1/2}^{(j)} \cos[j(\bar{\lambda}' - \bar{\lambda})] - \alpha^{-1} \cos(\bar{\lambda}' - \bar{\lambda}), \tag{9.26}$$

$$\mathcal{S}_1' = \frac{\alpha}{2} \sum_{j=-\infty,\infty} \left\{ k(-2j - \alpha D) b_{1/2}^{(j)} \cos[j\bar{\lambda}' + (1-j)\bar{\lambda}] \right.$$

$$+ h(-2j - \alpha D) b_{1/2}^{(j)} \sin[j\bar{\lambda}' + (1-j)\bar{\lambda}]$$

$$+ k'(-1 + 2j + \alpha D) b_{1/2}^{(j-1)} \cos[j\bar{\lambda}' + (1-j)\bar{\lambda}]$$

$$+ h'(-1 + 2j + \alpha D) b_{1/2}^{(j-1)} \sin[j\bar{\lambda}' + (1-j)\bar{\lambda}] \right\}$$

$$+ \frac{\alpha^{-1}}{2} \left\{ -k' \cos(2\bar{\lambda}' - \bar{\lambda}) - h' \sin(2\bar{\lambda}' - \bar{\lambda}) + 3k' \cos\bar{\lambda} + 3h' \sin\bar{\lambda} \right.$$

$$\left. - 4k \cos(\bar{\lambda}' - 2\bar{\lambda}) + 4h \sin(\bar{\lambda}' - 2\bar{\lambda}) \right\}, \tag{9.27}$$

and

$$\mathcal{S}_2' = \frac{1}{8}(h^2 + k^2 + h'^2 + k'^2)\alpha(2\alpha D + \alpha^2 D^2) b_{1/2}^{(0)} - \frac{1}{8}(p^2 + q^2 + p'^2 + q'^2)\alpha^2 b_{3/2}^{(1)}$$

$$+ \frac{1}{4}(kk' + hh')\alpha(2 - 2\alpha D - \alpha^2 D^2) b_{1/2}^{(1)}$$

$$+ \frac{1}{4}(pp' + qq')\alpha^2 b_{3/2}^{(1)}. \tag{9.28}$$

Here, $\epsilon = m/(M + m') = \tilde{\mu}/\mu' \ll 1$, and $n' = (\mu'/a'^3)^{1/2}$.

9.3 Secular evolution of planetary orbits

As a specific example of the use of orbital perturbation theory, let us determine the evolution of the osculating orbital elements of the two planets in our model solar system due to the secular terms in their disturbing functions (i.e., the terms that are independent of the mean longitudes $\bar{\lambda}$ and $\bar{\lambda}'$). This is equivalent to averaging the osculating elements over the relatively short timescales associated with the periodic terms in the disturbing functions (i.e., the terms that depend on $\bar{\lambda}$ and $\bar{\lambda}'$, and, therefore, oscillate on timescales similar to the orbital periods of the planets). From Equations (9.16)–(9.18), the secular part of the first planet's disturbing function takes the form

$$\mathcal{S}_s = \mathcal{S}_{0s} + \mathcal{S}_{2s}, \tag{9.29}$$

where

$$\mathcal{S}_{0s} = \frac{1}{2} b_{1/2}^{(0)}, \tag{9.30}$$

$$\mathcal{S}_{2s} = \frac{1}{8}(h^2 + k^2 + h'^2 + k'^2)\alpha b_{3/2}^{(1)} - \frac{1}{8}(p^2 + q^2 + p'^2 + q'^2)\alpha b_{3/2}^{(1)}$$

$$- \frac{1}{4}(kk' + hh')\alpha b_{3/2}^{(2)} + \frac{1}{4}(pp' + qq')\alpha b_{3/2}^{(1)}, \tag{9.31}$$

because, as can be demonstrated (Brouwer and Clemence 1961),

$$(2\,\alpha\,D + \alpha^2\,D^2)\,b_{1/2}^{(0)} \equiv \alpha\,b_{3/2}^{(1)} \tag{9.32}$$

and

$$(2 - 2\,\alpha\,D - \alpha^2\,D^2)\,b_{1/2}^{(1)} \equiv -\alpha\,b_{3/2}^{(2)}. \tag{9.33}$$

Evaluating the right-hand sides of Equations (9.10)–(9.15) to $\mathcal{O}(\epsilon'\,e\,n)$ (it is assumed that $\epsilon,\,\epsilon' \ll e,\,h,\,k,\,p,\,q,\,h',\,k',\,p',\,q' \ll 1,\,\alpha$), we find that

$$\frac{da^{(1)}}{dt} = 0, \tag{9.34}$$

$$\frac{d\bar{\lambda}^{(1)}}{dt} = \epsilon'\,n\left[-\frac{3}{2}\,a^{(1)} - 2\,\alpha^2\,\frac{\partial \mathcal{S}_{0\,s}}{\partial \alpha}\right], \tag{9.35}$$

$$\frac{dh}{dt} = \epsilon'\,n\left(\alpha\,\frac{\partial \mathcal{S}_{2\,s}}{\partial k}\right), \tag{9.36}$$

$$\frac{dk}{dt} = \epsilon'\,n\left(-\alpha\,\frac{\partial \mathcal{S}_{2\,s}}{\partial h}\right), \tag{9.37}$$

$$\frac{dp}{dt} = \epsilon'\,n\left(\alpha\,\frac{\partial \mathcal{S}_{2\,s}}{\partial q}\right), \tag{9.38}$$

and

$$\frac{dq}{dt} = \epsilon'\,n\left(-\alpha\,\frac{\partial \mathcal{S}_{2\,s}}{\partial p}\right), \tag{9.39}$$

as $\partial \mathcal{S}_{0\,s}/\partial \bar{\lambda} = 0$. It follows that $a^{(1)} = 0$, and

$$\bar{\lambda} = \bar{\lambda}_0 + n\left[1 - \epsilon'\,\alpha^2\,D\,b_{1/2}^{(0)}\right]t, \tag{9.40}$$

$$\frac{dh}{dt} = \epsilon'\,n\left[\frac{1}{4}\,k\,\alpha^2\,b_{3/2}^{(1)} - \frac{1}{4}\,k'\,\alpha^2\,b_{3/2}^{(2)}\right], \tag{9.41}$$

$$\frac{dk}{dt} = \epsilon'\,n\left[-\frac{1}{4}\,h\,\alpha^2\,b_{3/2}^{(1)} + \frac{1}{4}\,h'\,\alpha^2\,b_{3/2}^{(2)}\right], \tag{9.42}$$

$$\frac{dp}{dt} = \epsilon'\,n\left[-\frac{1}{4}\,q\,\alpha^2\,b_{3/2}^{(1)} + \frac{1}{4}\,q'\,\alpha^2\,b_{3/2}^{(1)}\right], \tag{9.43}$$

and

$$\frac{dq}{dt} = \epsilon'\,n\left[\frac{1}{4}\,p\,\alpha^2\,b_{3/2}^{(1)} - \frac{1}{4}\,p'\,\alpha^2\,b_{3/2}^{(1)}\right]. \tag{9.44}$$

Note that the first planet's mean angular velocity is slightly modified in the presence of the second planet, but that its major radius remains the same.

From Equations (9.26)–(9.28), the secular part of the second planet's disturbing function takes the form

$$\mathcal{S}'_s = \mathcal{S}'_{0\,s} + \mathcal{S}'_{2\,s'}, \tag{9.45}$$

where

$$\mathcal{S}'_{0\,s} = \frac{1}{2}\,\alpha\,b_{1/2}^{(0)} \tag{9.46}$$

and

$$
\begin{aligned}
\mathcal{S}'_{2s} = {} & \frac{1}{8}(h^2 + k^2 + h'^2 + k'^2)\alpha^2 b^{(1)}_{3/2} - \frac{1}{8}(p^2 + q^2 + p'^2 + q'^2)\alpha^2 b^{(1)}_{3/2} \\
& -\frac{1}{4}(k\,k' + h\,h')\alpha^2 b^{(2)}_{3/2} + \frac{1}{4}(p\,p' + q\,q')\alpha^2 b^{(1)}_{3/2}.
\end{aligned} \tag{9.47}
$$

Evaluating the right-hand sides of Equations (9.20)–(9.25) to $\mathcal{O}(\epsilon\,e\,n')$, we find that

$$
\frac{da^{(1)'}}{dt} = 0, \tag{9.48}
$$

$$
\frac{d\bar{\lambda}^{(1)'}}{dt} = \epsilon\,n'\left(-\frac{3}{2}a^{(1)'} + 2\frac{\partial\mathcal{S}'_{0s}}{\partial\alpha}\right), \tag{9.49}
$$

$$
\frac{dh'}{dt} = \epsilon\,n'\left(\alpha^{-1}\frac{\partial\mathcal{S}'_{2s}}{\partial k'}\right), \tag{9.50}
$$

$$
\frac{dk'}{dt} = \epsilon\,n'\left(-\alpha^{-1}\frac{\partial\mathcal{S}'_{2s}}{\partial h'}\right), \tag{9.51}
$$

$$
\frac{dp'}{dt} = \epsilon\,n'\left(\alpha^{-1}\frac{\partial\mathcal{S}'_{2s}}{\partial q'}\right), \tag{9.52}
$$

and

$$
\frac{dq'}{dt} = \epsilon\,n'\left(-\alpha^{-1}\frac{\partial\mathcal{S}'_{2s}}{\partial p'}\right). \tag{9.53}
$$

It follows that $a^{(1)'} = 0$, and

$$
\bar{\lambda}' = \bar{\lambda}'_0 + n'\left[1 - \epsilon\,D\left(\alpha\,b^{(0)}_{1/2}\right)\right]t, \tag{9.54}
$$

$$
\frac{dh'}{dt} = \epsilon\,n'\left[\frac{1}{4}k'\,\alpha\,b^{(1)}_{3/2} - \frac{1}{4}k\,\alpha\,b^{(2)}_{3/2}\right], \tag{9.55}
$$

$$
\frac{dk'}{dt} = \epsilon\,n'\left[-\frac{1}{4}h'\,\alpha\,b^{(1)}_{3/2} + \frac{1}{4}h\,\alpha\,b^{(2)}_{3/2}\right], \tag{9.56}
$$

$$
\frac{dp'}{dt} = \epsilon\,n'\left[-\frac{1}{4}q'\,\alpha\,b^{(1)}_{3/2} + \frac{1}{4}q\,\alpha\,b^{(1)}_{3/2}\right], \tag{9.57}
$$

and

$$
\frac{dq'}{dt} = \epsilon\,n'\left[\frac{1}{4}p'\,\alpha\,b^{(1)}_{3/2} - \frac{1}{4}p\,\alpha\,b^{(1)}_{3/2}\right]. \tag{9.58}
$$

The second planet's mean angular velocity is also slightly modified in the presence of the first planet, but its major radius remains the same.

Let us now generalize the preceding analysis to take all eight of the major planets in the solar system into account. Let planet i (where i runs from 1 to 8) have mass m_i, major radius a_i, eccentricity e_i, longitude of the perihelion ϖ_i, inclination I_i, and longitude of the ascending node Ω_i. As before, it is convenient to introduce the alternative orbital elements $h_i = e_i\sin\varpi_i$, $k_i = e_i\cos\varpi_i$, $p_i = \sin I_i\sin\Omega_i$, and $q_i = \sin I_i\cos\Omega_i$. It is also helpful to define the following parameters:

$$\alpha_{ij} = \begin{cases} a_i/a_j & a_j > a_i \\ a_j/a_i & a_j < a_i \end{cases}, \tag{9.59}$$

and

$$\bar{\alpha}_{ij} = \begin{cases} a_i/a_j & a_j > a_i \\ 1 & a_j < a_i \end{cases}, \tag{9.60}$$

as well as

$$\epsilon_{ij} = \frac{m_j}{M + m_i} \tag{9.61}$$

and

$$n_i = [G\,(M + m_i)/a_i^3]^{1/2}, \tag{9.62}$$

where M is the mass of the Sun. It then follows, from the preceding analysis, that the secular terms in the planetary disturbing functions cause the h_i, k_i, p_i, and q_i to vary in time as

$$\frac{dh_i}{dt} = \sum_{j=1,8} A_{ij}\,k_j, \tag{9.63}$$

$$\frac{dk_i}{dt} = -\sum_{j=1,8} A_{ij}\,h_j, \tag{9.64}$$

$$\frac{dp_i}{dt} = \sum_{j=1,8} B_{ij}\,q_j, \tag{9.65}$$

and

$$\frac{dq_i}{dt} = -\sum_{j=1,8} B_{ij}\,p_j, \tag{9.66}$$

where

$$A_{ii} = \sum_{j \neq i} \frac{n_i}{4}\,\epsilon_{ij}\,\alpha_{ij}\,\bar{\alpha}_{ij}\,b_{3/2}^{(1)}(\alpha_{ij}), \tag{9.67}$$

$$A_{ij} = -\frac{n_i}{4}\,\epsilon_{ij}\,\alpha_{ij}\,\bar{\alpha}_{ij}\,b_{3/2}^{(2)}(\alpha_{ij}), \tag{9.68}$$

$$B_{ii} = -\sum_{j \neq i} \frac{n_i}{4}\,\epsilon_{ij}\,\alpha_{ij}\,\bar{\alpha}_{ij}\,b_{3/2}^{(1)}(\alpha_{ij}), \tag{9.69}$$

and

$$B_{ij} = \frac{n_i}{4}\,\epsilon_{ij}\,\alpha_{ij}\,\bar{\alpha}_{ij}\,b_{3/2}^{(1)}(\alpha_{ij}), \tag{9.70}$$

for $j \neq i$. Here, Mercury is planet 1, Venus is planet 2, and so on, and Neptune is planet 8. Note that the time evolution of the h_i and the k_i, which determine the eccentricities of the planetary orbits, is decoupled from that of the p_i and the q_i, which determine the inclinations. Let us search for normal mode solutions to Equations (9.63)–(9.66) of the form

$$h_i = \sum_{l=1,8} e_{il} \, \sin(g_l \, t + \beta_l), \tag{9.71}$$

$$k_i = \sum_{l=1,8} e_{il} \, \cos(g_l \, t + \beta_l), \tag{9.72}$$

$$p_i = \sum_{l=1,8} I_{il} \, \sin(f_l \, t + \gamma_l), \tag{9.73}$$

and

$$q_i = \sum_{l=1,8} I_{il} \, \cos(f_l \, t + \gamma_l). \tag{9.74}$$

It follows that

$$\sum_{j=1,8} \left(A_{ij} - \delta_{ij} \, g_l \right) e_{jl} = 0 \tag{9.75}$$

and

$$\sum_{j=1,8} \left(B_{ij} - \delta_{ij} \, f_l \right) I_{jl} = 0. \tag{9.76}$$

At this stage, we have effectively reduced the problem of determining the secular evolution of the planetary orbits to a pair of matrix eigenvalue equations (?) that can be solved via standard numerical techniques (Press et al. 1992). Once we have determined the eigenfrequencies, g_l and f_l, and the corresponding eigenvectors, e_{il} and I_{il}, we find the phase angles β_l and γ_l by demanding that, at $t = 0$, Equations (9.71)–(9.74) lead to the values of e_i, I_i, ϖ_i, and Ω_i given in Table 3.1.

The theory outlined here is generally referred to as *Laplace-Lagrange secular evolution theory*. The eigenfrequencies, eigenvectors, and phase angles obtained from this theory are listed in Tables 9.1–9.3. Note that the largest eigenfrequency is of magnitude 25.90 arc seconds per year, which translates to an oscillation period of about 5×10^4 years. In other words, the typical timescale over which the secular evolution of the solar system predicted by Laplace-Lagrange theory takes place is at least 5×10^4 years, and is, therefore, much longer than the orbital period of any planet.

Figure 9.2 compares the observed perihelion and ascending node precession rates of the Planets at $t = 0$ (which corresponds to the epoch J2000) with those calculated from the theory described previously. It can be seen that, generally, there is good agreement between the theoretical and observed precession rates, which gives us some degree of confidence in the theory. On the whole, the degree of agreement exhibited in the left-hand panel of Figure 9.2 is better than that exhibited in Figure 4.1, indicating that the Laplace-Lagrange secular evolution theory described in this chapter is an improvement on the (highly simplified) Gaussian secular evolution theory outlined in Section 4.4.

Observe that one of the inclination eigenfrequencies, f_5, takes the value zero. This is a consequence of the conservation of angular momentum. Because the solar system is effectively an isolated dynamical system, its net angular momentum vector, $\mathbf{L}$, is constant in both magnitude and direction. The plane normal to $\mathbf{L}$ that passes through

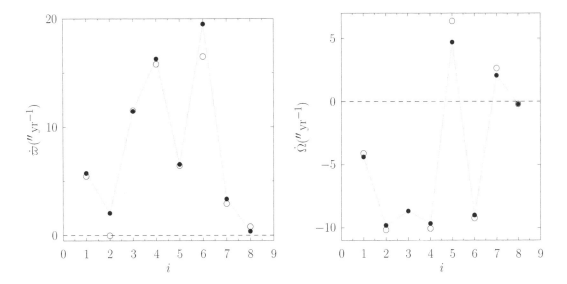

Fig. 9.2 The filled circles show the observed planetary perihelion precession rates (left-hand panel) and ascending node precession rates (right-hand panel) at J2000. All ascending nodes are measured relative to the mean ecliptic at J2000. The empty circles show the theoretical precession rates calculated from Laplace-Lagrange secular evolution theory. Source (for observational data): Standish and Williams 1992.

the center of mass of the solar system (which lies very close to the Sun) is known as the *invariable plane*. If all the planetary orbits were to lie in the invariable plane, the net angular velocity vector of the solar system would be parallel to its fixed net angular momentum vector. Moreover, the angular momentum vector would be parallel to one of the solar system's principal axes of rotation. In this situation, we would expect the Solar System to remain in the invariable plane. (See Chapter 7.) In other words, we would not expect any time evolution of the planetary inclinations. (Of course, lack of time variation implies an eigenfrequency of zero.) According to Equations (9.73) and (9.74), and the data shown in Tables 9.1 and 9.3, if the solar system were in the inclination eigenstate associated with the null eigenfrequency, f_5, then we would have

$$p_i = 2.751 \times 10^{-2} \sin \gamma_5 \tag{9.77}$$

and

$$q_i = 2.751 \times 10^{-2} \cos \gamma_5, \tag{9.78}$$

for $i = 1, 8$. Because $p_i = \sin I_i \sin \Omega_i$ and $q_i = \sin I_i \cos \Omega_i$, it follows that all the planetary orbits would lie in the same plane, and this plane—which is, of course, the invariable plane—is inclined at $I_5 = \sin^{-1}(2.751 \times 10^{-2}) = 1.576°$ to the ecliptic plane. Furthermore, the longitude of the ascending node of the invariable plane, with respect to the ecliptic plane, is $\Omega_5 = \gamma_5 = 107.5°$. Actually, it is generally more convenient to measure the inclinations of the planetary orbits with respect to the invariable plane, rather than the ecliptic plane, as the inclination of the latter plane varies in time. We can achieve this goal by simply omitting the fifth inclination eigenstate when calculating orbital inclinations from Equations (9.73) and (9.74).

Table 9.1 Eigenfrequencies and phase angles obtained from Laplace-Lagrange secular evolution theory				
l	$g_l('' \text{ yr}^{-1})$	$f_l('' \text{ yr}^{-1})$	$\beta_l(°)$	$\gamma_l(°)$
1	5.462	−5.201	89.65	20.23
2	7.346	−6.570	195.0	318.3
3	17.33	−18.74	336.1	255.6
4	18.00	−17.64	319.0	296.9
5	3.724	0.000	30.12	107.5
6	22.44	−25.90	131.0	127.3
7	2.708	−2.911	109.9	315.6
8	0.6345	−0.6788	67.98	202.8

Table 9.2 Components of eccentricity eigenvectors e_{il} obtained from Laplace-Lagrange secular evolution theory. All components multiplied by 10^5							
$i = 1$	2	3	4	5	6	7	8
$l = 1$	18128	629	404	66	0	0	0	0
2	−2331	1919	1497	265	−1	−1	0	0
3	154	−1262	1046	2979	0	0	0	0
4	−169	1489	−1485	7281	0	0	0	0
5	2446	1636	1634	1878	4331	3416	−4388	159
6	10	−51	242	1562	−1560	4830	−180	−13
7	59	58	62	82	207	189	2999	−322
8	0	1	1	2	6	6	144	954

Table 9.3 Components of inclination eigenvectors I_{il} obtained from Laplace-Lagrange secular evolution theory. All components multiplied by 10^5							
$i = 1$	2	3	4	5	6	7	8
$l = 1$	12548	1180	850	180	−2	−2	2	0
2	−3548	1006	811	180	−1	−1	0	0
3	409	−2684	2446	−3595	0	0	0	0
4	116	−685	451	5021	0	−1	0	0
5	2751	2751	2751	2751	2751	2751	2751	2751
6	27	14	279	954	−636	1587	−69	−7
7	−333	−191	−173	−125	−95	−77	1757	−206
8	−144	−132	−129	−122	−116	-112	109	1181

Planet	e_{min}	e_{max}	$\langle\varpi\rangle$	$I_{min}(°)$	$I_{max}(°)$	$\langle\dot{\Omega}\rangle$
Mercury	0.130	0.233	g_1	4.57	9.86	f_1
Venus	0.000	0.0705	–	0.000	3.38	–
Earth	0.000	0.0638	–	0.000	2.95	–
Mars	0.0444	0.141	g_4	0.000	5.84	–
Jupiter	0.0256	0.0611	g_5	0.241	0.489	f_6
Saturn	0.0121	0.0845	g_6	0.797	1.02	f_6
Uranus	0.0106	0.0771	g_5	0.902	1.11	f_7
Neptune	0.00460	0.0145	g_8	0.554	0.800	f_8

Table 9.4 Maximum/minimum eccentricities and inclinations of planetary orbits, and mean perihelion/nodal precession rates, from Laplace-Lagrange secular evolution theory. All inclinations relative to invariable plane

Consider the ith planet. Suppose one of the e_{il} coefficients—say, e_{ik}—is sufficiently large that

$$|e_{ik}| > \sum_{\substack{l=1,8 \\ l\neq k}} |e_{il}|. \tag{9.79}$$

This is known as the *Lagrange condition* (Hagihara 1971). As can be demonstrated, if the Lagrange condition is satisfied, the eccentricity of the ith planet's orbit varies between the minimum value,

$$e_{i\,min} = |e_{ik}| - \sum_{\substack{l=1,8 \\ l\neq k}} |e_{il}|, \tag{9.80}$$

and the maximum value,

$$e_{i\,max} = |e_{ik}| + \sum_{\substack{l=1,8 \\ l\neq k}} |e_{il}|. \tag{9.81}$$

Moreover, on average, the ith planet's perihelion point precesses at the associated eigenfrequency, g_k. The precession is prograde (i.e., in the same direction as the orbital motion) if the frequency is positive, and retrograde (i.e., in the opposite direction) if the frequency is negative. If the Lagrange condition is not satisfied, all we can say is that the maximum eccentricity is given by Equation (9.81), and there is no minimum eccentricity (i.e., the eccentricity can vary all the way down to zero). Furthermore, no mean precession rate of the perihelion point can be identified. It can be seen from Table 9.2 that the Lagrange condition for the orbital eccentricities is satisfied for all planets except Venus and Earth. The maximum and minimum eccentricities, and mean perihelion precession rates, of the planets (when they exist) are given in Table 9.4. Note that Jupiter and Uranus have the same mean perihelion precession rates, and that all planets that possess mean precession rates exhibit prograde precession.

There is also a Lagrange condition associated with the inclinations of the planetary orbits (Hagihara 1971). This condition is satisfied for the ith planet if one of the I_{il}—say,

I_{ik}—is sufficiently large that

$$|I_{ik}| > \sum_{l=1,8}^{l \neq k, l \neq 5} |I_{il}|. \tag{9.82}$$

The fifth inclination eigenmode is omitted from this summation because we are now measuring inclinations relative to the invariable plane. If the Lagrange condition is satisfied, the inclination of the ith planet's orbit with respect to the invariable plane varies between the minimum value

$$I_{i\,\min} = \sin^{-1}\left(|I_{ik}| - \sum_{l=1,8}^{l \neq k, l \neq 5} |I_{il}|\right) \tag{9.83}$$

and the maximum value

$$I_{i\,\max} = \sin^{-1}\left(|I_{ik}| + \sum_{l=1,8}^{l \neq k, l \neq 5} |I_{il}|\right). \tag{9.84}$$

Moreover, on average, the ascending node precesses at the associated eigenfrequency, f_k. The precession is prograde (i.e., in the same direction as the orbital motion) if the frequency is positive, and retrograde (i.e., in the opposite direction) if the frequency is negative. If the Lagrange condition is not satisfied, all we can say is that the maximum inclination is given by Equation (9.84), and there is no minimum inclination (i.e., the inclination can vary all the way down to zero). Furthermore, no mean precession rate of the ascending node can be identified. It can be seen from Table 9.3 that the Lagrange condition for the orbital inclinations is satisfied for all planets except Venus, Earth, and Mars. The maximum and minimum inclinations, and mean nodal precession rates, of the planets (when they exist) are given in Table 9.4. The four outer planets, which possess most of the mass of the solar system, all have orbits whose inclinations to the invariable plane remain small. On the other hand, the four relatively light inner planets have orbits whose inclinations to the invariable plane can become relatively large. Observe that Jupiter and Saturn have the same mean nodal precession rates, and that all planets that possess mean precession rates exhibit retrograde precession.

Figure 9.3 shows the time variation of the eccentricity, inclination, perihelion precession rate, and ascending node precession rate of Mercury, as predicted by the Laplace-Lagrange secular perturbation theory described earlier. It can be seen that the eccentricity and inclination do indeed oscillate between the upper and lower bounds specified in Table 9.4. Moreover, the perihelion and ascending node precession rates do appear to oscillate about the mean values (g_1 and f_1, respectively) specified in the same table.

According to Laplace-Lagrange secular perturbation theory, the mutual gravitational interactions between the various planets in the solar system cause their orbital eccentricities and inclinations to oscillate between *fixed bounds* on timescales that are long compared with their orbital periods. Recall, however, that these results depend on a great many approximations: the neglect of all nonsecular terms in the planetary disturbing functions, and the neglect of secular terms beyond first order in the planetary masses and beyond second order in the orbital eccentricities and inclinations. It turns out that when the neglected terms are included in the analysis, the largest correction to standard

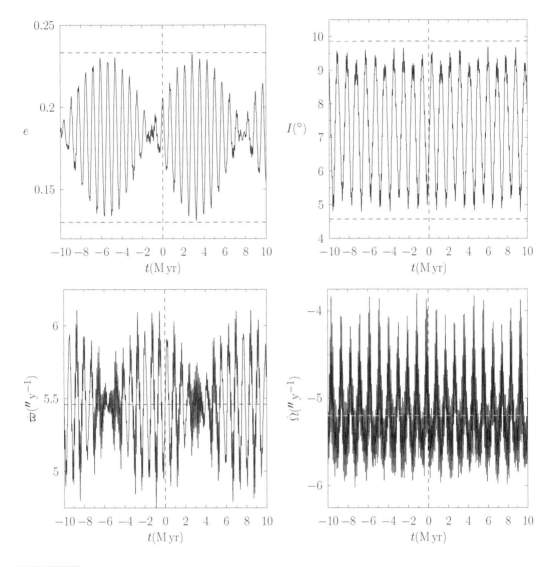

Fig. 9.3 Time variation of the eccentricity (top left), inclination (top right), perihelion precession rate (bottom left), and ascending node precession rate (bottom right) of Mercury, as predicted by Laplace-Lagrange secular perturbation theory. Time is measured in millions of years relative to J2000. All inclinations are relative to the invariable plane. The horizontal dashed lines in the top panels indicate the predicted minimum and maximum eccentricities and inclinations from Table 9.4. The horizontal dashed lines in the bottom panels indicate the predicted mean perihelion and ascending node precession rates from the same table.

Laplace-Lagrange theory is a second-order (in the planetary masses) effect caused by periodic terms in the disturbing functions of Jupiter and Saturn that oscillate on a relatively long timescale (i.e., almost 900 years), because the orbital periods of these two planets are almost commensurable (i.e., five times the orbital period of Jupiter is almost equal to two times the orbital period of Saturn). In 1950, Brouwer and van Woerkom worked out a modified version of Laplace-Lagrange secular perturbation theory that

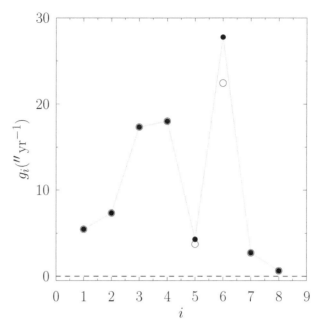

Fig. 9.4 The eccentricity eigenfrequencies obtained from Brouwer and van Woerkom's refinement of standard Laplace-Lagrange secular evolution theory (filled circles), compared with the corresponding eigenfrequencies from Table 9.1 (open circles).

takes the aforementioned correction into account (Brouwer and van Woerkom 1950). This refined secular evolution theory is described, in detail, in Murray and Dermott (1999). As is illustrated in Figure 9.4, the values of the eccentricity eigenfrequencies g_5 and g_6 in the Brouwer-van Woerkom theory differ somewhat from those specified in Table 9.1. The corresponding eigenvectors are also somewhat modified. The Brouwer-van Woerkom theory also contains two additional, relatively small-amplitude, eccentricity eigenmodes that oscillate at the eigenfrequencies $g_9 = 2\,g_5 - g_6$ and $g_{10} = 2\,g_6 - g_5$. On the other hand, the Brouwer-van Woerkom theory does not give rise to any significant modifications to the inclination eigenmodes. Figure 9.5 shows the maximum and minimum orbital eccentricities predicted by the Brouwer-van Woerkom theory, compared with the corresponding limits from Table 9.4. It can be seen that the refinements introduced by Brouwer and van Woerkom modify the oscillation limits for the orbital eccentricities of Mercury and Uranus somewhat, but do not significantly change the limits for the other planets. Of course, the oscillation limits for the orbital inclinations are unaffected by these refinements (because the inclination eigenmodes are unaffected).

It must be emphasized that the Brouwer-van Woerkom secular evolution theory is only approximate in nature. In fact, the theory is capable of predicting the secular evolution of the solar system with reasonable accuracy up to a million or so years into the future or the past. However, over longer timescales, it becomes inaccurate because the true long-term dynamics of the solar system contain *chaotic* elements. These elements originate from two secular resonances among the planets: $(g_4 - g_3) - (f_4 - f_3) \simeq 0$, which is

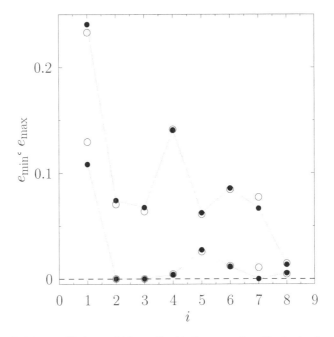

The maximum and minimum orbital eccentricities predicted by Brouwer and van Woerkom's refinement of standard Laplace-Lagrange secular evolution theory (filled circles), compared with the corresponding limits from Table 9.4 (open circles).

related to the gravitational interaction of Mars and the Earth; and $(g_1 - g_5) - (f_1 - f_2) \simeq 0$, which is related to the interaction of Mercury, Venus, and Jupiter (Laskar 1990).

9.4 Secular evolution of asteroid orbits

Let us now consider the perturbing influence of the planets on the orbit of an asteroid. Because asteroids have much smaller masses than planets, it is reasonable to suppose that the perturbing influence of the asteroid on the planetary orbits is negligible. Let the asteroid have the standard osculating orbital elements a, $\bar{\lambda}_0$, e, I, ϖ, Ω, and the alternative elements $h = e \sin \varpi$, $k = e \cos \varpi$, $p = \sin I \sin \Omega$, and $q = \sin I \cos \Omega$. Thus, the mean orbital angular velocity of the asteroid is $n = (G M/a^3)^{1/2}$, where M is the solar mass. Likewise, let the eight planets have the standard osculating orbital elements a_i, $\bar{\lambda}_{0i}$, e_i, I_i, ϖ_i, Ω_i, and the alternative elements $h_i = e_i \sin \varpi_i$, $k_i = e_i \cos \varpi_i$, $p_i = \sin I_i \sin \Omega_i$, and $q_i = \sin I_i \cos \Omega_i$, for $i = 1, 8$. It is helpful to define the following parameters:

$$\alpha_i = \begin{cases} a/a_i & a_i > a \\ a_i/a & a_i < a \end{cases} \tag{9.85}$$

and

$$\bar{\alpha}_i = \begin{cases} a/a_i & a_i > a \\ 1 & a_i < a \end{cases}, \tag{9.86}$$

as well as

$$\epsilon_i = \frac{m_i}{M}. \tag{9.87}$$

By analogy with the analysis in the previous section, the secular terms in the disturbing function of the asteroid, generated by the perturbing influence of the planets, cause the asteroid's osculating orbital elements to evolve in time as

$$\frac{dh}{dt} = A\,k + \sum_{i=1,8} A_i\,k_i, \tag{9.88}$$

$$\frac{dk}{dt} = -A\,h - \sum_{i=1,8} A_i\,h_i, \tag{9.89}$$

$$\frac{dp}{dt} = B\,q + \sum_{i=1,8} B_i\,q_i, \tag{9.90}$$

and

$$\frac{dq}{dt} = -B\,p - \sum_{i=1,8} B_i\,p_i, \tag{9.91}$$

where

$$A = \sum_{i=1,8} \frac{n}{4}\,\epsilon_i\,\alpha_i\,\bar{\alpha}_i\,b_{3/2}^{(1)}(\alpha_i), \tag{9.92}$$

$$A_i = -\frac{n}{4}\,\epsilon_i\,\alpha_i\,\bar{\alpha}_i\,b_{3/2}^{(2)}(\alpha_i), \tag{9.93}$$

$$B = -\sum_{i=1,8} \frac{n}{4}\,\epsilon_i\,\alpha_i\,\bar{\alpha}_i\,b_{3/2}^{(1)}(\alpha_i), \tag{9.94}$$

and

$$B_i = \frac{n}{4}\,\epsilon_i\,\alpha_i\,\bar{\alpha}_i\,b_{3/2}^{(1)}(\alpha_i). \tag{9.95}$$

However, as we have already seen, the planetary osculating elements themselves evolve in time as

$$h_i = \sum_{l=1,8} e_{il}\,\sin(g_l\,t + \beta_l), \tag{9.96}$$

$$k_i = \sum_{l=1,8} e_{il}\,\cos(g_l\,t + \beta_l), \tag{9.97}$$

$$p_i = \sum_{l=1,8} I_{il}\,\sin(f_l\,t + \gamma_l), \tag{9.98}$$

and

$$q_i = \sum_{l=1,8} I_{il}\,\cos(f_l\,t + \gamma_l). \tag{9.99}$$

Equations (9.88)–(9.91) can be solved to give

$$h(t) = e_{\text{free}} \sin(A\,t + \beta_{\text{free}}) + h_{\text{forced}}(t), \tag{9.100}$$

$$k(t) = e_{\text{free}} \cos(A\,t + \beta_{\text{free}}) + k_{\text{forced}}(t), \tag{9.101}$$

$$p(t) = \sin I_{\text{free}} \sin(B\,t + \gamma_{\text{free}}) + p_{\text{forced}}(t), \tag{9.102}$$

and

$$q(t) = \sin I_{\text{free}} \cos(B\,t + \gamma_{\text{free}}) + q_{\text{forced}}(t), \tag{9.103}$$

where

$$h_{\text{forced}} = -\sum_{l=1,8} \frac{v_l}{A - g_l} \sin(g_l\,t + \beta_l), \tag{9.104}$$

$$k_{\text{forced}} = -\sum_{l=1,8} \frac{v_l}{A - g_l} \cos(g_l\,t + \beta_l), \tag{9.105}$$

$$p_{\text{forced}} = -\sum_{l=1,8} \frac{\mu_l}{B - f_l} \sin(f_l\,t + \gamma_l), \tag{9.106}$$

and

$$q_{\text{forced}} = -\sum_{l=1,8} \frac{\mu_l}{B - f_l} \cos(f_l\,t + \gamma_l), \tag{9.107}$$

as well as

$$v_l = \sum_{i=1,8} A_i\,e_{il} \tag{9.108}$$

and

$$\mu_l = \sum_{i=1,8} B_i\,I_{il}. \tag{9.109}$$

The parameters e_{free} and I_{free} appearing in Equations (9.100)–(9.103) are the eccentricity and inclination, respectively, that the asteroid orbit would possess were it not for the perturbing influence of the planets. These parameters are usually called the *free*, or *proper*, eccentricity and inclination, respectively. Roughly speaking, the planetary perturbations cause the osculating eccentricity, $e = (h^2 + k^2)^{1/2}$, and inclination, $I = \sin^{-1}\{[p^2 + q^2]^{1/2}\}$, to oscillate about the corresponding free quantities, e_{free} and I_{free}, respectively.

Figure 9.6 shows the osculating eccentricity plotted against the sine of the osculating inclination for the orbits of the first 100,000 numbered asteroids (asteroids are numbered in order of their discovery). No particular patten is apparent. Figures 9.7 and 9.8 show the *free* eccentricity plotted against the sine of the *free* inclination for the same 100,000 orbits. In Figure 9.7, the free orbital elements are determined from standard Laplace-Lagrange secular evolution theory, whereas in Figure 9.8 they are determined from Brouwer and van Woerkom's refinement of this theory. It can be seen that many of the points representing the asteroid orbits have condensed into clumps. These clumps, which are somewhat clearer in Figure 9.8 than in Figure 9.7, are known as *Hirayama*

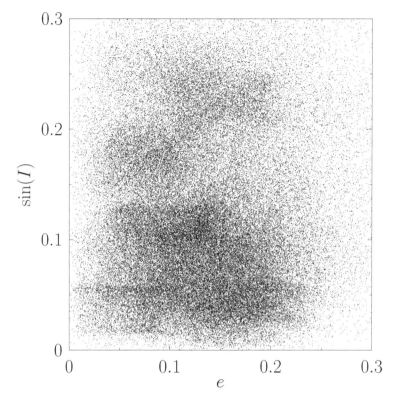

Fig. 9.6   The osculating eccentricity plotted against the sine of the osculating inclination (relative to J2000 equinox and ecliptic) for the orbits of the first 100,000 numbered asteroids at MJD 55400. Raw data from JPL Small-Body Database.

families after their discoverer, the Japanese astronomer Kiyotsugu Hirayama (1874–1943). It is thought that the asteroids making up a given family had a common origin—most likely due to the break up of some much larger body (Bertotti et al. 2003). As a consequence of this origin, the asteroids originally had similar orbital elements. However, as time progressed, these elements were jumbled by the perturbing influence of the planets. Thus, only when this influence is removed does the commonality of the orbits becomes apparent. Hirayama families are named after their largest member. The most prominent families are the (4) Vesta, (15) Eunomia, (24) Themis, (44) Nysa, (158) Koronis, (221) Eos, and (1272) Gefion families. (The number in parentheses is that of the corresponding asteroid.)

9.5 Secular evolution of artificial satellite orbits

Consider a nonrotating (with respect to the distant stars) frame of reference whose origin coincides with the center of the Earth. We can regard such a frame as approximately inertial when we consider orbital motion in the Earth's immediate vicinity. Let X, Y, Z be a Cartesian coordinate system in the said reference frame that is oriented such that

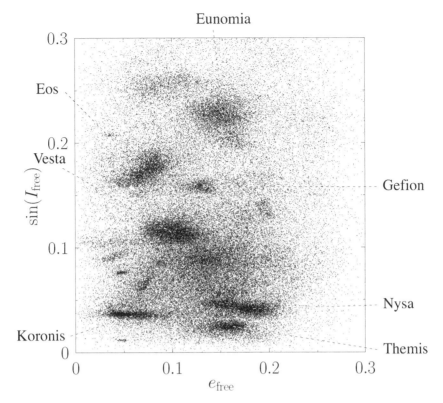

Fig. 9.7   The free eccentricity plotted against the sine of the free inclination (relative to J2000 equinox and ecliptic) for the
orbits of the first 100,000 numbered asteroids at MJD 55400. The free orbital elements are determined from
standard Laplace-Lagrange secular evolution theory. The most prominent Hirayama families are labeled.
Raw data from JPL Small-Body Database.

its Z-axis is aligned with the Earth's (approximately) constant axis of rotation (with the
terrestrial north pole lying at positive Z). As we saw in Section 7.9, the gravitational
potential in the immediate vicinity of the Earth can be written

$$\Phi(r, \vartheta) \simeq -\frac{G\,M}{r} + \frac{G\,(\mathcal{I}_{\parallel} - \mathcal{I}_{\perp})}{r^3}\,P_2(\cos\vartheta), \qquad (9.110)$$

where M is the Earth's mass, $\mathcal{I}_{\parallel}$ its moment of inertia about the Z-axis, and $\mathcal{I}_{\perp}$ its
moment of inertia about an axis lying in the X–Y plane. Here, $r = (X^2 + Y^2 + Z^2)^{1/2}$
and $\vartheta = \cos^{-1}(Z/r)$ are standard spherical coordinates. The first term on the right-
hand side of the preceding expression is the monopole gravitational potential that would
result were the Earth spherically symmetric. The second term is the small quadrupole
correction to this potential generated by the Earth's slight oblateness; see Section 5.5.
It is conventional to parameterize this correction in terms of the dimensionless quantity
(Yoder 1995)

$$J_2 = \frac{\mathcal{I}_{\parallel} - \mathcal{I}_{\perp}}{M\,R^2} = 1.083 \times 10^{-3}, \qquad (9.111)$$

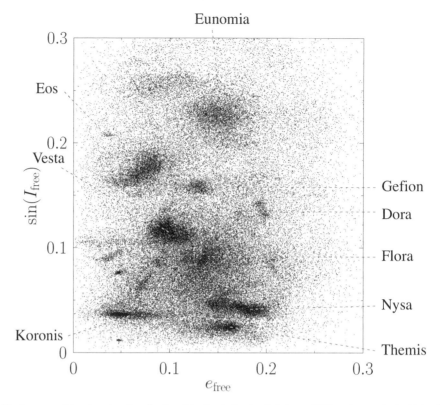

Fig. 9.8 The free eccentricity plotted against the sine of the free inclination (relative to J2000 equinox and ecliptic) for the orbits of the first 100,000 numbered asteroids at MJD 55400. The free orbital elements are determined from Brouwer and van Woerkom's improved secular evolution theory. The most prominent Hirayama families are labeled. Raw data from JPL Small-Body Database.

where R is the Earth's equatorial radius. Hence, Equation (9.110) can be written

$$\Phi(r, \vartheta) \simeq -\frac{G M}{r} + J_2 \frac{G M R^2}{r^3} P_2(\cos \vartheta). \tag{9.112}$$

Consider an artificial satellite in orbit around the Earth. The satellite's equation of motion in our approximately inertial geocentric reference frame takes the form

$$\ddot{\mathbf{r}} = -\nabla \Phi. \tag{9.113}$$

This can be combined with Equation (9.112) to give

$$\ddot{\mathbf{r}} + \mu \frac{\mathbf{r}}{r^3} = \nabla \mathcal{R}, \tag{9.114}$$

where $\mu = G M$, and

$$\mathcal{R}(r, \vartheta) = -J_2 \frac{\mu R^2}{r^3} P_2(\cos \vartheta). \tag{9.115}$$

Note that the preceding expression has exactly the same form as the canonical equation of motion, Equation (B.2), that is the starting point for orbital perturbation theory.

In particular, $\mathcal{R}(r, \vartheta)$ is the disturbing function that describes the perturbation to the Keplerian orbit of the satellite due to the Earth's small quadrupole gravitational field.

Let the satellite's osculating orbital elements be the major radius, a; the time of perigee passage, τ; the eccentricity, e; the inclination (to the Earth's equatorial plane), I; the argument of the perigee, ω; and the longitude of the ascending node, Ω. (See Section 3.12.) Actually, it is more convenient to replace τ by the mean anomaly, $\mathcal{M} = n(t - \tau)$, where $n = (\mu/a^3)^{1/2}$ is the (unperturbed) mean orbital angular velocity. According to standard orbital perturbation theory, the time evolution of the satellite's orbital elements is governed by the *Lagrange planetary equations*, which, for the particular set of elements under consideration, take the form (see Section B.6)

$$\frac{da}{dt} = \frac{2}{n\,a}\frac{\partial \mathcal{R}}{\partial \mathcal{M}}, \tag{9.116}$$

$$\frac{d\mathcal{M}}{dt} = n - \frac{2}{n\,a}\frac{\partial \mathcal{R}}{\partial a} - \frac{1-e^2}{n\,a^2\,e}\frac{\partial \mathcal{R}}{\partial e}, \tag{9.117}$$

$$\frac{de}{dt} = \frac{1-e^2}{n\,a^2\,e}\frac{\partial \mathcal{R}}{\partial \mathcal{M}} - \frac{(1-e^2)^{1/2}}{n\,a^2\,e}\frac{\partial \mathcal{R}}{\partial \omega}, \tag{9.118}$$

$$\frac{dI}{dt} = \frac{\cot I}{n\,a^2\,(1-e^2)^{1/2}}\frac{\partial \mathcal{R}}{\partial \omega} - \frac{(1-e^2)^{-1/2}}{n\,a^2\,\sin I}\frac{\partial \mathcal{R}}{\partial \Omega}, \tag{9.119}$$

$$\frac{d\omega}{dt} = \frac{(1-e^2)^{1/2}}{n\,a^2\,e}\frac{\partial \mathcal{R}}{\partial e} - \frac{\cot I}{n\,a^2\,(1-e^2)^{1/2}}\frac{\partial \mathcal{R}}{\partial I}, \tag{9.120}$$

and

$$\frac{d\Omega}{dt} = \frac{(1-e^2)^{-1/2}}{n\,a^2\,\sin I}\frac{\partial \mathcal{R}}{\partial I}. \tag{9.121}$$

According to Equation (3.74),

$$\cos \vartheta = \frac{Z}{r} = \sin(\omega + \theta)\,\sin I, \tag{9.122}$$

where θ is the satellite's true anomaly. Thus, Equation (9.115) can be written

$$\mathcal{R} = \frac{J_2}{2}\frac{\mu R^2}{r^3}\left[1 - \frac{3}{2}\,\sin^2 I + \frac{3}{2}\,\sin^2 I\,\cos(2\,\omega + 2\,\theta)\right]. \tag{9.123}$$

We are interested primarily in the *secular* evolution of the satellite's orbital elements, that is, the evolution on timescales much longer than the orbital period. We can concentrate on this evolution, and filter out any relatively short-term oscillations in the elements, by averaging the disturbing function over an orbital period. In other words, in Equations (9.116)–(9.121), we need to replace $\mathcal{R}$ by

$$\overline{\mathcal{R}} = \frac{1}{T}\int_0^T \mathcal{R}\,dt = \frac{n}{h}\oint r^2\,\mathcal{R}\,\frac{d\theta}{2\pi} = (1-e^2)^{-1/2}\oint \frac{r^2}{a^2}\,\mathcal{R}\,\frac{d\theta}{2\pi}, \tag{9.124}$$

where $T = 2\pi/n$. Here, we have made use of the fact that $r^2\,\dot{\theta} = h = (1-e^2)^{1/2}\,n\,a^2$ is a constant of the motion in a Keplerian orbit. (See Chapter 3.) A Keplerian orbit is also characterized by $r/a = (1-e^2)(1 + e\,\cos\theta)^{-1}$. Hence, the previous two equations can be combined to give

$$\overline{\mathcal{R}} = \frac{J_2}{2} \frac{\mu R^2}{a^3} (1 - e^2)^{-3/2} \oint (1 + e \cos\theta) \left[1 - \frac{3}{2} \sin^2 I + \frac{3}{2} \sin^2 I \cos(2\omega + 2\theta) \right] \frac{d\theta}{2\pi},$$
(9.125)

which evaluates to

$$\overline{\mathcal{R}} = \frac{J_2}{2} \frac{\mu R^2}{a^3} (1 - e^2)^{-3/2} \left(1 - \frac{3}{2} \sin^2 I \right).$$
(9.126)

Substitution of this expression into Equations (9.116), (9.118), and (9.119) (recalling that we are replacing $\mathcal{R}$ by $\overline{\mathcal{R}}$) reveals that there is no secular evolution of the satellite's orbital major radius, a, eccentricity, e, and inclination, I, due to the Earth's oblateness (because $\overline{\mathcal{R}}$ does not depend on $\mathcal{M}$, ω, or Ω). On the other hand, according to Equations (9.120) and (9.121), the oblateness causes the satellite's perigee and ascending node to process at the constant rates

$$\dot{\omega} = \frac{3 J_2}{4} n \frac{R^2}{a^2} \frac{(5 \cos^2 I - 1)}{(1 - e^2)^2}$$
(9.127)

and

$$\dot{\Omega} = -\frac{3 J_2}{2} n \frac{R^2}{a^2} \frac{\cos I}{(1 - e^2)^2},$$
(9.128)

respectively. These formulae suggest that the precession of the ascending node is always in the opposite sense to the orbital motion: that is, it is retrograde. Note, however, that the ascending node remains fixed in the special case of a so-called polar orbit that passes over the terrestrial poles (i.e., $I = \pi/2$). The formulae also suggest that the perigee precesses in a prograde fashion when $\cos I > 1/\sqrt{5}$, precesses in a retrograde fashion when $\cos I < 1/\sqrt{5}$, and remains fixed when $\cos I = 1/\sqrt{5}$. In other words, the perigee of an orbit lying in the Earth's equatorial plane precesses in the same direction as the orbital motion, the perigee of a polar orbit precesses in the opposite direction, and the perigee of an orbit that is inclined at the critical angle of $63.4°$ to the Earth's equatorial plane does not precess at all.

Exercises

9.1 Consider the secular evolution of two planets moving around a star in coplanar orbits of low eccentricity. Let a, e, and ϖ be the orbital major radius, eccentricity, and longitude of the periastron (i.e., the point of closest approach to the star) of the first planet, respectively, and let a', e', and ϖ' be the the corresponding parameters for the second planet. Suppose that $a' > a$. Let $h = e \sin\varpi$, $k = e \cos\varpi$, $h' = e' \sin\varpi'$, and $k' = e' \cos\varpi'$. Consider normal mode solutions of the two planets' secular evolution equations of the form $h(t) = \hat{e} \sin(g\,t + \beta)$, $k(t) = \hat{e} \cos(g\,t + \beta)$, $h'(t) = \hat{e}' \sin(g\,t + \beta)$, and $k'(t) = \hat{e}' \cos(g\,t + \beta)$, where $\hat{e}$, $\hat{e}'$, g, and β are constants.

Demonstrate that

$$\left(\begin{array}{cc} \hat{g} - q\,\alpha & q\,\alpha\,\beta \\ \alpha^{3/2}\,\beta & \hat{g} - \alpha^{3/2} \end{array} \right) \left(\begin{array}{c} \hat{e} \\ \hat{e}' \end{array} \right) = \left(\begin{array}{c} 0 \\ 0 \end{array} \right),$$

where $\hat{g} = g/[(1/4)\,\epsilon\,n\,\alpha\,b_{3/2}^{(1)}(\alpha)]$, $\epsilon = m/M$, $n = (G\,M/a^3)^{1/2}$, $\alpha = a/a'$, $q = m'/m$, and $\beta = b_{3/2}^{(2)}(\alpha)/b_{3/2}^{(1)}(\alpha)$. Here, m, m', and M are the masses of the first planet, second planet, and star, respectively. It is assumed that $M \gg m, m'$. Hence, deduce that the general time variation of the osculating orbital elements $h(t)$, $k(t)$, $h'(t)$, and $k'(t)$ is a linear combination of two normal modes of oscillation, which are characterized by

$$\hat{g} = \frac{1}{2}\left\{ q\,\alpha + \alpha^{3/2} \pm \left[(q\,\alpha - \alpha^{3/2})^2 + 4\,q\,\alpha^{5/2}\,\beta^2 \right]^{1/2} \right\},$$

and

$$\frac{\hat{e}'}{\hat{e}} = \frac{q\,\alpha - \hat{g}}{q\,\alpha\,\beta} = \frac{\alpha^{3/2}\,\beta}{\alpha^{3/2} - \hat{g}}.$$

Demonstrate that in the limit $\alpha \ll 1$, in which $b_{3/2}^{(1)}(\alpha) \simeq 3\,\alpha$ and $b_{3/2}^{(2)}(\alpha) \simeq (15/4)\,\alpha^2$, the first normal mode is such that $g \simeq (3/4)\,\epsilon'\,n\,\alpha^3$ and $\hat{e}'/\hat{e} \simeq -(5/4)\,\alpha^{3/2}/q$ (assuming that $q \gg \alpha^{1/2}$), whereas the second mode is such that $g \simeq (3/4)\,\epsilon\,n'\,\alpha^2$ and $\hat{e}'/\hat{e} \simeq (4/5)\,\alpha^{-1}$. Here, $\epsilon' = m'/M$ and $n' = (G\,M/a'^3)^{1/2}$. (Modified from Murray and Dermott 1999.)

9.2 The gravitational potential of the Sun in the vicinity of the planet Mercury can be written

$$\Phi(r) = -\frac{G\,M}{r} - \frac{G\,M\,h^2}{c^2\,r^3},$$

where M is the mass of the Sun, r the radial distance of Mercury from the center of the Sun, h the conserved angular momentum per unit mass of Mercury, and c the velocity of light in vacuum. The second term on the right-hand side of the preceding expression comes from a small general relativistic correction to Newtonian gravity (Rindler 1977). Show that Mercury's equation of motion can be written in the standard form

$$\ddot{\mathbf{r}} + \mu\,\frac{\mathbf{r}}{r^3} = \nabla\mathcal{R},$$

where $\mu = G\,M$, and

$$\mathcal{R} = \frac{\mu\,h^2}{c^2\,r^3}$$

is the disturbing function due to the general relativistic correction. Demonstrate that when the disturbing function is averaged over an orbital period it becomes

$$\overline{\mathcal{R}} = \frac{\mu\,h^2}{c^2\,a^3\,(1 - e^2)^{3/2}},$$

where a and e are the major radius and eccentricity, respectively, of Mercury's orbit. Hence, deduce from Lagrange's planetary equations that the general relativistic correction causes the argument of the perigee of Mercury's orbit to precess at the rate

$$\dot\omega = \frac{3\,\mu\,h^2}{c^2\,n\,a^5\,(1-e^2)^2} = \frac{3\,\mu^{3/2}}{c^2\,a^{5/2}\,(1-e^2)},$$

where n is Mercury's mean orbital angular velocity. Finally, show that the preceding expression evaluates to $0.43''\,\mathrm{yr}^{-1}$.

9.3 The gravitational potential in the immediate vicinity of the Earth can be written

$$\Phi(r,\vartheta) = -\frac{\mu}{r}\left[1 - J_2\left(\frac{R}{r}\right)^2 P_2(\cos\vartheta) - J_3\left(\frac{R}{r}\right)^3 P_3(\cos\vartheta) + \cdots\right],$$

where $\mu = GM$, M is the terrestrial mass, r, ϑ, ϕ are spherical coordinates that are centered on the Earth and aligned with its axis of rotation, R is the Earth's equatorial radius, and $J_2 = 1.083 \times 10^{-3}$, $J_3 = -2.112 \times 10^{-6}$ (Yoder 1995). In the preceding expression, the term involving J_2 is caused by the Earth's small oblateness, and the term involving J_3 is caused by the Earth's slightly asymmetric mass distribution between its northern and southern hemispheres. Consider an artificial satellite in orbit around the Earth. Let a, e, I, and ω be the orbital major radius, eccentricity, inclination (to the Earth's equatorial plane), and argument of the perigee, respectively. Furthermore, let $n = (\mu/a^3)^{1/2}$ be the unperturbed mean orbital angular velocity.

Demonstrate that, when averaged over an orbital period, the disturbing function due to the J_3 term takes the form

$$\overline{\mathcal{R}} = -\frac{3\,J_3}{2}\,\frac{\mu R^3}{a^4}\,\frac{e}{(1-e^2)^{5/2}}\left(\frac{5}{4}\sin^2 I - 1\right)\sin I\,\sin\omega.$$

Hence, deduce that the J_3 term causes the eccentricity and inclination of the satellite orbit to evolve in time as

$$\frac{de}{d(n\,t)} = -\frac{3\,J_3}{8}\left(\frac{R}{a}\right)^3\frac{(5\cos^2 I - 1)}{(1-e^2)^2}\sin I\,\cos\omega,$$

$$\frac{dI}{d(n\,t)} = \frac{3\,J_3}{8}\left(\frac{R}{a}\right)^3\frac{e\,(5\cos^2 I - 1)}{(1-e^2)^3}\cos I\,\cos\omega,$$

respectively. Given that the (much larger) J_2 term causes the argument of the perigee to precess at the approximately constant (assuming that the variations in e and I are small) rate

$$\frac{d\omega}{d(n\,t)} = \frac{3\,J_2}{4}\left(\frac{R}{a}\right)^2\frac{(5\cos^2 I - 1)}{(1-e^2)^2},$$

deduce that the variations in the orbital eccentricity and inclination induced by the J_3 term can be written

$$e \simeq e_0 - \frac{J_3}{2\,J_2}\frac{R}{a}\sin I_0\,\sin\omega,$$

$$I \simeq I_0 + \frac{J_3}{2\,J_2}\frac{R}{a}\frac{e_0}{1-e_0^2}\cos I_0\,\sin\omega,$$

respectively, where e_0 and I_0 are constants. (Modified from Murray and Dermott 1999.)

10 Lunar motion

10.1 Introduction

The orbital motion of the planets around the Sun is fairly accurately described by Kepler's laws. (See Chapter 3.) Similarly, to a first approximation, the orbital motion of the Moon around the Earth can also be accounted for via these laws. However, unlike the planetary orbits, the deviations of the lunar orbit from a Keplerian ellipse are sufficiently large that they are easily apparent to the naked eye. Indeed, the largest of these deviations, which is generally known as *evection*, was discovered in ancient times by the Alexandrian astronomer Claudius Ptolemy (90 CE–168 CE) (Pannekoek 2011). Moreover, the next largest deviation, which is called *variation*, was first observed by Tycho Brahe (1546–1601) without the aid of a telescope (Godfray 1853). Another non-Keplerian feature of the lunar orbit, which is sufficiently obvious that it was known to the ancient Greeks, is the fact that the lunar perigee (i.e., the point of closest approach to the Earth) precesses (i.e., orbits about the Earth in the same direction as the Moon) at such a rate that, on average, it completes a full circuit every 8.85 years.[1] The ancient Greeks also noticed that the lunar *ascending node* (i.e., the point at which the Moon passes through the fixed plane of the Earth's orbit around the Sun from south to north) regresses (i.e., orbits about the Earth in the opposite direction to the Moon) at such a rate that, on average, it completes a full circuit every 18.6 years (Pannekoek 2011). Of course, according to standard two-body orbit theory, the lunar perigee and ascending node should both be stationary. (See Chapter 3.)

Newton demonstrated, in Book III of his *Principia*, that the deviations of the lunar orbit from a Keplerian ellipse are due to the gravitational influence of the Sun, which is sufficiently large that it is not completely negligible compared with the mutual gravitational attraction of the Earth and the Moon. However, Newton was not able to give a full account of these deviations (in the *Principia*), because of the complexity of the equations of motion that arise in a system of three mutually gravitating bodies. (See Chapter 8.) In fact, Alexis Clairaut (1713–1765) is generally credited with the first reasonably accurate and complete theoretical explanation of the Moon's orbit to be published. His method of calculation makes use of an expansion of the lunar equations of motion in terms of various small parameters. Clairaut, however, initially experienced difficulty in accounting for the precession of the lunar perigee. Indeed, his first calculation overestimated the period of this precession by a factor of about two, leading him to question

---

[1] This precession rate is about 10^4 times greater than any of the planetary perihelion precession rates discussed in Sections 4.4 and 9.3.

Newton's inverse-square law of gravitation. Later, he realized that he could account for the precession in terms of standard Newtonian dynamics by continuing his expansion in small parameters to higher order. (See Section 10.6.) After Clairaut, the theory of lunar motion was further elaborated in major works by D'Alembert (1717–1783), Euler (1707–1783), Laplace (1749–1827), Damoiseau (1768–1846), Plana (1781–1864), Poisson (1781–1840), Hansen (1795–1874), De Pontécoulant (1795–1874), J. Herschel (1792–1871), Airy (1801–1892), Delaunay (1816–1872), G.W. Hill (1836–1914), and E.W. Brown (1836–1938) (Brown 1896). The fact that so many celebrated mathematicians and astronomers devoted so much time and effort to lunar theory is a tribute to its inherent difficulty, as well as its great theoretical and practical interest. Indeed, for a period of about one hundred years (between 1767 and about 1850) the so-called *method of lunar distance* was the principal means used by mariners to determine terrestrial longitude at sea (Cotter 1968). This method depends crucially on a precise knowledge of the position of the Moon in the sky as a function of time. Consequently, astronomers and mathematicians during the period in question were spurred to make ever more accurate observations of the Moon's orbit, and to develop lunar theory to greater and greater precision. An important outcome of these activities was the making of various tables of lunar motion (e.g., those of Mayer, Damoiseau, Plana, Hansen, and Brown), the majority of which were published at public expense.

   This chapter contains an introduction to lunar theory in which approximate expressions for evection, variation, the precession of the perigee, and the regression of the ascending node, are derived from the laws of Newtonian mechanics. Further information on lunar theory can be obtained from Godfray (1853), Brown (1896), Adams (1900), and Cook (1988).

10.2 Preliminary analysis

Let $\mathbf{r}_E$ and $\mathbf{r}_M$ denote the position vectors of the Earth and Moon, respectively, in a non-rotating reference frame in which the Sun is at rest at the origin. Treating this reference frame as inertial (which is an excellent approximation, given that the mass of the Sun is very much greater than that of the Earth or the Moon), the Earth's equation of motion becomes (see Chapter 3)

$$\ddot{\mathbf{r}}_E + n'^2 a'^3 \frac{\mathbf{r}_E}{|\mathbf{r}_E|^3} = 0, \tag{10.1}$$

where $n' = 0.98560912°$ per day and $a' = 149{,}598{,}261$ km are the mean angular velocity and major radius, respectively, of the terrestrial orbit about the Sun (Yoder 1995). Here, $\ddot{} \equiv d^2/dt^2$. On the other hand, the Moon's equation of motion takes the form

$$\ddot{\mathbf{r}}_M + n'^2 a'^3 \frac{\mathbf{r}_M}{|\mathbf{r}_M|^3} = -n^2 a^3 \frac{\mathbf{r}_M - \mathbf{r}_E}{|\mathbf{r}_M - \mathbf{r}_E|^3}, \tag{10.2}$$

where $n = 13.176359°$ per day and $a = 384{,}399$ km are the mean angular velocity and major radius, respectively, of the lunar orbit about the Earth (Yoder 1995). We have retained the acceleration due to the Earth in the lunar equation of motion, Equation (10.2), while neglecting the acceleration due to the Moon in the terrestrial equation of motion, Equation (10.1), because the former acceleration is significantly greater [by a factor $M_E/M_M \simeq 81.3$, where M_E is the mass of the Earth, and M_M the mass of the Moon (Yoder 1995)] than the latter.

Let

$$\mathbf{r} = \mathbf{r}_M - \mathbf{r}_E \tag{10.3}$$

and

$$\mathbf{r}' = -\mathbf{r}_E \tag{10.4}$$

be the position vectors of the Moon and Sun, respectively, relative to the Earth. It follows, from Equations (10.1)–(10.4), that in a noninertial reference frame, S (say), in which the Earth is at rest at the origin but the coordinate axes point in fixed directions, the lunar and solar equations of motion take the form

$$\ddot{\mathbf{r}} + n^2 a^3 \frac{\mathbf{r}}{|\mathbf{r}|^3} = n'^2 a'^3 \left(\frac{\mathbf{r}' - \mathbf{r}}{|\mathbf{r}' - \mathbf{r}|^3} - \frac{\mathbf{r}'}{|\mathbf{r}'|^3} \right) \tag{10.5}$$

and

$$\ddot{\mathbf{r}}' + n'^2 a'^3 \frac{\mathbf{r}'}{|\mathbf{r}'|^3} = 0, \tag{10.6}$$

respectively. One obvious way of proceeding would be to express the right-hand side of the lunar equation of motion, Equation (10.5), as the gradient of a disturbing function (see Exercise 10.1), and then to use this function to determine the time evolution of the Moon's osculating orbital elements from Lagrange's planetary equations. Unfortunately, this approach is fraught with mathematical difficulties. (See Brouwer and Clemence 1961.) It is actually more straightforward to solve Equation (10.5) in a Cartesian coordinate system, centered on the Earth, that rotates about an axis normal to the ecliptic plane at the Moon's mean orbital angular velocity. This method of solution is outlined as follows.

Let us set up a conventional Cartesian coordinate system in S that is such that the (apparent) orbit of the Sun about the Earth lies in the x–y plane. This implies that the x–y plane corresponds to the so-called ecliptic plane. Accordingly, in S, the Sun appears to orbit the Earth at the mean angular velocity $\boldsymbol{\omega}' = n'\, \mathbf{e}_z$ (assuming that the z-axis points toward the so-called north ecliptic pole), whereas the projection of the Moon onto the ecliptic plane orbits the Earth at the mean angular velocity $\boldsymbol{\omega} = n\, \mathbf{e}_z$.

In the following, for the sake of simplicity, we shall neglect the small eccentricity, $e' = 0.016711$, of the Sun's apparent orbit about the Earth (which is actually the eccentricity of the Earth's orbit about the Sun), and approximate the solar orbit as a circle, centered on the Earth. Thus, if x', y', z' are the Cartesian coordinates of the Sun in S, then an appropriate solution of the solar equation of motion, Equation (10.6), is

$$x' = a' \cos(n' t), \tag{10.7}$$

$$y' = a' \sin(n' t), \tag{10.8}$$

and

$$z' = 0. \tag{10.9}$$

10.3 Lunar equations of motion

As we have already mentioned, it is convenient to solve the lunar equation of motion, Equation (10.5), in a geocentric frame of reference, S_1 (say), that rotates with respect to S at the fixed angular velocity ω. Thus, if the lunar orbit were a circle, centered on the Earth and lying in the ecliptic plane, then the Moon would appear stationary in S_1. In fact, the small eccentricity of the lunar orbit, $e = 0.05488$, combined with its slight inclination to the ecliptic plane, $I = 5.16°$, causes the Moon to execute a small periodic orbit about the stationary point (Yoder 1995).

Let x, y, z and x_1, y_1, z_1 be the Cartesian coordinates of the Moon in S and S_1, respectively. It is easily demonstrated that (see Section A.6)

$$x = \cos(n\,t)\,x_1 - \sin(n\,t)\,y_1, \tag{10.10}$$

$$y = \sin(n\,t)\,x_1 + \cos(n\,t)\,y_1, \tag{10.11}$$

and

$$z = z_1. \tag{10.12}$$

Moreover, if x'_1, y'_1, z'_1 are the Cartesian components of the Sun in S_1, then (see Section A.6)

$$x'_1 = \cos(n\,t)\,x' + \sin(n\,t)\,y', \tag{10.13}$$

$$y'_1 = -\sin(n\,t)\,x' + \cos(n\,t)\,y', \tag{10.14}$$

and

$$z'_1 = z', \tag{10.15}$$

giving

$$x'_1 = a'\,\cos[(n - n')\,t], \tag{10.16}$$

$$y'_1 = -a'\,\sin[(n - n')\,t], \tag{10.17}$$

and

$$z'_1 = 0, \tag{10.18}$$

where use has been made of Equations (10.7)–(10.9).

In the rotating frame S_1, the lunar equation of motion, Equation (10.5), transforms to (see Section 5.2)

$$\ddot{\mathbf{r}} + 2\,\omega \times \dot{\mathbf{r}} + \omega \times (\omega \times \mathbf{r}) + n^2\,a^3\,\frac{\mathbf{r}}{|\mathbf{r}|^3} = n'^2\,a'^3\left(\frac{\mathbf{r}' - \mathbf{r}}{|\mathbf{r}' - \mathbf{r}|^3} - \frac{\mathbf{r}'}{|\mathbf{r}'|^3}\right), \tag{10.19}$$

where $\dot{} \equiv d/dt$. Furthermore, expanding the final term on the right-hand side of Equation (10.19) to lowest order in the small parameter $a/a' = 0.00257$, we obtain

$$\ddot{\mathbf{r}} + 2\,\omega \times \dot{\mathbf{r}} + \omega \times (\omega \times \mathbf{r}) + n^2\,a^3\,\frac{\mathbf{r}}{|\mathbf{r}|^3} \simeq \frac{n'^2\,a'^3}{|\mathbf{r}'|^3}\left[\frac{(3\,\mathbf{r}\cdot\mathbf{r}')\,\mathbf{r}'}{|\mathbf{r}'|^2} - \mathbf{r}\right]. \qquad (10.20)$$

When written in terms of Cartesian coordinates, this equation yields

$$\ddot{x}_1 - 2\,n\,\dot{y}_1 - \left(n^2 + n'^2/2\right)x_1 + n^2\,a^3\,\frac{x_1}{r^3} \simeq \frac{3}{2}\,n'^2\,\cos[2\,(n-n')\,t]\,x_1$$
$$- \frac{3}{2}\,n'^2\,\sin[2\,(n-n')\,t]\,y_1, \quad (10.21)$$

$$\ddot{y}_1 + 2\,n\,\dot{x}_1 - \left(n^2 + n'^2/2\right)y_1 + n^2\,a^3\,\frac{y_1}{r^3} \simeq -\frac{3}{2}\,n'^2\,\sin[2\,(n-n')\,t]\,x_1$$
$$- \frac{3}{2}\,n'^2\,\cos[2\,(n-n')\,t]\,y_1, \quad (10.22)$$

and

$$\ddot{z}_1 + n'^2\,z_1 + n^2\,a^3\,\frac{z_1}{r^3} \simeq 0, \qquad (10.23)$$

where $r = (x_1^2 + y_1^2 + z_1^2)^{1/2}$ and use has been made of Equations (10.16)–(10.18).

It is convenient, at this stage, to normalize all lengths to a, and all times to n^{-1}. Accordingly, let

$$X = x_1/a, \qquad (10.24)$$
$$Y = y_1/a, \qquad (10.25)$$

and

$$Z = z_1/a; \qquad (10.26)$$

let $r/a = R = (X^2 + Y^2 + Z^2)^{1/2}$, and $T = n\,t$. In normalized form, Equations (10.21)–(10.23) become

$$\ddot{X} - 2\,\dot{Y} - (1 + m^2/2)\,X + \frac{X}{R^3} \simeq \frac{3}{2}\,m^2\,\cos[2\,(1-m)\,T]\,X$$
$$- \frac{3}{2}\,m^2\,\sin[2\,(1-m)\,T]\,Y, \qquad (10.27)$$

$$\ddot{Y} + 2\,\dot{X} - (1 + m^2/2)\,Y + \frac{Y}{R^3} \simeq -\frac{3}{2}\,m^2\,\sin[2\,(1-m)\,T]\,X$$
$$- \frac{3}{2}\,m^2\,\cos[2\,(1-m)\,T]\,Y, \qquad (10.28)$$

and

$$\ddot{Z} + m^2\,Z + \frac{Z}{R^3} \simeq 0, \qquad (10.29)$$

respectively, where $m = n'/n = 0.07480$ is a measure of the perturbing influence of the Sun on the lunar orbit. Here, $\ddot{} \equiv d^2/dT^2$ and $\dot{} \equiv d/dT$.

Finally, let us write

$$X = X_0 + \delta X, \qquad (10.30)$$
$$Y = \delta Y, \qquad (10.31)$$

and

$$Z = \delta Z, \tag{10.32}$$

where $X_0 = (1 + m^2/2)^{-1/3}$, and $|\delta X|, |\delta Y|, |\delta Z| \ll X_0$. Thus, if the lunar orbit were a circle, centered on the Earth, and lying in the ecliptic plane, then, in the rotating frame S_1, the Moon would appear stationary at the point $(X_0, 0, 0)$. Expanding Equations (10.27)–(10.29) to second order in δX, δY, δZ, and neglecting terms of order m^4 and $m^2 \delta X^2$, and so on, we obtain

$$\delta \ddot{X} - 2\,\delta \dot{Y} - 3\,(1 + m^2/2)\,\delta X \simeq \frac{3}{2}\,m^2\,\cos[2\,(1-m)\,T] + \frac{3}{2}\,m^2\,\cos[2\,(1-m)\,T]\,\delta X$$

$$- \frac{3}{2}\,m^2\,\sin[2\,(1-m)\,T]\,\delta Y - 3\,\delta X^2$$

$$+ \frac{3}{2}\,(\delta Y^2 + \delta Z^2), \tag{10.33}$$

$$\delta \ddot{Y} + 2\,\delta \dot{X} \simeq -\frac{3}{2}\,m^2\,\sin[2\,(1-m)\,T]$$

$$- \frac{3}{2}\,m^2\,\sin[2\,(1-m)\,T]\,\delta X$$

$$- \frac{3}{2}\,m^2\,\cos[2\,(1-m)\,T]\,\delta Y + 3\,\delta X\,\delta Y, \tag{10.34}$$

and

$$\delta \ddot{Z} + (1 + 3\,m^2/2)\,\delta Z \simeq 3\,\delta X\,\delta Z. \tag{10.35}$$

After the preceding three equations have been solved for δX, δY, and δZ, the Cartesian coordinates, x, y, z, of the Moon in the nonrotating geocentric frame S are obtained from Equations (10.10)–(10.12), (10.24)–(10.26), and (10.30)–(10.32). However, it is more convenient to write $x = r\cos\lambda$, $y = r\sin\lambda$, and $z = r\sin\beta$, where r is the radial distance between the Earth and Moon and λ and β are termed the Moon's *geocentric* (i.e., centered on the Earth) *ecliptic longitude* and *ecliptic latitude*, respectively. Moreover, it is easily seen that, to second order in δX, δY, δZ, and neglecting terms of order m^3 and $m^2 \delta X$, and so on,

$$\frac{r}{a} - 1 + \frac{m^2}{6} \simeq \delta X + \frac{1}{2}\,\delta Y^2 + \frac{1}{2}\,\delta Z^2, \tag{10.36}$$

$$\lambda - n\,t \simeq \delta Y - \delta X\,\delta Y, \tag{10.37}$$

and

$$\beta \simeq \delta Z - \delta X\,\delta Z. \tag{10.38}$$

10.4 Unperturbed lunar motion

Let us, first of all, neglect the perturbing influence of the Sun on the Moon's orbit by setting $m = 0$ in the lunar equations of motion, Equations (10.33)–(10.35). For the sake of simplicity, let us also neglect nonlinear effects in these equations by setting $\delta X^2 = \delta Y^2 = \delta Z^2 = \delta X \, \delta Y = \delta X \, \delta Z = 0$. In this case, the equations reduce to

$$\delta \ddot{X} - 2 \, \delta \dot{Y} - 3 \, \delta X \simeq 0, \tag{10.39}$$

$$\delta \ddot{Y} + 2 \, \delta \dot{X} \simeq 0, \tag{10.40}$$

and

$$\delta \ddot{Z} + \delta Z \simeq 0. \tag{10.41}$$

By inspection, appropriate solutions are

$$\delta X \simeq -e \, \cos(T - \alpha_0), \tag{10.42}$$

$$\delta Y \simeq 2 \, e \, \sin(T - \alpha_0), \tag{10.43}$$

$$\delta Z \simeq I \, \sin(T - \gamma_0), \tag{10.44}$$

where e, α_0, I, and γ_0 are arbitrary constants. Recalling that $T = nt$, it follows from Equations (10.36)–(10.38) that

$$r \simeq a \, [1 - e \, \cos(n \, t - \alpha_0)], \tag{10.45}$$

$$\lambda \simeq n \, t + 2 \, e \, \sin(n \, t - \alpha_0), \tag{10.46}$$

and

$$\beta \simeq I \, \sin(n \, t - \gamma_0). \tag{10.47}$$

However, Equations (10.45) and (10.46) are simply first-order (in e) approximations to the familiar Keplerian laws (see Chapter 3)

$$r = \frac{a \, (1 - e^2)}{1 + e \, \cos \theta} \tag{10.48}$$

and

$$r^2 \, \dot{\theta} = (1 - e^2)^{1/2} \, n \, a^2, \tag{10.49}$$

where $\theta = \lambda - \alpha_0$ is the Moon's true anomaly and $\dot{} \equiv d/dt$. (See Section 3.13.) Of course, these two laws describe a body that executes an elliptical orbit, confocal with the Earth, of major radius a, mean angular velocity n, and eccentricity e, such that the radius vector connecting the body to the Earth sweeps out equal areas in equal time intervals. We conclude, unsurprisingly, that the unperturbed lunar orbit is a Keplerian ellipse. Note that the lunar perigee lies at the fixed ecliptic longitude $\lambda = \alpha_0$. Equation (10.47) is the first-order approximation to

$$\beta = I \, \sin(\lambda - \gamma_0). \tag{10.50}$$

(See Section 3.13.) This expression implies that the unperturbed lunar orbit is coplanar but is inclined at an angle I to the ecliptic plane. Moreover, the ascending node lies at the fixed ecliptic longitude $\lambda = \gamma_0$. Incidentally, the neglect of nonlinear terms in Equations (10.39)–(10.41) is valid only as long as $e, I \ll 1$, in other words, provided the unperturbed lunar orbit is only slightly elliptical and slightly inclined to the ecliptic plane. In fact, the observed values of e and I are 0.05488 and 0.09008 radians, respectively (Yoder 1995; Standish and Williams 1992), so this is a good approximation.

10.5 Perturbed lunar motion

The perturbed nonlinear lunar equations of motion, Equations (10.33)–(10.35), take the general form

$$\delta\ddot{X} - 2\,\delta\dot{Y} - 3\,(1 + m^2/2)\,\delta X \simeq R_X,\tag{10.51}$$

$$\delta\ddot{Y} + 2\,\delta\dot{X} \simeq R_Y,\tag{10.52}$$

and

$$\delta\ddot{Z} + (1 + 3\,m^2/2)\,\delta Z \simeq R_Z,\tag{10.53}$$

where (see Tables 10.1 and 10.2)

$$R_X = a_0 + \sum_{j=1,5} a_j\,\cos(\omega_j\,T - \alpha_j),\tag{10.54}$$

$$R_Y = \sum_{j=1,5} b_j\,\sin(\omega_j\,T - \alpha_j),\tag{10.55}$$

and

$$R_Z = \sum_{j=1,5}^{j\neq 4} c_j\,\sin(\Omega_j\,T - \gamma_j).\tag{10.56}$$

Let us search for solutions of the general form

$$\delta X = x_0 + \sum_{j=1,5} x_j\,\cos(\omega_j\,T - \alpha_j),\tag{10.57}$$

$$\delta Y = \sum_{j=1,5} y_j\,\sin(\omega_j\,T - \alpha_j),\tag{10.58}$$

and

$$\delta Z = \sum_{j=1,5}^{j\neq 4} z_j\,\sin(\Omega_j\,T - \gamma_j).\tag{10.59}$$

Substituting Equations (10.54)–(10.59) into Equations (10.51)–(10.53), we can easily demonstrate that

$$x_0 = -\frac{a_0}{3\,(1 + m^2/2)},\tag{10.60}$$

$$x_j = \frac{\omega_j\,a_j - 2\,b_j}{\omega_j\,(1 - 3\,m^2/2 - \omega_j^2)},\tag{10.61}$$

$$y_j = \frac{(\omega_j^2 + 3 + 3\,m^2/2)\,b_j - 2\,\omega_j\,a_j}{\omega_j^2\,(1 - 3\,m^2/2 - \omega_j^2)},\tag{10.62}$$

and

$$z_j = \frac{c_j}{1 + 3\,m^2/2 - \Omega_j^2},\tag{10.63}$$

where $j = 1, 5$.

Table 10.1 Angular frequencies and phase shifts associated with principal periodic driving terms appearing in perturbed nonlinear lunar equations of motion

j	ω_j	α_j	Ω_j	γ_j
1	$1 + c\,m^2$	α_0	$1 + g\,m^2$	γ_0
2	$2(1 + c\,m^2)$	$2\,\alpha_0$	$(c - g)\,m^2$	$\alpha_0 - \gamma_0$
3	$2(1 + g\,m^2)$	$2\,\gamma_0$	$2 + (c + g)\,m^2$	$\alpha_0 + \gamma_0$
4	$2 - 2\,m$	0		
5	$1 - 2\,m - c\,m^2$	$-\alpha_0$	$1 - 2\,m - g\,m^2$	$-\gamma_0$

The angular frequencies, ω_j, Ω_j, and phase shifts, α_j, γ_j, of the principal periodic driving terms that appear on the right-hand sides of the perturbed nonlinear lunar equations of motion, Equations (10.51)–(10.53), are specified in Table 10.1. Here, c and g are, as yet, unspecified $\mathcal{O}(1)$ constants associated with the precession of the lunar perigee and the regression of the ascending node, respectively. Note that ω_1 and Ω_1 are the frequencies of the Moon's unforced motion in ecliptic longitude and latitude, respectively. Moreover, ω_4 is the forcing frequency associated with the perturbing influence of the Sun. All other frequencies appearing in Table 10.1 are combinations of these three fundamental frequencies. In fact, $\omega_2 = 2\,\omega_1$, $\omega_3 = 2\,\Omega_1$, $\omega_5 = \omega_4 - \omega_1$, $\Omega_2 = \omega_1 - \Omega_1$, $\Omega_3 = \omega_1 + \Omega_1$, and $\Omega_5 = \omega_4 - \Omega_1$. Note that there is no Ω_4.

A comparison of Equations (10.33)–(10.35), (10.51)–(10.53), and Table 10.1 reveals that

$$
\begin{aligned}
R_X = {} & \frac{3}{2}\,m^2\,\cos(\omega_4\,T - \alpha_4) + \frac{3}{2}\,m^2\,\cos(\omega_4\,T - \alpha_4)\,\delta X \\
& - \frac{3}{2}\,m^2\,\sin(\omega_4\,T - \alpha_4)\,\delta Y - 3\,\delta X^2 + \frac{3}{2}\,(\delta Y^2 + \delta Z^2),
\end{aligned} \tag{10.64}
$$

$$
\begin{aligned}
R_Y = {} & -\frac{3}{2}\,m^2\,\sin(\omega_4\,T - \alpha_4) - \frac{3}{2}\,m^2\,\sin(\omega_4\,T - \alpha_4)\,\delta X \\
& - \frac{3}{2}\,m^2\,\cos(\omega_4\,T - \alpha_4)\,\delta Y + 3\,\delta X\,\delta Y,
\end{aligned} \tag{10.65}
$$

and

$$
R_Z = 3\,\delta X\,\delta Z. \tag{10.66}
$$

Substitution of Equations (10.57)–(10.59) into these equations, followed by a comparison with Equations (10.54)–(10.56), yields the amplitudes a_j, b_j, and c_j specified in Table 10.2. In calculating these amplitudes, we have neglected all contributions to the periodic driving terms appearing in Equations (10.51)–(10.53) that involve cubic, or higher order, combinations of e, I, m^2, x_j, y_j, and z_j, because we expanded Equations (10.33)–(10.35) only to *second order* in δX, δY, and δZ. We have also assumed that $x_1 = -e$, $y_1 = 2e$, $z_1 = I$, in accordance with Equations (10.42)–(10.44).

For $j = 0$, it follows from Equation (10.60) and Table 10.2 that

$$
x_0 \simeq -\frac{1}{2}\,e^2 - \frac{1}{4}\,I^2. \tag{10.67}
$$

j	a_j	b_j	c_j
	Table 10.2 Amplitudes of periodic driving terms appearing in perturbed nonlinear lunar equations of motion		
0	$\frac{3}{2} e^2 + \frac{3}{4} I^2$		
1	$\frac{3}{4} m^2 x_5 - \frac{3}{4} m^2 y_5 - 3 x_4 x_5 + \frac{3}{2} y_4 y_5$	$-\frac{3}{4} m^2 x_5 + \frac{3}{4} m^2 y_5 + \frac{3}{2} y_4 x_5 - \frac{3}{2} y_5 x_4$	$-\frac{3}{2} x_4 z_5$
2	$-\frac{9}{2} e^2$	$-3 e^2$	$\frac{3}{2} e I$
3	$-\frac{3}{4} I^2$	0	$-\frac{3}{2} e I$
4	$\frac{3}{2} m^2$	$-\frac{3}{2} m^2$	
5	$-\frac{9}{4} m^2 e + 3 e x_4 + 3 e y_4$	$\frac{9}{4} m^2 e - 3 e x_4 - \frac{3}{2} e y_4$	$-\frac{3}{2} I x_4$

For $j = 2$, making the approximation $\omega_2 \simeq 2$ (see Table 10.1), we see from Equations (10.61) and (10.62) and Table 10.2 that

$$x_2 \simeq \frac{1}{2} e^2 \tag{10.68}$$

and

$$y_2 \simeq \frac{1}{4} e^2. \tag{10.69}$$

Likewise, making the approximation $\Omega_2 \simeq 0$ (see Table 10.1), we see from Equation (10.63) and Table 10.2 that

$$z_2 \simeq \frac{3}{2} e I. \tag{10.70}$$

For $j = 3$, making the approximation $\omega_3 \simeq 2$ (see Table 10.1), we see from Equations (10.61) and (10.62) and Table 10.2 that

$$x_3 \simeq \frac{1}{4} I^2 \tag{10.71}$$

and

$$y_3 \simeq -\frac{1}{4} I^2. \tag{10.72}$$

Likewise, making the approximation $\Omega_3 \simeq 2$ (see Table 10.1), we see from Equation (10.63) and Table 10.2 that

$$z_3 \simeq \frac{1}{2} e I. \tag{10.73}$$

For $j = 4$, making the approximation $\omega_4 \simeq 2$ (see Table 10.1), we see from Equations (10.61) and (10.62) and Table 10.2 that

$$x_4 \simeq -m^2 \tag{10.74}$$

and

$$y_4 \simeq \frac{11}{8} m^2. \tag{10.75}$$

Thus, according to Table 10.2,

$$a_5 \simeq -\frac{9}{8} m^2 e, \tag{10.76}$$

$$b_5 \simeq \frac{51}{16} m^2 e, \tag{10.77}$$

and

$$c_5 \simeq \frac{3}{2} m^2 I. \tag{10.78}$$

For $j = 5$, making the approximation $\omega_5 \simeq 1 - 2m$ (see Table 10.1), we see from Equations (10.61), (10.62), (10.76), and (10.77) that

$$x_5 \simeq -\frac{15}{8} m e \tag{10.79}$$

and

$$y_5 \simeq \frac{15}{4} m e. \tag{10.80}$$

Likewise, making the approximation $\Omega_5 \simeq 1 - 2m$ (see Table 10.1), we see from Equations (10.63) and (10.78) that

$$z_5 \simeq \frac{3}{8} m I. \tag{10.81}$$

Thus, according to Table 10.2,

$$a_1 \simeq -\frac{135}{64} m^3 e, \tag{10.82}$$

$$b_1 \simeq \frac{765}{128} m^3 e, \tag{10.83}$$

and

$$c_1 \simeq \frac{9}{16} m^3 I. \tag{10.84}$$

Finally, for $j = 1$, by analogy with Equations (10.42)–(10.44), we expect

$$x_1 \simeq -e, \tag{10.85}$$

$$y_1 \simeq 2e, \tag{10.86}$$

and

$$z_1 \simeq I. \tag{10.87}$$

Thus, because $\omega_1 = 1 + c m^2$ (see Table 10.1), it follows from Equations (10.61), (10.82), (10.83), and (10.85) that

$$-e \simeq \frac{-(225/16) m^3 e}{-(3/2) m^2 - 2 c m^2}, \tag{10.88}$$

which yields

$$c \simeq -\frac{3}{4} - \frac{225}{32} m + \mathcal{O}(m^2). \tag{10.89}$$

Likewise, because $\Omega_1 = 1 + g\,m^2$ (see Table 10.1), it follows from Equations (10.63), (10.84), and (10.87) that

$$I \simeq \frac{(9/16)\,m^3\,I}{(3/2)\,m^2 - 2\,g\,m^2},\tag{10.90}$$

which yields

$$g \simeq \frac{3}{4} - \frac{9}{32}\,m + \mathcal{O}(m^2).\tag{10.91}$$

According to this analysis, our final expressions for δX, δY, and δZ are

$$\begin{aligned}
\delta X \simeq {}& -\frac{1}{2}\,e^2 - \frac{1}{4}\,I^2 - e\,\cos[(1 + c\,m^2)\,T - \alpha_0] \\
& + \frac{1}{2}\,e^2\,\cos[2\,(1 + c\,m^2)\,T - 2\,\alpha_0] \\
& + \frac{1}{4}\,I^2\,\cos[2\,(1 + g\,m^2)\,T - 2\,\gamma_0] - m^2\,\cos[2\,(1 - m)\,T] \\
& - \frac{15}{8}\,m\,e\,\cos[(1 - 2\,m - c\,m^2)\,T + \alpha_0],
\end{aligned}\tag{10.92}$$

$$\begin{aligned}
\delta Y \simeq {}& 2\,e\,\sin[(1 + c\,m^2)\,T - \alpha_0] + \frac{1}{4}\,e^2\,\sin[2\,(1 + c\,m^2)\,T - 2\,\alpha_0] \\
& - \frac{1}{4}\,I^2\,\sin[2\,(1 + g\,m^2)\,T - 2\,\gamma_0] + \frac{11}{8}\,m^2\,\cos[2\,(1 - m)\,T] \\
& + \frac{15}{4}\,m\,e\,\sin[(1 - 2\,m - c\,m^2)\,T + \alpha_0],
\end{aligned}\tag{10.93}$$

and

$$\begin{aligned}
\delta Z \simeq {}& I\,\sin[(1 + g\,m^2)\,T - \gamma_0] + \frac{3}{2}\,e\,I\,\sin[(c - g)\,m^2\,T - \alpha_0 + \gamma_0] \\
& + \frac{1}{2}\,e\,I\,\sin[(2 + c\,m^2 + g\,m^2)\,T - \alpha_0 - \gamma_0] \\
& + \frac{3}{8}\,m\,I\,\sin[(1 - 2\,m - g\,m^2)\,T + \gamma_0].
\end{aligned}\tag{10.94}$$

Thus, making use of Equations (10.36)–(10.38), we find that

$$\begin{aligned}
\frac{r}{a} \simeq {}& 1 - e\,\cos[(1 + c\,m^2)\,T - \alpha_0] \\
& + \frac{1}{2}\,e^2 - \frac{1}{6}\,m^2 - \frac{1}{2}\,e^2\,\cos[2\,(1 + c\,m^2)\,T - 2\,\alpha_0] \\
& - m^2\,\cos[2\,(1 - m)\,T] - \frac{15}{8}\,m\,e\,\cos[(1 - 2\,m - c\,m^2)\,T + \alpha_0],
\end{aligned}\tag{10.95}$$

$$\begin{aligned}
\lambda \simeq {}& T + 2\,e\,\sin[(1 + c\,m^2)\,T - \alpha_0] + \frac{5}{4}\,e^2\,\sin[2\,(1 + c\,m^2)\,T - 2\,\alpha_0] \\
& - \frac{1}{4}\,I^2\,\sin[2\,(1 + g\,m^2)\,T - 2\,\gamma_0] + \frac{11}{8}\,m^2\,\sin[2\,(1 - m)\,T] \\
& + \frac{15}{4}\,m\,e\,\sin[(1 - 2\,m - c\,m^2)\,T + \alpha_0],
\end{aligned}\tag{10.96}$$

and

$$
\begin{aligned}
\beta \simeq \; & I \sin[(1 + g\,m^2)\,T - \gamma_0] + e\,I \, \sin[(c - g)\,m^2\,T - \alpha_0 + \gamma_0] \\
& + e\,I \, \sin[(2 + c\,m^2 + g\,m^2)\,T - \alpha_0 - \gamma_0] \\
& + \frac{3}{8}\,m\,I \, \sin[(1 - 2\,m - g\,m^2)\,T + \gamma_0].
\end{aligned}
\tag{10.97}
$$

These expressions are accurate up to second order in the small parameters e, I, and m.

10.6 Description of lunar motion

To better understand the expressions for perturbed lunar motion derived in the previous section, it is helpful to introduce the concept of the *mean moon*. This is an imaginary body that orbits the Earth, in the ecliptic plane, at a steady angular velocity equal to the Moon's mean orbital angular velocity, n. Likewise, the *mean sun* is a second imaginary body that orbits the Earth, in the ecliptic plane, at a steady angular velocity equal to the Sun's mean (apparent) orbital angular velocity, n'. Thus, the geocentric ecliptic longitudes of the mean moon and the mean sun are

$$
\bar{\lambda} = n\,t
\tag{10.98}
$$

and

$$
\bar{\lambda}' = n'\,t,
\tag{10.99}
$$

respectively. Here, for the sake of simplicity, and also for the sake of consistency with our previous analysis, we have assumed that both objects are located at ecliptic longitude 0 at time $t = 0$.

From Equation (10.95), to first order in small parameters, the lunar perigee corresponds to $(1 + c\,m^2)\,n\,t - \alpha_0 = j\,2\pi$, where j is an integer. However, this condition can also be written $\bar{\lambda} = \alpha$, where

$$
\alpha = \alpha_0 + \alpha'\,n'\,t,
\tag{10.100}
$$

and, making use of Equation (10.89), together with the definition $m = n'/n$,

$$
\alpha' = \frac{3}{4}\,m + \frac{225}{32}\,m^2 + \mathcal{O}(m^3).
\tag{10.101}
$$

Thus, we can identify α as the mean ecliptic longitude of the perigee. Moreover, according to Equation (10.100), the perigee precesses (i.e., its longitude increases in time) at the mean rate of $360\,\alpha'$ degrees per year. (Of course, a year corresponds to $\Delta t = 2\pi/n'$.) Furthermore, it is clear that this precession is entirely due to the perturbing influence of the Sun, because it depends only on the parameter m, which is a measure of this influence. Given that $m = 0.07480$, we find that the perigee advances by $34.36°$ per year. Hence, we predict that the perigee completes a full circuit about the Earth every $1/\alpha' = 10.5$ years. In fact, the lunar perigee completes a full circuit every 8.85 years. Our prediction is somewhat inaccurate because our previous analysis neglected $\mathcal{O}(m^2)$, and smaller, contributions to the parameter c [see Equation (10.89)], and these turn out to be significant.

From Equation (10.97), to first order in small parameters, the Moon passes through its ascending node when $(1 + g\,m^2)\,n\,t - \gamma_0 = j\,2\pi$, where j is an integer. However, this condition can also be written $\bar{\lambda} = \gamma$, where

$$\gamma = \gamma_0 - \gamma'\,n'\,t, \tag{10.102}$$

and, making use of Equation (10.91),

$$\gamma' = \frac{3}{4}\,m - \frac{9}{32}\,m^2 + \mathcal{O}(m^3). \tag{10.103}$$

Thus, we can identify γ as the mean ecliptic longitude of the ascending node. Moreover, according to Equation (10.102), the ascending node regresses (i.e., its longitude decreases in time) at the mean rate of $360\,\gamma'$ per year. As before, it is clear that this regression is entirely due to the perturbing influence of the Sun. Moreover, we find that the ascending node retreats by $19.63°$ per year. Hence, we predict that the ascending node completes a full circuit about the Earth every $1/\gamma' = 18.3$ years. In fact, the lunar ascending node completes a full circuit every 18.6 years, so our prediction is fairly accurate.

It is interesting to note that Clairaut's initial lunar theory, produced in 1747, neglected the $\mathcal{O}(m^2)$ contributions to α' and γ'—see Equations (10.101) and (10.103), respectively—leading to the prediction that the lunar perigee should precess at the same rate at which the ascending node regresses, and that both the perigee and the node should complete full circuits around the Earth every 17.8 years (Taton and Wilson 1995). Of course, this prediction was in serious disagreement with observations, according to which the lunar perigee precesses at about twice the rate at which the ascending node regresses. This discrepancy between theory and observations lead Clairaut to briefly doubt the inverse-square nature of Newtonian gravity. Fortunately, Clairaut realized in 1748 that the discrepancy could be resolved by carrying his expansion to higher order in m. Indeed, as is clear from Equations (10.101) and (10.103), the $\mathcal{O}(m^2)$ contributions to α' and γ' give rise to a significant increase in the precession rate of the lunar perigee relative to the regression rate of the ascending node. In fact, the complete expression for α' (Delaunay 1867),

$$\alpha' = \frac{3}{4}\,m + \frac{225}{32}\,m^2 + \frac{4071}{128}\,m^3 + \frac{265493}{2048}\,m^4 + \frac{12822631}{24576}\,m^5 + \cdots, \tag{10.104}$$

takes the form of a power series in m. Despite the fact that, for the case of the Moon, m takes the relatively small value 0.07480, this series is slowly converging, and many terms must be retained to get an accurate value for the precession rate of the perigee. The complete expression for γ' (Delaunay 1867),

$$\gamma' = \frac{3}{4}\,m - \frac{9}{32}\,m^2 - \frac{273}{128}\,m^3 - \frac{9797}{2048}\,m^4 + \cdots, \tag{10.105}$$

also takes the form of a power series in m. Fortunately, for the case of the Moon, this series converges relatively quickly, which accounts for the fact that our prediction for the regression rate of the lunar ascending node is considerably more accurate than that for the precession rate of the perigee.

It is helpful to introduce the lunar *mean anomaly*,

$$\mathcal{M} = \bar{\lambda} - \alpha, \tag{10.106}$$

which is defined as the angular distance (in geocentric ecliptic longitude) between the mean Moon and the perigee. It is also helpful to introduce the lunar *mean argument of latitude*,

$$F = \bar{\lambda} - \gamma, \tag{10.107}$$

which is defined as the angular distance (in geocentric ecliptic longitude) between the mean Moon and the ascending node. Finally, it is helpful to introduce the *mean elongation* of the Moon,

$$D = \bar{\lambda} - \bar{\lambda}', \tag{10.108}$$

which is defined as the difference between the geocentric ecliptic longitudes of the mean Moon and the mean Sun.

When expressed in terms of $\mathcal{M}$, F, and D, our previous expression for the true geocentric ecliptic longitude of the Moon, Equation (10.96), becomes

$$\lambda = \bar{\lambda} + \delta\lambda, \tag{10.109}$$

where

$$\delta\lambda = 2e \sin\mathcal{M} + \frac{5}{4}e^2 \sin 2\mathcal{M} - \frac{1}{4}I^2 \sin 2F + \frac{11}{8}m^2 \sin 2D + \frac{15}{4}me \sin(2D - \mathcal{M})$$

$$\tag{10.110}$$

is the angular distance (in geocentric ecliptic longitude) between the Moon and the mean moon.

The first three terms on the right-hand side of Equation (10.110) are Keplerian in origin (i.e., they are independent of the perturbing action of the Sun). In fact, the first is due to the eccentricity of the lunar orbit (i.e., the fact that the geometric center of the orbit is slightly shifted from the center of the Earth), the second is due to the ellipticity of the orbit (i.e., the fact that the orbit is slightly noncircular), and the third is due to the slight inclination of the orbit to the ecliptic plane. The first and third terms are usually called the *major inequality* and the *reduction to the ecliptic*, respectively.

The fourth term on the right-hand side of Equation (10.110) corresponds to variation; it is clearly due to the perturbing influence of the Sun (because it depends only on the parameter m, which is a measure of this influence). Variation attains its maximal amplitude around the so-called *octant points*, at which the Moon's disk is either one-quarter or three-quarters illuminated (i.e., when $D = 45°$, $135°$, $225°$, or $315°$). Conversely, the amplitude of variation is zero around the so-called *quadrant points*, at which the Moon's disk is either fully illuminated, half illuminated, or not illuminated at all (i.e., when $D = 0°, 90°, 180°$, or $270°$). Variation generates a perturbation in the lunar ecliptic longitude that oscillates sinusoidally with a period of half a synodic month.[2] This oscillation period is in good agreement with observations. However, the amplitude of the oscillation (calculated using $m = 0.07480$) is 1,630 arc seconds, which is considerably

---

[2] A synodic month, which is 29.53 days, is the mean period between successive new moons.

less than the observed amplitude of 2,370 arc seconds (Chapront-Touzé and Chapront 1988). This discrepancy between theory and observation is due to the fact that, for the sake of simplicity, we have calculated only the lowest order (in m) contribution to variation.

The fifth term on the right-hand side of Equation (10.110) corresponds to evection and is due to the combined action of the Sun and the eccentricity of the lunar orbit. In fact, evection can be thought of as causing a slight reduction in the eccentricity of the lunar orbit around the times of the new moon and the full moon (i.e., $D = 0°$ and $D = 180°$), and a corresponding slight increase in the eccentricity around the times of the first and last quarter moons (i.e., $D = 90°$ and $D = 270°$). (See Exercise 10.2.) This follows because the evection term in Equation (10.110) augments the eccentricity term, $2e \sin \mathcal{M}$, when $\cos 2D = -1$, and reduces the term when $\cos 2D = +1$. Evection generates a perturbation in the lunar ecliptic longitude that oscillates sinusoidally with a period of 31.8 days. This oscillation period is in good agreement with observations. However, the amplitude of the oscillation (calculated using $m = 0.07480$ and $e = 0.05488$) is 3,218 arc seconds, which is considerably less than the observed amplitude of 4,586 arc seconds (Chapront-Touzé and Chapront 1988). Again, this discrepancy between theory and observation exists because we have calculated only the lowest order (in m and e) contribution to evection.

Recall that we previously neglected the slight eccentricity, $e' = 0.016711$, of the Sun's apparent orbit about the Earth in our calculation. In fact, the eccentricity of the solar orbit gives rise to a small addition term on the right-hand side of Equation (10.110), which, to lowest order, takes the form $-3\,m\,e'\,\sin \mathcal{M}'$. Here, $\mathcal{M}'$ is the Sun's mean anomaly. This term, which is known as the *annual inequality*, generates a perturbation in the lunar ecliptic longitude that oscillates with a period of a solar year and has an amplitude of 772 arc seconds. As before, the oscillation period is in good agreement with observations, whereas the amplitude is inaccurate [it should be 666 arc seconds (Chapront-Touzé and Chapront 1988)] because of the omission of higher order (in m and e') contributions.

When written in terms of D and F, our previous expression for the geocentric ecliptic latitude of the Moon, Equation (10.97) becomes

$$\beta = I \sin(F + \delta\lambda) + \frac{3}{8} m I \sin(2D - F). \tag{10.111}$$

The first term on the right-hand side of Equation (10.111) is Keplerian in origin (i.e., it is essentially independent of the perturbing influence of the Sun). The second term, which is known as *evection in latitude*, is due to the combined action of the Sun and the inclination of the lunar orbit to the ecliptic. Evection in latitude can be thought of as causing a slight increase in the inclination of the lunar orbit at the times of the first and last quarter moons, and a slight decrease at the times of the new moon and the full moon. (See Exercise 10.2.) Evection in latitude generates a perturbation in the lunar ecliptic latitude that oscillates sinusoidally with a period of 32.3 days; it has an amplitude of 521 arc seconds. As before, the oscillation period is in good agreement with observations, but the amplitude is inaccurate (it should be 624 arc seconds; Chapront-Touzé and Chapront 1988) due to the omission of higher-order (in m and I) contributions.

Exercises

10.1 Demonstrate that the lunar equation of motion, Equation (10.5), can be written in the canonical form

$$\ddot{\mathbf{r}} + n^2 a^3 \frac{\mathbf{r}}{r^3} = \nabla \mathcal{R},$$

where

$$\mathcal{R} = n'^2 a'^3 \left(\frac{1}{|\mathbf{r}' - \mathbf{r}|} - \frac{\mathbf{r} \cdot \mathbf{r}'}{r'^3} \right)$$

is the disturbing function due to the gravitational influence of the Sun. Show that if the Earth's orbit about the Sun is approximated as a circle of radius a', then, to lowest order in a/a', the solar disturbing function can be written

$$\mathcal{R} \simeq n'^2 r^2 P_2(\cos \psi),$$

where ψ is the angle subtended between the vectors $\mathbf{r}$ and $\mathbf{r}'$.

10.2 Demonstrate that the evection term

$$\frac{15}{4} m e \sin(2D - \mathcal{M}),$$

appearing in Equation (10.110), can be represented as the combined effect of periodic variations in the eccentricity, e, of the lunar orbit, and the mean longitude, α, of the lunar perigee; in other words, $e \to e[1 - (15/8)m \cos 2D]$ and $\alpha \to \alpha - (15/8)m \sin 2D$. Likewise, show that the evection term

$$\frac{3}{8} m I \sin(2D - F),$$

appearing in Equation (10.111), can be represented as the combined effect of periodic variations in the inclination, I, of the lunar orbit, and the mean longitude, γ, of the lunar ascending node; in other words, $I \to I[1 - (3/8)m \cos 2D]$ and $\gamma \to \gamma - (3/8)m \sin 2D$.

10.3 Suppose that the major radius of the lunar orbit were reduced by a multiplicative factor ζ: that is, $a \to \zeta a$, where $0 < \zeta < 1$. Assuming that the masses of the Earth and Sun and the major radius of the terrestrial orbit remain constant, demonstrate that the parameter m, which measures the perturbing influence of the Sun on the lunar orbit, would be reduced by a factor $\zeta^{3/2}$: that is, $m \to \zeta^{3/2} m$. Given that $m = 0.07480$ for the true lunar orbit, how small would ζ have to be before the (theoretical) precession rate of the lunar perigee became equal to the regression rate of the ascending node to within 1 percent? What is the corresponding major radius of the lunar orbit in units of mean Earth radii? (The true major radius of the lunar orbit is 60.9 mean Earth radii.)

10.4 An artificial satellite orbits the Moon in a low-eccentricity orbit whose major radius is twice the lunar radius. The plane of the satellite orbit is slightly inclined to the plane of the Moon's orbit about the Earth. Given that the mass of the Earth is 81.3 times that of the Moon, and the major radius of the lunar orbit is 221.3 times the lunar radius, estimate the precession period of the satellite orbit's perilune (i.e., its point of closest approach to the Moon) in months due to the perturbing influence of the Earth. Likewise, estimate the regression rate of the satellite orbit's ascending node (with respect to the plane of the lunar orbit) in months. (Assume that the Moon is a perfect sphere.)

10.5 The mean ecliptic longitudes (measured with respect to the vernal equinox at a fixed epoch) of the Moon and the Sun increase at the rates $13.176359°$ per day and $0.98560912°$ per day, respectively. However, the vernal equinox regresses in such a manner that, on average, it completes a full circuit every 25,772 years. Furthermore, the lunar perigee precesses in such a manner that, on average, it completes a full circuit every 8.848 years, whereas the lunar ascending mode regresses in such a manner that, on average, it completes a full circuit every 18.615 years (Yoder 1995). A *sidereal month* is the mean period of the Moon's orbit with respect to the fixed stars, a *tropical month* is the mean time required for the Moon's ecliptic longitude (with respect to the true vernal equinox) to increase by $360°$, a *synodic month* is the mean period between successive new moons, an *anomalistic month* is the mean period between successive passages of the Moon through its perigee, and a *draconic month* is the mean period between successive passages of the Moon through its ascending node. Use the preceding information to demonstrate than the lengths of a sidereal, tropical, synodic, anomalistic, and draconic month are 27.32166, 27.32158, 29.5306, 27.5546, and 27.2123 days, respectively.

10.6 To first order in the Moon's orbital eccentricity and inclination, the geocentric ecliptic longitude and latitude of the Moon, relative to the Sun, are written

$$\Delta\lambda \simeq (n - n')\,t + 2\,e\,\sin[(n - n_p)\,t]$$

and

$$\Delta\beta \simeq I\,\sin[(n - n_n)\,t],$$

respectively, where $e = 0.05488$ and $I = 0.9008$ radians. Here, n and n' are the mean geocentric orbital angular velocities of the Moon and Sun, respectively, n_p is the mean orbital angular velocity of the lunar perigee, and n_n is the mean orbital angular velocity of the lunar ascending node. Note that $2\pi/(n - n') = T_{\text{synodic}}$, $2\pi/(n - n_p) = T_{\text{anomalistic}}$, and $2\pi/(n - n_n) = T_{\text{draconic}}$, where T_{synodic}, $T_{\text{anomalistic}}$, and T_{draconic} are the lengths of a synodic, anomalistic, and draconic month, respectively. At $t = 0$, we have $\Delta\lambda = \Delta\beta = 0°$. In other words, at $t = 0$, the Moon and Sun have exactly the same geocentric ecliptic longitudes and latitudes, which implies that a solar eclipse occurs at this time. Suppose we can find some time period T that satisfies $T = j_1\,T_{\text{synodic}} = j_2\,T_{\text{anomalistic}} = j_3\,T_{\text{draconic}}$, where j_1, j_2, j_3 are positive integers. Demonstrate that $\Delta\lambda = \Delta\beta = 0°$ at $t = T$. Thus, if the period T, which is known as the *saros*, existed, then solar (and lunar) eclipses would occur in infinite sequences spaced j_1 synodic months apart (Roy 2005). Show that for $0 < j_1, j_2, j_3 < 1,000$, the closest approximation to the saros is obtained when $j_1 = 223$, $j_2 = 239$, and $j_3 = 242$. Demonstrate that if $\Delta\lambda = \Delta\beta = 0°$ at $t = 0$ (i.e., if there is a solar eclipse at $t = 0$) then, exactly 223 synodic months later, $\Delta\lambda = -0.3°$ and $\Delta\beta = -0.1°$. It turns out that these values of $\Delta\lambda$

and $\Delta\beta$ are sufficiently small that the eclipse recurs. In fact, because 223 synodic months almost satisfies the saros condition, solar (and lunar) eclipses occur in series of about 70 eclipses spaced 223 synodic months, or 18 years and 11 days, apart.

10.7 Let the x-, y-, and z-axes be the lunar principal axes of rotation passing through the lunar center of mass. Because the Moon is not quite spherically symmetric, its principal moments of inertia are not exactly equal to one another. Let us label the principal axes such that $\mathcal{I}_{zz} > \mathcal{I}_{yy} > \mathcal{I}_{xx}$. To a first approximation, the Moon is spinning about the z-axis, which is oriented normal to its orbital plane. Moreover, the Moon spins in such a manner that the x-axis always points approximately in the direction of the Earth. Let η be the (small) angle subtended between the x-axis and the line joining the centers of the Moon and the Earth. A slight generalization of the analysis in Section 7.11 reveals that

$$\eta = \eta_o + \eta_p,$$

where $\eta_o = \delta\lambda$ and η_p are the Moon's optical and physical libration (in ecliptic longitude), respectively, $\delta\lambda$ is defined in Equation (10.110), and

$$\ddot{\eta}_p + n_0^2\,\eta_p = -n_0^2\,\eta_o.$$

Here, $n_0 = [3\,(\mathcal{I}_{yy} - \mathcal{I}_{xx})/\mathcal{I}_{zz}]^{1/2}\,n = 0.3446°$ per day is the Moon's free libration rate, whereas $n = 13.1764°$ per day is the lunar mean sidereal orbital angular velocity.

We can write

$$\delta\lambda \simeq \sum_{i=1,5} A_i\,\sin(n_i\,t - \gamma_i),$$

where the $i = 1, 2, 3, 4,$ and 5 terms correspond to the major inequality, the reduction to the ecliptic, variation, evection, and the annual inequality, respectively. Furthermore, $A_1 = 22640''$, $A_2 = 418''$, $A_3 = 2370''$, $A_4 = 4586''$, $A_5 = 666''$, and $n_1 = n$, $n_2 = 2\,n_{dr}$, $n_3 = 2\,n_{sy}$, $n_4 = 2\,n_{sy} - n_{an}$, $n_5 = n'$. Here, $n_{dr} = 13.2293°$ per day, $n_{sy} = 12.1908°$ per day, and $n_{an} = 13.0650°$ per day are the lunar mean draconic, synodic, and anomalistic orbital angular velocities, respectively, and $n' = 0.9856°$ per day is the Earth's mean sidereal orbital angular velocity. Demonstrate that

$$\eta_p \simeq \sum_{i=1,5} A_i'\,\sin(n_i\,t - \gamma_i),$$

where $A_1' = 15.7''$, $A_2' = 0.07''$, $A_3' = 0.48''$, $A_4' = 4.3''$, and $A_5' = 94.0''$ are the forced libration amplitudes associated with the major inequality, reduction to the ecliptic, variation, evection, and the annual inequality, respectively. [The observed values of A_1', A_3', A_4', and A_5' are $16.8''$, $0.50''$, $4.1''$, and $90.7''$, respectively. A_2' is too small to measure. (Meeus 2005).]

Appendix A Useful mathematics

A.1 Calculus

$$\frac{d}{dx}e^x = e^x \tag{A.1}$$

$$\frac{d}{dx}\ln x = \frac{1}{x} \tag{A.2}$$

$$\frac{d}{dx}\sin x = \cos x \tag{A.3}$$

$$\frac{d}{dx}\cos x = -\sin x \tag{A.4}$$

$$\frac{d}{dx}\tan x = \frac{1}{\cos^2 x} \tag{A.5}$$

$$\frac{d}{dx}\sin^{-1} x = \frac{1}{\sqrt{1-x^2}} \tag{A.6}$$

$$\frac{d}{dx}\cos^{-1} x = -\frac{1}{\sqrt{1-x^2}} \tag{A.7}$$

$$\frac{d}{dx}\tan^{-1} x = \frac{1}{1+x^2} \tag{A.8}$$

$$\frac{d}{dx}\sinh x = \cosh x \tag{A.9}$$

$$\frac{d}{dx}\cosh x = \sinh x \tag{A.10}$$

$$\frac{d}{dx}\tanh x = \frac{1}{\cosh^2 x} \tag{A.11}$$

$$\frac{d}{dx}\sinh^{-1} x = \frac{1}{\sqrt{1+x^2}} \tag{A.12}$$

$$\frac{d}{dx}\cosh^{-1} x = \frac{1}{\sqrt{x^2-1}} \tag{A.13}$$

$$\frac{d}{dx}\tanh^{-1} x = \frac{1}{1-x^2} \tag{A.14}$$

A.2 Series expansions

Notation: $k! = k\,(k-1)\,(k-2)\dots2.1$, $f^{(n)}(x) = d^n f(x)/dx^n$.

$$f(x) = f(a) + \frac{(x-a)}{1!}\,f^{(1)}(a) + \frac{(x-a)^2}{2!}\,f^{(2)}(a) + \cdots \frac{(x-a)^n}{n!}\,f^{(n)}(a) + \cdots$$

(A.15)

$$(1+x)^\alpha = 1 + \alpha\,x + \frac{\alpha\,(\alpha-1)}{2!}\,x^2 + \frac{\alpha\,(\alpha-1)\,(\alpha-2)}{3!}\,x^3 + \cdots$$

(A.16)

$$e^x = 1 + x + \frac{x^2}{2!} + \frac{x^3}{3!} + \cdots$$

(A.17)

$$\ln(1+x) = x - \frac{x^2}{2} + \frac{x^3}{3} - \frac{x^4}{4} + \cdots$$

(A.18)

$$\sin x = x - \frac{x^3}{3!} + \frac{x^5}{5!} - \frac{x^7}{7!} + \cdots$$

(A.19)

$$\cos x = 1 - \frac{x^2}{2!} + \frac{x^4}{4!} - \frac{x^6}{6!} + \cdots$$

(A.20)

$$\tan x = x + \frac{x^3}{3} + \frac{2\,x^5}{15} + \frac{17\,x^7}{315} + \cdots$$

(A.21)

$$\sinh x = x + \frac{x^3}{3!} + \frac{x^5}{5!} + \frac{x^7}{7!} + \cdots$$

(A.22)

$$\cosh x = 1 + \frac{x^2}{2!} + \frac{x^4}{4!} + \frac{x^6}{6!} + \cdots$$

(A.23)

$$\tanh x = x - \frac{x^3}{3} + \frac{2\,x^5}{15} - \frac{17\,x^7}{315} + \cdots$$

(A.24)

A.3 Trigonometric identities

$$\sin(-\alpha) = -\sin\alpha$$

(A.25)

$$\cos(-\alpha) = +\cos\alpha$$

(A.26)

$$\tan(-\alpha) = -\tan\alpha$$

(A.27)

$$\sin^2\alpha + \cos^2\alpha = 1$$

(A.28)

$$\sin(\alpha\pm\beta) = \sin\alpha\,\cos\beta \pm \cos\alpha\,\sin\beta$$

(A.29)

$$\cos(\alpha\pm\beta) = \cos\alpha\,\cos\beta \mp \sin\alpha\,\sin\beta$$

(A.30)

$$\tan(\alpha\pm\beta) = = \frac{\tan\alpha \pm \tan\beta}{1 \mp \tan\alpha\,\tan\beta}$$

(A.31)

$$\sin\alpha + \sin\beta = 2\,\sin\left(\frac{\alpha+\beta}{2}\right)\cos\left(\frac{\alpha-\beta}{2}\right)$$

(A.32)

$$\sin\alpha - \sin\beta = 2 \cos\left(\frac{\alpha+\beta}{2}\right) \sin\left(\frac{\alpha-\beta}{2}\right) \tag{A.33}$$

$$\cos\alpha + \cos\beta = 2 \cos\left(\frac{\alpha+\beta}{2}\right) \cos\left(\frac{\alpha-\beta}{2}\right) \tag{A.34}$$

$$\cos\alpha - \cos\beta = -2 \sin\left(\frac{\alpha+\beta}{2}\right) \sin\left(\frac{\alpha-\beta}{2}\right) \tag{A.35}$$

$$\sin\alpha \, \sin\beta = \frac{1}{2}\left[\cos(\alpha-\beta) - \cos(\alpha+\beta)\right] \tag{A.36}$$

$$\cos\alpha \, \cos\beta = \frac{1}{2}\left[\cos(\alpha-\beta) + \cos(\alpha+\beta)\right] \tag{A.37}$$

$$\sin\alpha \, \cos\beta = \frac{1}{2}\left[\sin(\alpha-\beta) + \sin(\alpha+\beta)\right] \tag{A.38}$$

$$\sin(\alpha/2) = \pm\left(\frac{1-\cos\alpha}{2}\right)^{1/2} \tag{A.39}$$

$$\cos(\alpha/2) = \pm\left(\frac{1+\cos\alpha}{2}\right)^{1/2} \tag{A.40}$$

$$\tan(\alpha/2) = \pm\left(\frac{1-\cos\alpha}{1+\cos\alpha}\right)^{1/2} = \frac{1-\cos\alpha}{\sin\alpha} = \frac{\sin\alpha}{1+\cos\alpha} \tag{A.41}$$

$$\sin(2\alpha) = 2 \sin\alpha \, \cos\alpha \tag{A.42}$$

$$\cos(2\alpha) = \cos^2\alpha - \sin^2\alpha = 2\cos^2\alpha - 1 = 1 - 2\sin^2\alpha \tag{A.43}$$

$$\sin(3\alpha) = -4 \sin^3\alpha + 3 \sin\alpha \tag{A.44}$$

$$\cos(3\alpha) = 4 \cos^3\alpha - 3 \cos\alpha \tag{A.45}$$

$$\sin(4\alpha) = (-8 \sin^3\alpha + 4 \sin\alpha) \cos\alpha \tag{A.46}$$

$$\cos(4\alpha) = 8 \cos^4\alpha - 8 \cos^2\alpha + 1 \tag{A.47}$$

$$\sin^2\alpha = \frac{1}{2}(1 - \cos 2\alpha) \tag{A.48}$$

$$\cos^2\alpha = \frac{1}{2}(1 + \cos 2\alpha) \tag{A.49}$$

$$\sin^3\alpha = \frac{1}{4}(3 \sin\alpha - \sin 3\alpha) \tag{A.50}$$

$$\cos^3\alpha = \frac{1}{4}(3 \cos\alpha + \cos 3\alpha) \tag{A.51}$$

$$\sin^4\alpha = \frac{1}{8}(3 - 4 \cos 2\alpha + \cos 4\alpha) \tag{A.52}$$

$$\cos^4\alpha = \frac{1}{8}(3 + 4 \cos 2\alpha + \cos 4\alpha) \tag{A.53}$$

$$\sinh(-\alpha) = -\sinh\alpha \tag{A.54}$$

$$\cosh(-\alpha) = +\cosh\alpha \tag{A.55}$$

$$\tanh(-\alpha) = -\tanh\alpha \tag{A.56}$$

$$\cosh^2 \alpha - \sinh^2 \alpha = 1 \tag{A.57}$$

$$\sinh(\alpha \pm \beta) = \sinh \alpha \, \cosh \beta \pm \cosh \alpha \, \sinh \beta \tag{A.58}$$

$$\cosh(\alpha \pm \beta) = \cosh \alpha \, \cosh \beta \pm \sinh \alpha \, \sinh \beta \tag{A.59}$$

$$\tanh(\alpha \pm \beta) = \frac{\tanh \alpha \pm \tanh \beta}{1 \pm \tanh \alpha \, \tanh \beta} \tag{A.60}$$

$$\sinh \alpha + \sinh \beta = 2 \sinh \left(\frac{\alpha + \beta}{2} \right) \cosh \left(\frac{\alpha - \beta}{2} \right) \tag{A.61}$$

$$\sinh \alpha - \sinh \beta = 2 \cosh \left(\frac{\alpha + \beta}{2} \right) \sinh \left(\frac{\alpha - \beta}{2} \right) \tag{A.62}$$

$$\cosh \alpha + \cosh \beta = 2 \cosh \left(\frac{\alpha + \beta}{2} \right) \cosh \left(\frac{\alpha - \beta}{2} \right) \tag{A.63}$$

$$\cosh \alpha - \cosh \beta = 2 \sinh \left(\frac{\alpha + \beta}{2} \right) \sinh \left(\frac{\alpha - \beta}{2} \right) \tag{A.64}$$

$$\sinh \alpha \, \sinh \beta = \frac{1}{2} \left[\cosh(\alpha + \beta) - \cosh(\alpha - \beta) \right] \tag{A.65}$$

$$\cosh \alpha \, \cosh \beta = \frac{1}{2} \left[\cosh(\alpha + \beta) + \cosh(\alpha - \beta) \right] \tag{A.66}$$

$$\sinh \alpha \, \cosh \beta = \frac{1}{2} \left[\sinh(\alpha + \beta) + \sinh(\alpha - \beta) \right] \tag{A.67}$$

$$\sinh(\alpha/2) = \left(\frac{\cosh \alpha - 1}{2} \right)^{1/2} \tag{A.68}$$

$$\cosh(\alpha/2) = \left(\frac{\cosh \alpha + 1}{2} \right)^{1/2} \tag{A.69}$$

$$\tanh(\alpha/2) = \left(\frac{\cosh \alpha - 1}{\cosh \alpha + 1} \right)^{1/2} = \frac{\cosh \alpha - 1}{\sinh \alpha} = \frac{\sinh \alpha}{\cosh \alpha + 1} \tag{A.70}$$

$$\sinh(2\alpha) = 2 \sinh \alpha \, \cosh \alpha \tag{A.71}$$

$$\cosh(2\alpha) = \cosh^2 \alpha + \sinh^2 \alpha = 2 \cosh^2 \alpha - 1 = 2 \sinh^2 \alpha + 1 \tag{A.72}$$

A.4   Vector identities

Notation: $\mathbf{a}$, $\mathbf{b}$, $\mathbf{c}$, $\mathbf{d}$ are general vectors; ϕ, ψ are general scalar fields; $\mathbf{a} = a_x \, \mathbf{e}_x + a_y \, \mathbf{e}_y + a_x \, \mathbf{e}_z = (a_x, a_y, a_z)$, and so on, where $\mathbf{e}_x$, $\mathbf{e}_y$, and $\mathbf{e}_z$ are right-handed Cartesian basis vectors.

$$|\mathbf{a}| = \left(a_x^2 + a_y^2 + a_z^2 \right)^{1/2} \tag{A.73}$$

$$\mathbf{a} \cdot \mathbf{b} = a_x \, b_x + a_y \, b_y + a_z \, b_z \tag{A.74}$$

$$\mathbf{a} \times \mathbf{b} = \begin{vmatrix} \mathbf{e}_x & \mathbf{e}_y & \mathbf{e}_z \\ a_x & a_y & a_z \\ b_x & b_y & b_z \end{vmatrix}$$

$$= (a_y \, b_z - a_z \, b_y) \, \mathbf{e}_x + (a_z \, b_x - a_x \, b_z) \, \mathbf{e}_y + (a_x \, b_y - a_y \, b_x) \, \mathbf{e}_z \tag{A.75}$$

$$\mathbf{a} \times (\mathbf{b} \times \mathbf{c}) = (\mathbf{a} \cdot \mathbf{c})\,\mathbf{b} - (\mathbf{a} \cdot \mathbf{b})\,\mathbf{c} \tag{A.76}$$

$$(\mathbf{a} \times \mathbf{b}) \times \mathbf{c} = (\mathbf{c} \cdot \mathbf{a})\,\mathbf{b} - (\mathbf{c} \cdot \mathbf{b})\,\mathbf{a} \tag{A.77}$$

$$(\mathbf{a} \times \mathbf{b}) \cdot (\mathbf{c} \times \mathbf{d}) = (\mathbf{a} \cdot \mathbf{c})(\mathbf{b} \cdot \mathbf{d}) - (\mathbf{a} \cdot \mathbf{d})(\mathbf{b} \cdot \mathbf{c}) \tag{A.78}$$

$$(\mathbf{a} \times \mathbf{b}) \times (\mathbf{c} \times \mathbf{d}) = (\mathbf{a} \times \mathbf{b} \cdot \mathbf{d})\,\mathbf{c} - (\mathbf{a} \times \mathbf{b} \cdot \mathbf{c})\,\mathbf{d} \tag{A.79}$$

$$\nabla \phi = \frac{\partial \phi}{\partial x}\,\mathbf{e}_x + \frac{\partial \phi}{\partial y}\,\mathbf{e}_y + \frac{\partial \phi}{\partial z}\,\mathbf{e}_z \tag{A.80}$$

$$\nabla(\phi\,\psi) = \phi\,\nabla\psi + \psi\,\nabla\phi \tag{A.81}$$

A.5 Conservative fields

Consider a vector field $\mathbf{A}(\mathbf{r})$. In general, the line integral $\int_P^Q \mathbf{A} \cdot d\mathbf{r}$ depends on the path taken between the end points, P and Q. However, for some special vector fields, the integral is path independent. Such fields are called *conservative* fields. It can be shown that if $\mathbf{A}$ is a conservative field, then $\mathbf{A} = \nabla V$ for some scalar field $V(\mathbf{r})$. The proof of this is straightforward. Keeping P fixed, we have

$$\int_P^Q \mathbf{A} \cdot d\mathbf{r} = V(Q), \tag{A.82}$$

where $V(Q)$ is a well-defined function, owing to the path-independent nature of the line integral. Consider moving the position of the end point by an infinitesimal amount dx in the x-direction. We have

$$V(Q + dx) = V(Q) + \int_Q^{Q+dx} \mathbf{A} \cdot d\mathbf{r} = V(Q) + A_x\,dx. \tag{A.83}$$

Hence,

$$\frac{\partial V}{\partial x} = A_x, \tag{A.84}$$

with analogous relations for the other components of $\mathbf{A}$. It follows that

$$\mathbf{A} = \left(\frac{\partial V}{\partial x}, \frac{\partial V}{\partial y}, \frac{\partial V}{\partial z} \right) \equiv \nabla V. \tag{A.85}$$

A.6 Rotational coordinate transformations

Consider a conventional right-handed Cartesian coordinate system, x, y, z. Suppose that we transform to a new coordinate system, x', y', z', that is obtained from the x, y, z system by rotating the coordinate axes through an angle θ about the z-axis. (See Figure A.1.) Let the coordinates of a general point P be (x, y, z) in the first coordinate system, and (x', y', z') in the second. According to simple trigonometry, these two sets

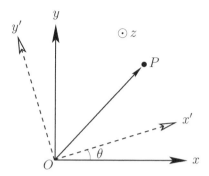

Rotation of the coordinate axes about the z-axis.

of coordinates are related to one another via the transformation

$$x' = \cos\theta\, x + \sin\theta\, y, \tag{A.86}$$

$$y' = -\sin\theta\, x + \cos\theta\, y, \tag{A.87}$$

and

$$z' = z. \tag{A.88}$$

When expressed in matrix form, this transformation becomes

$$\begin{pmatrix} x' \\ y' \\ z' \end{pmatrix} = \begin{pmatrix} \cos\theta & \sin\theta & 0 \\ -\sin\theta & \cos\theta & 0 \\ 0 & 0 & 1 \end{pmatrix} \begin{pmatrix} x \\ y \\ z \end{pmatrix}. \tag{A.89}$$

The reverse transformation is accomplished by rotating the coordinate axes through an angle $-\theta$ about the z'-axis:

$$\begin{pmatrix} x \\ y \\ z \end{pmatrix} = \begin{pmatrix} \cos\theta & -\sin\theta & 0 \\ \sin\theta & \cos\theta & 0 \\ 0 & 0 & 1 \end{pmatrix} \begin{pmatrix} x' \\ y' \\ z' \end{pmatrix}. \tag{A.90}$$

It follows that the matrix appearing in Equation (A.89) is the inverse of that appearing in Equation (A.90), and vice versa. However, because these two matrices are clearly also the transposes of each other, we deduce that both matrices are unitary. In fact, it is easily demonstrated that all rotation matrices must be unitary; otherwise, they would not preserve the lengths of the vectors on which they act.

A rotation through an angle ϕ about the x'-axis transforms the x', y', z' coordinate system into the x'', y'', z'' system, where, by analogy with the previous analysis,

$$\begin{pmatrix} x'' \\ y'' \\ z'' \end{pmatrix} = \begin{pmatrix} 1 & 0 & 0 \\ 0 & \cos\phi & \sin\phi \\ 0 & -\sin\phi & \cos\phi \end{pmatrix} \begin{pmatrix} x' \\ y' \\ z' \end{pmatrix}. \tag{A.91}$$

Thus, from Equations (A.89) and (A.91), a rotation through an angle θ about the z-axis, followed by a rotation through an angle ϕ about the x'-axis, transforms the x, y, z

coordinate system into the x'', y'', z'' system, where

$$
\begin{pmatrix} x'' \\ y'' \\ z'' \end{pmatrix} = \begin{pmatrix} 1 & 0 & 0 \\ 0 & \cos\phi & \sin\phi \\ 0 & -\sin\phi & \cos\phi \end{pmatrix} \begin{pmatrix} \cos\theta & \sin\theta & 0 \\ -\sin\theta & \cos\theta & 0 \\ 0 & 0 & 1 \end{pmatrix} \begin{pmatrix} x \\ y \\ z \end{pmatrix}. \tag{A.92}
$$

A.7 Precession

Suppose that some position vector $\mathbf{r}$ precesses (i.e., rotates) about the z-axis at the angular velocity Ω. If $x(t)$, $y(t)$, $z(t)$ are the Cartesian components of $\mathbf{r}$ at time t then it follows from the analysis in the previous section that

$$
\begin{pmatrix} x(t) \\ y(t) \\ z(t) \end{pmatrix} = \begin{pmatrix} \cos(\Omega t) & \sin(\Omega t) & 0 \\ -\sin(\Omega t) & \cos(\Omega t) & 0 \\ 0 & 0 & 1 \end{pmatrix} \begin{pmatrix} x(0) \\ y(0) \\ z(0) \end{pmatrix}. \tag{A.93}
$$

Hence, making use of the small angle approximations to the sine and cosine functions, we obtain

$$
\begin{pmatrix} x(\delta t) \\ y(\delta t) \\ z(\delta t) \end{pmatrix} - \begin{pmatrix} x(0) \\ y(0) \\ z(0) \end{pmatrix} \simeq \begin{pmatrix} 0 & \Omega\,\delta t & 0 \\ -\Omega\,\delta t & 0 & 0 \\ 0 & 0 & 0 \end{pmatrix} \begin{pmatrix} x(0) \\ y(0) \\ z(0) \end{pmatrix}, \tag{A.94}
$$

which immediately implies that

$$
\begin{pmatrix} \dot{x} \\ \dot{y} \\ \dot{z} \end{pmatrix} = \begin{pmatrix} 0 & \Omega & 0 \\ -\Omega & 0 & 0 \\ 0 & 0 & 0 \end{pmatrix} \begin{pmatrix} x \\ y \\ z \end{pmatrix} \tag{A.95}
$$

or

$$
\frac{d\mathbf{r}}{dt} = \mathbf{\Omega} \times \mathbf{r}, \tag{A.96}
$$

where $\mathbf{\Omega} = \Omega\,\mathbf{e}_z$ is the angular velocity of precession. Because vector equations are coordinate independent, we deduce that the preceding expression is the general equation for the time evolution of a position vector $\mathbf{r}$ that precesses at the angular velocity $\mathbf{\Omega}$.

A.8 Curvilinear coordinates

In the *cylindrical* coordinate system, the standard Cartesian coordinates x and y are replaced by $r = (x^2 + y^2)^{1/2}$ and $\theta = \tan^{-1}(y/x)$. Here, r is the perpendicular distance from the z-axis, and θ the angle subtended between the perpendicular radius vector and the x-axis. (See Figure A.2.) A general vector $\mathbf{A}$ is thus written

$$
\mathbf{A} = A_r\,\mathbf{e}_r + A_\theta\,\mathbf{e}_\theta + A_z\,\mathbf{e}_z, \tag{A.97}
$$

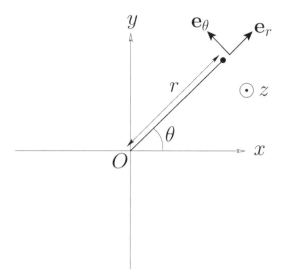

Cylindrical coordinates.

where $\mathbf{e}_r = \nabla r/|\nabla r|$ and $\mathbf{e}_\theta = \nabla\theta/|\nabla\theta|$. (See Figure A.2.) The unit vectors $\mathbf{e}_r$, $\mathbf{e}_\theta$, and $\mathbf{e}_z$ are mutually orthogonal. Hence, $A_r = \mathbf{A} \cdot \mathbf{e}_r$, etc. The volume element in this coordinate system is $d^3\mathbf{r} = r\,dr\,d\theta\,dz$. Moreover, the gradient of a general scalar field $V(\mathbf{r})$ takes the form

$$\nabla V = \frac{\partial V}{\partial r}\,\mathbf{e}_r + \frac{1}{r}\frac{\partial V}{\partial\theta}\,\mathbf{e}_\theta + \frac{\partial V}{\partial z}\,\mathbf{e}_z. \tag{A.98}$$

In the *spherical* coordinate system, the Cartesian coordinates x, y, and z are replaced by $r = (x^2 + y^2 + z^2)^{1/2}$, $\theta = \cos^{-1}(z/r)$, and $\phi = \tan^{-1}(y/x)$. Here, r is the radial distance from the origin, θ the angle subtended between the radius vector and the z-axis, and ϕ the angle subtended between the projection of the radius vector onto the x–y plane and the x-axis. (See Figure A.3.) Note that r and θ in the spherical system are *not* the same as their counterparts in the cylindrical system. A general vector $\mathbf{A}$ is written

$$\mathbf{A} = A_r\,\mathbf{e}_r + A_\theta\,\mathbf{e}_\theta + A_\phi\,\mathbf{e}_\phi, \tag{A.99}$$

where $\mathbf{e}_r = \nabla r/|\nabla r|$, $\mathbf{e}_\theta = \nabla\theta/|\nabla\theta|$, and $\mathbf{e}_\phi = \nabla\phi/|\nabla\phi|$. The unit vectors $\mathbf{e}_r$, $\mathbf{e}_\theta$, and $\mathbf{e}_\phi$ are mutually orthogonal. Hence, $A_r = \mathbf{A} \cdot \mathbf{e}_r$, and so on. The volume element in this coordinate system is $d^3\mathbf{r} = r^2\,\sin\theta\,dr\,d\theta\,d\phi$. Moreover, the gradient of a general scalar field $V(\mathbf{r})$ takes the form

$$\nabla V = \frac{\partial V}{\partial r}\,\mathbf{e}_r + \frac{1}{r}\frac{\partial V}{\partial\theta}\,\mathbf{e}_\theta + \frac{1}{r\sin\theta}\frac{\partial V}{\partial\phi}\,\mathbf{e}_\phi. \tag{A.100}$$

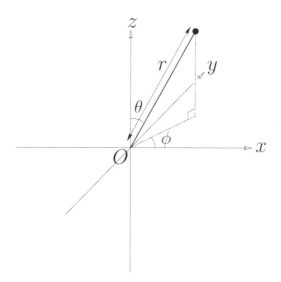

Spherical coordinates.

A.9 Conic sections

The ellipse, the parabola, and the hyperbola are collectively known as *conic sections*, as these three types of curve can be obtained by taking various different plane sections of a right cone.

An *ellipse*, centered on the origin, of major radius a and minor radius b, which are aligned along the x- and y-axes, respectively (see Figure A.4), satisfies the following

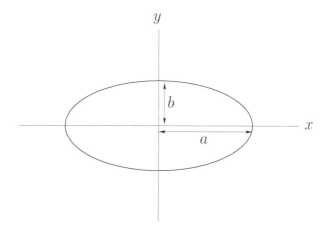

An ellipse.

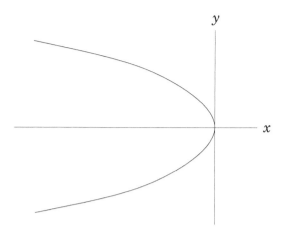

A parabola.

well-known equation:

$$\frac{x^2}{a^2} + \frac{y^2}{b^2} = 1, \tag{A.101}$$

where $0 < b \leq a$.

Likewise, a parabola, which is aligned along the $-x$-axis and passes through the origin (see Figure A.5), satisfies

$$y^2 + b\,x = 0, \tag{A.102}$$

where $b > 0$.

Finally, a hyperbola, which is aligned along the $-x$-axis and whose asymptotes intersect at the origin (see Figure A.6), satisfies

$$\frac{x^2}{a^2} - \frac{y^2}{b^2} = 1, \tag{A.103}$$

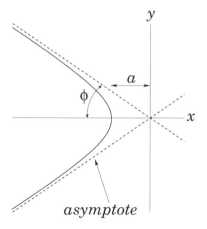

A hyperbola.

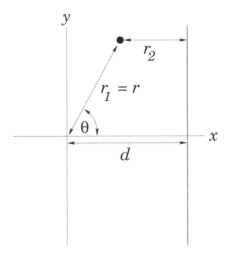

Conic sections in polar coordinates.

where $a, b > 0$. Here, a is the distance of closest approach to the origin. The asymptotes subtend an angle $\phi = \tan^{-1}(b/a)$ with the $-x$-axis.

It is not obvious, from the preceding formulae, what the ellipse, the parabola, and the hyperbola have in common. It turns out, in fact, that these three curves can all be represented as the locus of a movable point whose distance from a fixed point is in a constant ratio to its perpendicular distance to some fixed straight line. Let the fixed point—which is termed the *focus*—lie at the origin, and let the fixed line—which is termed the *directrix*—correspond to $x = d$ (with $d > 0$). Thus, the distance of a general point (x, y) (which lies to the left of the directrix) from the focus is $r_1 = (x^2 + y^2)^{1/2}$, whereas the perpendicular distance of the point from the directrix is $r_2 = d - x$. (See Figure A.7.) In polar coordinates, $r_1 = r$ and $r_2 = d - r\cos\theta$. Hence, the locus of a point for which r_1 and r_2 are in a fixed ratio satisfies the following equation:

$$\frac{r_1}{r_2} = \frac{(x^2 + y^2)^{1/2}}{d - x} = \frac{r}{d - r\cos\theta} = e, \tag{A.104}$$

where $e \geq 0$ is a constant. When expressed in terms of polar coordinates, the preceding equation can be rearranged to give

$$r = \frac{r_c}{1 + e\cos\theta}, \tag{A.105}$$

where $r_c = e\,d$.

When written in terms of Cartesian coordinates, Equation (A.104) can be rearranged to give

$$\frac{(x - x_c)^2}{a^2} + \frac{y^2}{b^2} = 1 \tag{A.106}$$

for $e < 1$. Here,

$$a = \frac{r_c}{1 - e^2}, \tag{A.107}$$

$$b = \frac{r_c}{\sqrt{1 - e^2}} = \sqrt{1 - e^2}\, a, \tag{A.108}$$

and

$$x_c = -\frac{e\, r_c}{1 - e^2} = -e\, a. \tag{A.109}$$

Equation (A.106) can be recognized as the equation of an ellipse whose center lies at $(x_c, 0)$, and whose major and minor radii, a and b, are aligned along the x- and y-axes, respectively [see Equation (A.101)]. Note, incidentally, that an ellipse actually possesses two foci located on the major axis ($y = 0$) a distance $e\, a$ on either side of the geometric center (i.e., at $x = 0$ and $x = -2\, e\, a$). Likewise, an ellipse possesses two directrices located at $x = a\,(1 \pm e^2)/e$.

When again written in terms of Cartesian coordinates, Equation (A.104) can be rearranged to give

$$y^2 + 2\, r_c\, (x - x_c) = 0 \tag{A.110}$$

for $e = 1$. Here, $x_c = r_c/2$. This is the equation of a *parabola* that passes through the point $(x_c, 0)$, and which is aligned along the $-x$-direction [see Equation (A.102)].

Finally, when written in terms of Cartesian coordinates, Equation (A.104) can be rearranged to give

$$\frac{(x - x_c)^2}{a^2} - \frac{y^2}{b^2} = 1 \tag{A.111}$$

for $e > 1$. Here,

$$a = \frac{r_c}{e^2 - 1}, \tag{A.112}$$

$$b = \frac{r_c}{\sqrt{e^2 - 1}} = \sqrt{e^2 - 1}\, a, \tag{A.113}$$

and

$$x_c = \frac{e\, r_c}{e^2 - 1} = e\, a. \tag{A.114}$$

Equation (A.111) can be recognized as the equation of a *hyperbola* whose asymptotes intersect at $(x_c, 0)$, and which is aligned along the $-x$-direction. The asymptotes subtend an angle

$$\phi = \tan^{-1}\left(\frac{b}{a}\right) = \tan^{-1}\left(\sqrt{e^2 - 1}\right) \tag{A.115}$$

with the $-x$-axis [see Equation (A.103)].

In conclusion, Equation (A.105) is the polar equation of a general conic section that is confocal with the origin (i.e., the origin lies at a focus). For $e < 1$, the conic section is an ellipse. For $e = 1$, the conic section is a parabola. Finally, for $e > 1$, the conic section is a hyperbola.

A.10 Elliptic expansions

The well-known *Bessel functions of the first kind*, $J_n(x)$, where n is an integer, are defined as the Fourier coefficients in the expansion of $\exp(\mathrm{i}\,x\sin\phi)$:

$$\mathrm{e}^{\mathrm{i}\,x\sin\phi} \equiv \sum_{n=-\infty,\infty} J_n(x)\,\mathrm{e}^{\mathrm{i}\,n\phi}. \qquad (A.116)$$

It follows that

$$J_n(x) = \frac{1}{2\pi}\oint \mathrm{e}^{-\mathrm{i}\,(n\phi-x\sin\phi)}\,d\phi = \frac{1}{\pi}\int_0^\pi \cos\left(x\sin\phi - n\phi\right)d\phi \qquad (A.117)$$

(Gradshteyn and Ryzhik 1980a). The Taylor expansion of $J_n(x)$ about $x = 0$ is

$$J_n(x) = \left(\frac{x}{2}\right)^n \sum_{k=0,\infty} \frac{(-x^2/4)^k}{k!\,(n+k)!} \qquad (A.118)$$

for $n \geq 0$ (Gradshteyn and Ryzhik 1980b). Moreover,

$$J_{-n}(x) = (-1)^n\,J_n(x), \qquad (A.119)$$

$$J_{-n}(-x) = J_n(x). \qquad (A.120)$$

In particular,

$$J_0(x) = 1 - \frac{x^2}{4} + \mathcal{O}(x^4), \qquad (A.121)$$

$$J_1(x) = \frac{x}{2} - \frac{x^3}{16} + \mathcal{O}(x^5), \qquad (A.122)$$

$$J_2(x) = \frac{x^2}{8} + \mathcal{O}(x^4), \qquad (A.123)$$

and

$$J_3(x) = \frac{x^3}{48} + \mathcal{O}(x^5). \qquad (A.124)$$

Let us write

$$\mathrm{e}^{\mathrm{i}\,E} = \sum_{n=-\infty,\infty} A_n\,\mathrm{e}^{\mathrm{i}\,n\,\mathcal{M}}, \qquad (A.125)$$

where E is the *eccentric anomaly*, and $\mathcal{M}$ the *mean anomaly*, of a Keplerian elliptic orbit. (See Section 3.11.) It follows that

$$A_n = \frac{1}{2\pi}\oint \mathrm{e}^{\mathrm{i}\,(E-n\,\mathcal{M})}\,d\mathcal{M}. \qquad (A.126)$$

Integrating by parts, we obtain

$$A_n = \frac{1}{2\pi\,n}\oint \mathrm{e}^{\mathrm{i}\,(E-n\,\mathcal{M})}\,dE. \qquad (A.127)$$

However, according to Equation (3.59), the relationship between the eccentric and the mean anomalies is

$$E - e\sin E = \mathcal{M}, \qquad (A.128)$$

where e is the orbital eccentricity. Hence,

$$A_n = \frac{1}{2\pi n} \oint e^{-i[(n-1)E - n e \sin E]} dE. \tag{A.129}$$

Comparison with Equation (A.117) reveals that

$$A_n = \frac{J_{n-1}(n e)}{n}. \tag{A.130}$$

For the special case $n = 0$, L'Hôpital's rule, together with Equations (A.119) and (A.122), yields

$$A_0 = e J'_{-1}(0) = -e J'_1(0) = -\frac{e}{2}, \tag{A.131}$$

where $'$ denotes a derivative.

The real part of Equation (A.125) gives

$$\cos E = A_0 + \sum_{n=1,\infty} (A_n + A_{-n}) \cos(n \mathcal{M})$$

$$= -\frac{e}{2} + \sum_{n=1,\infty} \left[\frac{J_{n-1}(n e) - J_{-n-1}(-n e)}{n} \right] \cos(n \mathcal{M})$$

$$= -\frac{e}{2} + \sum_{n=1,\infty} \left[\frac{J_{n-1}(n e) - J_{n+1}(n e)}{n} \right] \cos(n \mathcal{M}), \tag{A.132}$$

where use has been made of Equations (A.120), (A.130), and (A.131). Likewise, the imaginary part of (A.125) yields

$$\sin E = \sum_{n=1,\infty} \left[\frac{J_{n-1}(n e) + J_{n+1}(n e)}{n} \right] \sin(n \mathcal{M}). \tag{A.133}$$

It follows from Equations (A.121)–(A.124) that

$$\sin E = \left(1 - \frac{e^2}{8}\right) \sin \mathcal{M} + \frac{e}{2}\left(1 - \frac{e^2}{3}\right) \sin 2\mathcal{M} + \frac{3 e^2}{8} \sin 3\mathcal{M}$$

$$+ \frac{e^3}{3} \sin 4\mathcal{M} + \mathcal{O}(e^4), \tag{A.134}$$

and

$$\cos E = -\frac{e}{2} + \left(1 - \frac{3 e^2}{8}\right) \cos \mathcal{M} + \frac{e}{2}\left(1 - \frac{2 e^2}{3}\right) \cos 2\mathcal{M}$$

$$+ \frac{3 e^2}{8} \cos 3\mathcal{M} + \frac{e^3}{3} \cos 4\mathcal{M} + \mathcal{O}(e^4). \tag{A.135}$$

Hence, from (A.128),

$$E = \mathcal{M} + e \sin \mathcal{M} + \frac{e^2}{2} \sin 2\mathcal{M} + \frac{e^3}{8}(3 \sin 3\mathcal{M} - \sin \mathcal{M})$$

$$+ \frac{e^4}{6}(2 \sin 4\mathcal{M} - \sin 2\mathcal{M}) + \mathcal{O}(e^5). \tag{A.136}$$

According to Equation (3.69),

$$\frac{r}{a} = 1 - e \cos E, \tag{A.137}$$

where r is is the radial distance from the focus of the orbit and a is the orbital major radius. Thus,

$$\frac{r}{a} = 1 - e \cos \mathcal{M} - \frac{e^2}{2} (\cos 2\mathcal{M} - 1) - \frac{3\,e^3}{8} (\cos 3\mathcal{M} - \cos \mathcal{M})$$

$$- \frac{e^4}{3} (\cos 4\mathcal{M} - \cos 2\mathcal{M}) + \mathcal{O}(e^5). \tag{A.138}$$

Equations (3.39) and (3.67) imply that

$$\frac{d\theta}{d\mathcal{M}} = (1 - e^2)^{1/2} \left(\frac{a}{r}\right)^2, \tag{A.139}$$

where θ is the true anomaly. Hence, it follows from Equations (A.128) and (A.137), and the fact that $\theta = 0$ when $\mathcal{M} = 0$, that

$$\theta = (1 - e^2)^{1/2} \int_0^{\mathcal{M}} \left(\frac{dE}{d\mathcal{M}}\right)^2 d\mathcal{M}. \tag{A.140}$$

From Equation (A.136),

$$\frac{dE}{d\mathcal{M}} = 1 + e\left(1 - \frac{e^2}{8}\right) \cos \mathcal{M} + e^2\left(1 - \frac{e^2}{3}\right) \cos 2\mathcal{M} + \frac{9\,e^3}{8} \cos 3\mathcal{M}$$

$$+ \frac{4\,e^4}{3} \cos 4\mathcal{M} + \mathcal{O}(e^5). \tag{A.141}$$

Thus,

$$\left(\frac{dE}{d\mathcal{M}}\right)^2 = 1 + \frac{e^2}{2} + \frac{3\,e^4}{8} + 2\,e\left(1 + \frac{3\,e^2}{8}\right) \cos \mathcal{M} + \frac{5\,e^2}{2}\left(1 + \frac{2\,e^2}{15}\right) \cos 2\mathcal{M}$$

$$+ \frac{13\,e^3}{4} \cos 3\mathcal{M} + \frac{103\,e^4}{24} \cos 4\mathcal{M} + \mathcal{O}(e^5) \tag{A.142}$$

and

$$\theta = \mathcal{M} + 2\,e \sin \mathcal{M} + \frac{5\,e^2}{4} \sin 2\mathcal{M} + e^3\left(\frac{13}{12} \sin 3\mathcal{M} - \frac{1}{4} \sin \mathcal{M}\right)$$

$$+ e^4\left(\frac{103}{96} \sin 4\mathcal{M} - \frac{11}{24} \sin 2\mathcal{M}\right) + \mathcal{O}(e^5). \tag{A.143}$$

A.11 Matrix eigenvalue theory

Suppose that $\mathbf{A}$ is a *real symmetric* square matrix of dimension n. If follows that $\mathbf{A}^* = \mathbf{A}$ and $\mathbf{A}^T = \mathbf{A}$, where * denotes a complex conjugate, and ^{T} denotes a transpose. Consider the matrix equation

$$\mathbf{A}\mathbf{x} = \lambda\,\mathbf{x}. \tag{A.144}$$

Any column vector $\mathbf{x}$ that satisfies this equation is called an *eigenvector* of $\mathbf{A}$. Likewise, the associated number λ is called an *eigenvalue* of $\mathbf{A}$ (Gradshteyn and Ryzhik 1980c).

Let us investigate the properties of the eigenvectors and eigenvalues of a real symmetric matrix.

Equation (A.144) can be rearranged to give

$$(\mathbf{A} - \lambda \mathbf{1})\mathbf{x} = \mathbf{0}, \tag{A.145}$$

where $\mathbf{1}$ is the unit matrix. The preceding matrix equation is essentially a set of n homogeneous simultaneous algebraic equations for the n components of $\mathbf{x}$. A well-known property of such a set of equations is that it has a nontrivial solution only when the determinant of the associated matrix is set to zero (Gradshteyn and Ryzhik 1980c). Hence, a necessary condition for the preceding set of equations to have a nontrivial solution is that

$$|\mathbf{A} - \lambda \mathbf{1}| = 0, \tag{A.146}$$

where $|\ |$ denotes a *determinant*. This formula is essentially an nth-order *polynomial* equation for λ. We know that such an equation has n (possibly complex) roots. Hence, we conclude that there are n eigenvalues, and n associated eigenvectors, of the n-dimensional matrix $\mathbf{A}$.

Let us now demonstrate that the n eigenvalues and eigenvectors of the real symmetric matrix $\mathbf{A}$ are all *real*. We have

$$\mathbf{A}\,\mathbf{x}_i = \lambda_i\,\mathbf{x}_i, \tag{A.147}$$

and, taking the transpose and complex conjugate,

$$\mathbf{x}_i^{*T}\,\mathbf{A} = \lambda_i^*\,\mathbf{x}_i^{*T}, \tag{A.148}$$

where $\mathbf{x}_i$ and λ_i are the ith eigenvector and eigenvalue of $\mathbf{A}$, respectively. Left multiplying Equation (A.147) by $\mathbf{x}_i^{*T}$, we obtain

$$\mathbf{x}_i^{*T}\mathbf{A}\,\mathbf{x}_i = \lambda_i\,\mathbf{x}_i^{*T}\mathbf{x}_i. \tag{A.149}$$

Likewise, right multiplying Equation (A.148) by $\mathbf{x}_i$, we get

$$\mathbf{x}_i^{*T}\,\mathbf{A}\,\mathbf{x}_i = \lambda_i^*\,\mathbf{x}_i^{*T}\mathbf{x}_i. \tag{A.150}$$

The difference of the previous two equations yields

$$(\lambda_i - \lambda_i^*)\,\mathbf{x}_i^{*T}\mathbf{x}_i = 0. \tag{A.151}$$

It follows that $\lambda_i = \lambda_i^*$, because $\mathbf{x}_i^{*T}\mathbf{x}_i$ (which is $\mathbf{x}_i^* \cdot \mathbf{x}_i$ in vector notation) is real and positive definite. Hence, λ_i is real. It immediately follows that $\mathbf{x}_i$ is real.

Next, let us show that two eigenvectors corresponding to two *different* eigenvalues are *mutually orthogonal*. Let

$$\mathbf{A}\,\mathbf{x}_i = \lambda_i\,\mathbf{x}_i \tag{A.152}$$

and

$$\mathbf{A}\,\mathbf{x}_j = \lambda_j\,\mathbf{x}_j, \tag{A.153}$$

where $\lambda_i \neq \lambda_j$. Taking the transpose of the first equation and right multiplying by $\mathbf{x}_j$, and left multiplying the second equation by $\mathbf{x}_i^T$, we obtain

$$\mathbf{x}_i^T \mathbf{A} \mathbf{x}_j = \lambda_i \mathbf{x}_i^T \mathbf{x}_j \tag{A.154}$$

and

$$\mathbf{x}_i^T \mathbf{A} \mathbf{x}_j = \lambda_j \mathbf{x}_i^T \mathbf{x}_j. \tag{A.155}$$

Taking the difference of these two equations, we get

$$(\lambda_i - \lambda_j) \mathbf{x}_i^T \mathbf{x}_j = 0. \tag{A.156}$$

Because, by hypothesis, $\lambda_i \neq \lambda_j$, it follows that $\mathbf{x}_i^T \mathbf{x}_j = 0$. In vector notation, this is the same as $\mathbf{x}_i \cdot \mathbf{x}_j = 0$. Hence, the eigenvectors $\mathbf{x}_i$ and $\mathbf{x}_j$ are mutually orthogonal.

Suppose that $\lambda_i = \lambda_j = \lambda$. In this case, we cannot conclude that $\mathbf{x}_i^T \mathbf{x}_j = 0$ by the preceding argument. However, it is easily seen that any linear combination of $\mathbf{x}_i$ and $\mathbf{x}_j$ is an eigenvector of $\mathbf{A}$ with eigenvalue λ. Hence, it is possible to define two new eigenvectors of $\mathbf{A}$, with the eigenvalue λ, which are mutually orthogonal. For instance,

$$\mathbf{x}_i' = \mathbf{x}_i \tag{A.157}$$

and

$$\mathbf{x}_j' = \mathbf{x}_j - \left(\frac{\mathbf{x}_i^T \mathbf{x}_j}{\mathbf{x}_i^T \mathbf{x}_i} \right) \mathbf{x}_i. \tag{A.158}$$

It should be clear that this argument can be generalized to deal with any number of eigenvalues that take the same value.

In conclusion, a real symmetric n-dimensional matrix possesses n real eigenvalues, with n associated real eigenvectors, which are, or can be chosen to be, mutually orthogonal.

Appendix B Derivation of Lagrange planetary equations

B.1 Introduction

Consider a planet of mass m and relative position vector $\mathbf{r}$ that is orbiting around the Sun, whose mass is M. The planet's equation of motion is written (see Section 3.16)

$$\ddot{\mathbf{r}} + \mu\,\frac{\mathbf{r}}{r^3} = \mathbf{0}, \tag{B.1}$$

where $\mu = G\,(M + m)$. As described in Chapter 3, the solution to this equation is a Keplerian ellipse whose properties are fully determined after six integrals of the motion, known as *orbital elements*, are specified.

Now, suppose that the aforementioned Keplerian orbit is slightly perturbed—for example, by the presence of a second planet orbiting the Sun. In this case, the planet's modified equation of motion takes the general form

$$\ddot{\mathbf{r}} + \mu\,\frac{\mathbf{r}}{r^3} = \nabla\mathcal{R}, \tag{B.2}$$

where $\mathcal{R}(\mathbf{r})$ is a so-called *disturbing function* that fully describes the perturbation. Adopting the standard Cartesian coordinate system X, Y, Z, described in Section 3.12, we see that the preceding equation yields

$$\ddot{X} + \mu\,\frac{X}{r^3} = \frac{\partial\mathcal{R}}{\partial X}, \tag{B.3}$$

$$\ddot{Y} + \mu\,\frac{Y}{r^3} = \frac{\partial\mathcal{R}}{\partial Y}, \tag{B.4}$$

and

$$\ddot{Z} + \mu\,\frac{Z}{r^3} = \frac{\partial\mathcal{R}}{\partial Z}, \tag{B.5}$$

where $r = (X^2 + Y^2 + Z^2)^{1/2}$.

If the right-hand sides of Equations (B.3)–(B.5) are set to zero (i.e., if there is no perturbation), we obtain a Keplerian orbit of the general form

$$X = f_1(c_1, c_2, c_3, c_4, c_5, c_6, t), \tag{B.6}$$

$$Y = f_2(c_1, c_2, c_3, c_4, c_5, c_6, t), \tag{B.7}$$

$$Z = f_3(c_1, c_2, c_3, c_4, c_5, c_6, t), \tag{B.8}$$

$$\dot{X} = g_1(c_1, c_2, c_3, c_4, c_5, c_6, t), \tag{B.9}$$

$$\dot{Y} = g_2(c_1, c_2, c_3, c_4, c_5, c_6, t), \tag{B.10}$$

and

$$\dot{Z} = g_3(c_1, c_2, c_3, c_4, c_5, c_6, t). \tag{B.11}$$

Here, $c_1, \ldots, c_6$ are the six constant elements that determine the orbit. (See Section 3.12.) It follows that

$$g_k = \frac{\partial f_k}{\partial t} \tag{B.12}$$

for $k = 1, 2, 3$.

Let us now take the right-hand sides of Equations (B.3)–(B.5) into account. In this case, the orbital elements, $c_1, \ldots, c_6$, are no longer constants of the motion. However, provided the perturbation is sufficiently small, we would expect the elements to be relatively *slowly varying* functions of time. The purpose of this appendix is to derive evolution equations for these so-called *osculating orbital elements*. Our approach is largely based on that of Brouwer and Clemence (1961).

B.2 Preliminary analysis

According to Equation (B.6), we have

$$\frac{dX}{dt} = \frac{\partial f_1}{\partial t} + \sum_{k=1,6} \frac{\partial f_1}{\partial c_k} \frac{dc_k}{dt}. \tag{B.13}$$

If this expression, and the analogous expressions for dY/dt and dZ/dt, were differentiated with respect to time, and the results substituted into Equations (B.3)–(B.5), then we would obtain three time evolution equations for the six variables $c_1, \ldots, c_6$. To make the problem definite, three additional conditions must be introduced into the problem. It is convenient to choose

$$\sum_{k=1,6} \frac{\partial f_l}{\partial c_k} \frac{dc_k}{dt} = 0 \tag{B.14}$$

for $l = 1, 2, 3$. Hence, it follows from Equations (B.12) and (B.13) that

$$\frac{dX}{dt} = \frac{\partial f_1}{\partial t} = g_1, \tag{B.15}$$

$$\frac{dY}{dt} = \frac{\partial f_2}{\partial t} = g_2, \tag{B.16}$$

and

$$\frac{dZ}{dt} = \frac{\partial f_3}{\partial t} = g_3. \tag{B.17}$$

Differentiation of the these equations with respect to time yields

$$\frac{d^2X}{dt^2} = \frac{\partial^2 f_1}{\partial t^2} + \sum_{k=1,6} \frac{\partial g_1}{\partial c_k} \frac{dc_k}{dt}, \tag{B.18}$$

$$\frac{d^2Y}{dt^2} = \frac{\partial^2 f_2}{\partial t^2} + \sum_{k=1,6} \frac{\partial g_2}{\partial c_k} \frac{dc_k}{dt}, \tag{B.19}$$

and

$$\frac{d^2 Z}{dt^2} = \frac{\partial^2 f_3}{\partial t^2} + \sum_{k=1,6} \frac{\partial g_3}{\partial c_k} \frac{dc_k}{dt}. \tag{B.20}$$

Substitution into Equations (B.3)–(B.5) gives

$$\frac{\partial^2 f_1}{\partial t^2} + \mu \frac{f_1}{r^3} + \sum_{k=1,6} \frac{\partial g_1}{\partial c_k} \frac{dc_k}{dt} = \frac{\partial \mathcal{R}}{\partial X}, \tag{B.21}$$

$$\frac{\partial^2 f_2}{\partial t^2} + \mu \frac{f_2}{r^3} + \sum_{k=1,6} \frac{\partial g_2}{\partial c_k} \frac{dc_k}{dt} = \frac{\partial \mathcal{R}}{\partial Y}, \tag{B.22}$$

and

$$\frac{\partial^2 f_3}{\partial t^2} + \mu \frac{f_3}{r^3} + \sum_{k=1,6} \frac{\partial g_3}{\partial c_k} \frac{dc_k}{dt} = \frac{\partial \mathcal{R}}{\partial Z}, \tag{B.23}$$

where $r = (f_1^2 + f_2^2 + f_3^2)^{1/2}$. Because f_1, f_2, and f_3 are the respective solutions to Equation (B.3)–(B.5) when the right-hand sides are zero, and the orbital elements are thus constants, it follows that the first two terms in each of the preceding three equations cancel one another. Hence, writing f_1 as X, and g_1 as $\dot{X}$, and so on, we see that Equations (B.14) and (B.21)–(B.23) yield

$$\sum_{k=1,6} \frac{\partial X}{\partial c_k} \frac{dc_k}{dt} = 0, \tag{B.24}$$

$$\sum_{k=1,6} \frac{\partial Y}{\partial c_k} \frac{dc_k}{dt} = 0, \tag{B.25}$$

$$\sum_{k=1,6} \frac{\partial Z}{\partial c_k} \frac{dc_k}{dt} = 0, \tag{B.26}$$

$$\sum_{k=1,6} \frac{\partial \dot{X}}{\partial c_k} \frac{dc_k}{dt} = \frac{\partial \mathcal{R}}{\partial X}, \tag{B.27}$$

$$\sum_{k=1,6} \frac{\partial \dot{Y}}{\partial c_k} \frac{dc_k}{dt} = \frac{\partial \mathcal{R}}{\partial Y}, \tag{B.28}$$

and

$$\sum_{k=1,6} \frac{\partial \dot{Z}}{\partial c_k} \frac{dc_k}{dt} = \frac{\partial \mathcal{R}}{\partial Z}. \tag{B.29}$$

These six equations are equivalent to the three original equations of motion [(B.3)–(B.5)].

B.3 Lagrange brackets

Six new equations can be derived from Equations (B.24)–(B.29) by multiplying them successively by $-\partial \dot{X}/\partial c_j$, $-\partial \dot{Y}/\partial c_j$, $-\partial \dot{Z}/\partial c_j$, $\partial X/\partial c_j$, $\partial Y/\partial c_j$, and $\partial Z/\partial c_j$, and then summing the resulting equations. The right-hand sides of the new equations are

$$\frac{\partial \mathcal{R}}{\partial X}\frac{\partial X}{\partial c_j} + \frac{\partial \mathcal{R}}{\partial Y}\frac{\partial Y}{\partial c_j} + \frac{\partial \mathcal{R}}{\partial Z}\frac{\partial Z}{\partial c_j} \equiv \frac{\partial \mathcal{R}}{\partial c_j}. \tag{B.30}$$

The new equations can be written in a more compact form via the introduction of *Lagrange brackets*, which are defined as

$$[c_j, c_k] \equiv \sum_{l=1,3}\left(\frac{\partial X_l}{\partial c_j}\frac{\partial \dot{X}_l}{\partial c_k} - \frac{\partial X_l}{\partial c_k}\frac{\partial \dot{X}_l}{\partial c_j}\right), \tag{B.31}$$

where $X_1 \equiv X$, $X_2 \equiv Y$, and $X_3 \equiv Z$. Thus, the new equations become

$$\sum_{k=1,6}[c_j, c_k]\frac{dc_k}{dt} = \frac{\partial \mathcal{R}}{\partial c_j} \tag{B.32}$$

for $j = 1, 6$. Note, incidentally, that

$$[c_j, c_j] = 0 \tag{B.33}$$

and

$$[c_j, c_k] = -[c_k, c_j]. \tag{B.34}$$

Let

$$[p, q] = \sum_{l=1,3}\left(\frac{\partial X_l}{\partial p}\frac{\partial \dot{X}_l}{\partial q} - \frac{\partial X_l}{\partial q}\frac{\partial \dot{X}_l}{\partial p}\right), \tag{B.35}$$

where p and q are any two orbital elements. It follows that

$$\frac{\partial}{\partial t}[p, q] = \sum_{l=1,3}\left(\frac{\partial^2 X_l}{\partial p\,\partial t}\frac{\partial \dot{X}_l}{\partial q} + \frac{\partial X_l}{\partial p}\frac{\partial^2 \dot{X}_l}{\partial q\,\partial t} - \frac{\partial^2 X_l}{\partial q\,\partial t}\frac{\partial \dot{X}_l}{\partial p} - \frac{\partial X_l}{\partial q}\frac{\partial^2 \dot{X}_l}{\partial p\,\partial t}\right) \tag{B.36}$$

or

$$\frac{\partial}{\partial t}[p, q] = \sum_{l=1,3}\left[\frac{\partial}{\partial p}\left(\frac{\partial X_l}{\partial t}\frac{\partial \dot{X}_l}{\partial q} - \frac{\partial X_l}{\partial q}\frac{\partial \dot{X}_l}{\partial t}\right) - \frac{\partial}{\partial q}\left(\frac{\partial X_l}{\partial t}\frac{\partial \dot{X}_l}{\partial p} - \frac{\partial X_l}{\partial p}\frac{\partial \dot{X}_l}{\partial t}\right)\right]. \tag{B.37}$$

However, in the preceding expression, X_l and $\dot{X}_l$ stand for coordinates and velocities of Keplerian orbits calculated with $c_1, \ldots, c_6$ treated as constants. Thus, we can write $\partial X_l/\partial t \equiv \dot{X}_l$ and $\partial \dot{X}_l/\partial t \equiv \ddot{X}_l$, giving

$$\frac{\partial}{\partial t}[p, q] = \sum_{l=1,3}\left[\frac{\partial}{\partial p}\left(\frac{1}{2}\frac{\partial \dot{X}_l^2}{\partial q} - \frac{\partial F_0}{\partial X_l}\frac{\partial X_l}{\partial q}\right) - \frac{\partial}{\partial q}\left(\frac{1}{2}\frac{\partial \dot{X}_l^2}{\partial p} - \frac{\partial F_0}{\partial X_l}\frac{\partial X_l}{\partial p}\right)\right], \tag{B.38}$$

because

$$\ddot{X}_l = \frac{\partial F_0}{\partial X_l}, \tag{B.39}$$

where $F_0 = \mu/r$. Equation (B.38) reduces to

$$\frac{\partial}{\partial t}[p, q] = \frac{1}{2}\frac{\partial^2 v^2}{\partial p\,\partial q} - \frac{\partial^2 F_0}{\partial p\,\partial q} - \frac{1}{2}\frac{\partial^2 v^2}{\partial q\,\partial p} + \frac{\partial^2 F_0}{\partial q\,\partial p} = 0, \tag{B.40}$$

where $v^2 = \sum_{l=1,3}\dot{X}_l^2$. Hence, we conclude that Lagrange brackets are functions of the osculating orbital elements, $c_1, \ldots, c_6$, but are not explicit functions of t. It follows that we can evaluate these brackets at any convenient point in the orbit.

B.4 Transformation of Lagrange brackets

The most common set of orbital elements used to parameterize Keplerian orbits consists of the *major radius*, a; the *mean longitude at epoch*, $\bar{\lambda}_0$; the *eccentricity*, e; the *inclination* (relative to some reference plane), I; the *longitude of the perihelion*, ϖ; and *the longitude of the ascending node*, Ω. (See Section 3.12.) The mean orbital angular velocity is $n = (\mu/a^3)^{1/2}$ [see Equation (3.116)].

Consider how a particular Lagrange bracket transforms under a rotation of the coordinate system X, Y, Z about the Z-axis (if we look along the axis). We can write

$$[p,q] = \frac{\partial(X,\dot{X})}{\partial(p,q)} + \frac{\partial(Y,\dot{Y})}{\partial(p,q)} + \frac{\partial(Z,\dot{Z})}{\partial(p,q)}, \tag{B.41}$$

where

$$\frac{\partial(a,b)}{\partial(c,d)} \equiv \frac{\partial a}{\partial c}\frac{\partial b}{\partial d} - \frac{\partial a}{\partial d}\frac{\partial b}{\partial c}. \tag{B.42}$$

Let the new coordinate system be x', y', z'. A rotation about the Z-axis through an angle Ω brings the ascending node to the x'-axis. (See Figure 3.6.) The relation between the old and new coordinates is (see Section A.6)

$$X = \cos\Omega\, x' - \sin\Omega\, y', \tag{B.43}$$

$$Y = \sin\Omega\, x' + \cos\Omega\, y', \tag{B.44}$$

and

$$Z = z'. \tag{B.45}$$

The partial derivatives with respect to p can be written

$$\frac{\partial X}{\partial p} = A_1\,\cos\Omega - B_1\,\sin\Omega, \tag{B.46}$$

$$\frac{\partial Y}{\partial p} = B_1\,\cos\Omega + A_1\,\sin\Omega, \tag{B.47}$$

$$\frac{\partial \dot{X}}{\partial p} = C_1\,\cos\Omega - D_1\,\sin\Omega, \tag{B.48}$$

and

$$\frac{\partial \dot{Y}}{\partial p} = D_1\,\cos\Omega + C_1\,\sin\Omega, \tag{B.49}$$

where

$$A_1 = \frac{\partial x'}{\partial p} - y'\frac{\partial\Omega}{\partial p}, \tag{B.50}$$

$$B_1 = \frac{\partial y'}{\partial p} + x'\frac{\partial\Omega}{\partial p}, \tag{B.51}$$

$$C_1 = \frac{\partial \dot{x}'}{\partial p} - \dot{y}'\frac{\partial\Omega}{\partial p}, \tag{B.52}$$

and

$$D_1 = \frac{\partial \dot{y}'}{\partial p} + \dot{x}' \frac{\partial \Omega}{\partial p}. \tag{B.53}$$

Let A_2, B_2, C_2, and D_2 be the equivalent quantities obtained by replacing p by q in the above equations. It thus follows that

$$\frac{\partial(X, \dot{X})}{\partial(p, q)} = (A_1 C_2 - A_2 C_1) \cos^2 \Omega + (B_1 D_2 - B_2 D_1) \sin^2 \Omega$$

$$+ (-A_1 D_2 - B_1 C_2 + A_2 D_1 + B_2 C_1) \sin \Omega \cos \Omega, \tag{B.54}$$

and

$$\frac{\partial(Y, \dot{Y})}{\partial(p, q)} = (B_1 D_2 - B_2 D_1) \cos^2 \Omega + (A_1 C_2 - A_2 C_1) \sin^2 \Omega$$

$$+ (A_1 D_2 + B_1 C_2 - A_2 D_1 - B_2 C_1) \sin \Omega \cos \Omega. \tag{B.55}$$

Hence,

$$[p, q] = A_1 C_2 - A_2 C_1 + B_1 D_2 - B_2 D_1 + \frac{\partial(Z, \dot{Z})}{\partial(p, q)}. \tag{B.56}$$

Now,

$$A_1 C_2 - A_2 C_1 = \left(\frac{\partial x'}{\partial p} - y' \frac{\partial \Omega}{\partial p} \right) \left(\frac{\partial \dot{x}'}{\partial q} - \dot{y}' \frac{\partial \Omega}{\partial q} \right) - \left(\frac{\partial x'}{\partial q} - y' \frac{\partial \Omega}{\partial q} \right) \left(\frac{\partial \dot{x}'}{\partial p} - \dot{y}' \frac{\partial \Omega}{\partial p} \right)$$

$$= \frac{\partial(x', \dot{x}')}{\partial(p, q)}$$

$$+ \left(-y' \frac{\partial \dot{x}'}{\partial q} + \dot{y}' \frac{\partial x'}{\partial q} \right) \frac{\partial \Omega}{\partial p} + \left(-\dot{y}' \frac{\partial x'}{\partial p} + y' \frac{\partial \dot{x}'}{\partial p} \right) \frac{\partial \Omega}{\partial q}. \tag{B.57}$$

Similarly,

$$B_1 D_2 - B_2 D_1 = \left(\frac{\partial y'}{\partial p} + x' \frac{\partial \Omega}{\partial p} \right) \left(\frac{\partial \dot{y}'}{\partial q} + \dot{x}' \frac{\partial \Omega}{\partial q} \right) - \left(\frac{\partial y'}{\partial q} + x' \frac{\partial \Omega}{\partial q} \right) \left(\frac{\partial \dot{y}'}{\partial p} + \dot{x}' \frac{\partial \Omega}{\partial p} \right)$$

$$= \frac{\partial(y', \dot{y}')}{\partial(p, q)} + \left(x' \frac{\partial \dot{y}'}{\partial q} - \dot{x}' \frac{\partial y'}{\partial q} \right) \frac{\partial \Omega}{\partial p} + \left(\dot{x}' \frac{\partial y'}{\partial p} - x' \frac{\partial \dot{y}'}{\partial p} \right) \frac{\partial \Omega}{\partial q}. \tag{B.58}$$

Let

$$[p, q]' = \frac{\partial(x', \dot{x}')}{\partial(p, q)} + \frac{\partial(y', \dot{y}')}{\partial(p, q)} + \frac{\partial(z', \dot{z}')}{\partial(p, q)}. \tag{B.59}$$

Because $Z = z'$ and $\dot{Z} = \dot{z}'$, it follows that

$$[p, q] = [p, q]' + \left(x' \frac{\partial \dot{y}'}{\partial q} + \dot{y}' \frac{\partial x'}{\partial q} - y' \frac{\partial \dot{x}'}{\partial q} - \dot{x}' \frac{\partial y'}{\partial q} \right) \frac{\partial \Omega}{\partial p}$$

$$- \left(x' \frac{\partial \dot{y}'}{\partial p} + \dot{y}' \frac{\partial x'}{\partial p} - y' \frac{\partial \dot{x}'}{\partial p} - \dot{x}' \frac{\partial y'}{\partial p} \right) \frac{\partial \Omega}{\partial q}$$

$$= [p, q]' + \frac{\partial(\Omega, x' \dot{y}' - y' \dot{x}')}{\partial(p, q)}. \tag{B.60}$$

However,

$$x'\,\dot{y}' - y'\,\dot{x}' = h\,\cos I = [\mu\,a\,(1-e^2)]^{1/2}\,\cos I \equiv \mathcal{G}, \tag{B.61}$$

because the left-hand side is the component of the angular momentum per unit mass parallel to the z'-axis. Of course, this axis is inclined at an angle I to the z-axis, which is parallel to the angular momentum vector. Thus, we obtain

$$[p,q] = [p,q]' + \frac{\partial(\Omega,\mathcal{G})}{\partial(p,q)}. \tag{B.62}$$

Consider a rotation of the coordinate system about the x'-axis. Let the new coordinate system be x'', y'', z''. A rotation through an angle I brings the orbit into the x''–y'' plane. (See Figure 3.6.) Let

$$[p,q]'' = \frac{\partial(x'',\dot{x}'')}{\partial(p,q)} + \frac{\partial(y'',\dot{y}'')}{\partial(p,q)} + \frac{\partial(z'',\dot{z}'')}{\partial(p,q)}. \tag{B.63}$$

By analogy with the previous analysis,

$$[p,q]' = [p,q]'' + \frac{\partial(I,y''\,\dot{z}'' - z''\,\dot{y}'')}{\partial(p,q)}. \tag{B.64}$$

However, z'' and $\dot{z}''$ are both zero, as the orbit lies in the x''–y'' plane. Hence,

$$[p,q]' = [p,q]''. \tag{B.65}$$

Consider, finally, a rotation of the coordinate system about the z''-axis. Let the final coordinate system be x, y, z. A rotation through an angle $\varpi - \Omega$ brings the perihelion to the x-axis. (See Figure 3.6.) Let

$$[p,q]''' = \frac{\partial(x,\dot{x})}{\partial(p,q)} + \frac{\partial(y,\dot{y})}{\partial(p,q)}. \tag{B.66}$$

By analogy with the previous analysis,

$$[p,q]'' = [p,q]''' + \frac{\partial(\varpi - \Omega, x\,\dot{y} - y\,\dot{x})}{\partial(p,q)}. \tag{B.67}$$

However,

$$x\,\dot{y} - y\,\dot{x} = h = [\mu\,a\,(1-e^2)]^{1/2} \equiv H, \tag{B.68}$$

so, from Equations (B.62) and (B.65),

$$[p,q] = [p,q]''' + \frac{\partial(\varpi - \Omega, H)}{\partial(p,q)} + \frac{\partial(\Omega,\mathcal{G})}{\partial(p,q)}. \tag{B.69}$$

It thus remains to calculate $[p,q]'''$.

The coordinates $x = r\,\cos\theta$ and $y = r\,\sin\theta$—where r represents radial distance from the Sun, and θ is the true anomaly—are functions of the major radius, a, the eccentricity, e, and the mean anomaly, $\mathcal{M} = \bar{\lambda}_0 - \varpi + n\,t$. Because the Lagrange brackets are independent of time, it is sufficient to evaluate them at $\mathcal{M} = 0$, that is, at the perihelion point. It is easily demonstrated from Equations (3.85) and (3.86) that

$$x = a(1 - e) + \mathcal{O}(\mathcal{M}^2), \tag{B.70}$$

$$y = a\mathcal{M}\left(\frac{1 + e}{1 - e}\right)^{1/2} + \mathcal{O}(\mathcal{M}^3), \tag{B.71}$$

$$\dot{x} = -an\frac{\mathcal{M}}{(1 - e)^2} + \mathcal{O}(\mathcal{M}^3), \tag{B.72}$$

and

$$\dot{y} = an\left(\frac{1 + e}{1 - e}\right)^{1/2} + \mathcal{O}(\mathcal{M}^2) \tag{B.73}$$

at small $\mathcal{M}$. Hence, at $\mathcal{M} = 0$,

$$\frac{\partial x}{\partial a} = 1 - e, \tag{B.74}$$

$$\frac{\partial x}{\partial e} = -a, \tag{B.75}$$

$$\frac{\partial y}{\partial(\bar{\lambda}_0 - \varpi)} = a\left(\frac{1 + e}{1 - e}\right)^{1/2}, \tag{B.76}$$

$$\frac{\partial \dot{x}}{\partial(\bar{\lambda}_0 - \varpi)} = -\frac{an}{(1 - e)^2}, \tag{B.77}$$

$$\frac{\partial \dot{y}}{\partial a} = -\frac{n}{2}\left(\frac{1 + e}{1 - e}\right)^{1/2}, \tag{B.78}$$

and

$$\frac{\partial \dot{y}}{\partial e} = an(1 + e)^{-1/2}(1 - e)^{-3/2}, \tag{B.79}$$

because $n \propto a^{-3/2}$. All other partial derivatives are zero. Because the orbit in the x, y, z coordinate system depends only on the elements a, e, and $\bar{\lambda}_0 - \varpi$, we can write

$$[p, q]''' = \frac{\partial(a, e)}{\partial(p, q)}\left[\frac{\partial(x, \dot{x})}{\partial(a, e)} + \frac{\partial(y, \dot{y})}{\partial(a, e)}\right]$$

$$+ \frac{\partial(e, \bar{\lambda}_0 - \varpi)}{\partial(p, q)}\left[\frac{\partial(x, \dot{x})}{\partial(e, \bar{\lambda}_0 - \varpi)} + \frac{\partial(y, \dot{y})}{\partial(e, \bar{\lambda}_0 - \varpi)}\right]$$

$$+ \frac{\partial(\bar{\lambda}_0 - \varpi, a)}{\partial(p, q)}\left[\frac{\partial(x, \dot{x})}{\partial(\bar{\lambda}_0 - \varpi, a)} + \frac{\partial(y, \dot{y})}{\partial(\bar{\lambda}_0 - \varpi, a)}\right]. \tag{B.80}$$

Substitution of the values of the derivatives evaluated at $\mathcal{M} = 0$ into this expression yields

$$\frac{\partial(x, \dot{x})}{\partial(a, e)} + \frac{\partial(y, \dot{y})}{\partial(a, e)} = 0, \tag{B.81}$$

$$\frac{\partial(x, \dot{x})}{\partial(e, \bar{\lambda}_0 - \varpi)} + \frac{\partial(y, \dot{y})}{\partial(e, \bar{\lambda}_0 - \varpi)} = 0, \tag{B.82}$$

$$\frac{\partial(x, \dot{x})}{\partial(\bar{\lambda}_0 - \varpi, a)} + \frac{\partial(y, \dot{y})}{\partial(\bar{\lambda}_0 - \varpi, a)} = \frac{an}{2}, \tag{B.83}$$

and

$$[p, q]''' = \frac{\partial(\bar{\lambda}_0 - \varpi, a)}{\partial(p, q)} \frac{n\, a}{2} = \frac{\partial(\bar{\lambda}_0 - \varpi, a)}{\partial(p, q)} \frac{\mu^{1/2}}{2\, a^{1/2}} = \frac{\partial(\bar{\lambda}_0 - \varpi, L)}{\partial(p, q)}, \tag{B.84}$$

where $L = (\mu\, a)^{1/2}$. Hence, from Equation (B.69), we obtain

$$[p, q] = \frac{\partial(\bar{\lambda}_0 - \varpi, L)}{\partial(p, q)} + \frac{\partial(\varpi - \Omega, H)}{\partial(p, q)} + \frac{\partial(\Omega, \mathcal{G})}{\partial(p, q)}. \tag{B.85}$$

B.5 Lagrange planetary equations

Now,

$$L = (\mu\, a)^{1/2}, \tag{B.86}$$

$$H = [\mu\, a\, (1 - e^2)]^{1/2}, \tag{B.87}$$

$$\mathcal{G} = [\mu\, a\, (1 - e^2)]^{1/2} \cos I, \tag{B.88}$$

and $n\, a = (\mu/a)^{1/2}$. Hence,

$$\frac{\partial L}{\partial a} = \frac{n\, a}{2}, \tag{B.89}$$

$$\frac{\partial H}{\partial a} = \frac{n\, a}{2} (1 - e^2)^{1/2}, \tag{B.90}$$

$$\frac{\partial H}{\partial e} = -n\, a^2\, e\, (1 - e^2)^{-1/2}, \tag{B.91}$$

$$\frac{\partial \mathcal{G}}{\partial a} = \frac{n\, a}{2} (1 - e^2)^{1/2} \cos I, \tag{B.92}$$

$$\frac{\partial \mathcal{G}}{\partial e} = -n\, a^2\, e\, (1 - e^2)^{-1/2} \cos I, \tag{B.93}$$

and

$$\frac{\partial \mathcal{G}}{\partial I} = -n\, a^2\, (1 - e^2)^{1/2} \sin I, \tag{B.94}$$

with all other partial derivatives zero. Thus, from Equation (B.85), the only nonzero Lagrange brackets are

$$[\bar{\lambda}_0, a] = -[a, \bar{\lambda}_0] = \frac{n\, a}{2}, \tag{B.95}$$

$$[\varpi, a] = -[a, \varpi] = -\frac{n\, a}{2} [1 - (1 - e^2)^{1/2}], \tag{B.96}$$

$$[\Omega, a] = -[a, \Omega] = -\frac{n\, a}{2} (1 - e^2)^{1/2} (1 - \cos I), \tag{B.97}$$

$$[\varpi, e] = -[e, \varpi] = -n\, a^2\, e\, (1 - e^2)^{-1/2}, \tag{B.98}$$

$$[\Omega, e] = -[e, \Omega] = n\, a^2\, e\, (1 - e^2)^{-1/2} (1 - \cos I), \tag{B.99}$$

and

$$[\Omega, I] = -[I, \Omega] = -n\,a^2\,(1 - e^2)^{1/2}\,\sin I. \tag{B.100}$$

Hence, Equations (B.32) yield

$$[a, \bar{\lambda}_0]\,\frac{d\bar{\lambda}_0}{dt} + [a, \varpi]\,\frac{d\varpi}{dt} + [a, \Omega]\,\frac{d\Omega}{dt} = \frac{\partial\mathcal{R}}{\partial a}, \tag{B.101}$$

$$[e, \varpi]\,\frac{d\varpi}{dt} + [e, \Omega]\,\frac{d\Omega}{dt} = \frac{\partial\mathcal{R}}{\partial e}, \tag{B.102}$$

$$[\bar{\lambda}_0, a]\,\frac{da}{dt} = \frac{\partial\mathcal{R}}{\partial\bar{\lambda}_0}, \tag{B.103}$$

$$[I, \Omega]\,\frac{d\Omega}{dt} = \frac{\partial\mathcal{R}}{\partial I}, \tag{B.104}$$

$$[\Omega, a]\,\frac{da}{dt} + [\Omega, e]\,\frac{de}{dt} + [\Omega, I]\,\frac{dI}{dt} = \frac{\partial\mathcal{R}}{\partial\Omega}, \tag{B.105}$$

and

$$[\varpi, a]\,\frac{da}{dt} + [\varpi, e]\,\frac{de}{dt} = \frac{\partial\mathcal{R}}{\partial\varpi}. \tag{B.106}$$

Finally, Equations (B.95)–(B.106) can be rearranged to give

$$\frac{da}{dt} = \frac{2}{n\,a}\,\frac{\partial\mathcal{R}}{\partial\bar{\lambda}_0}, \tag{B.107}$$

$$\frac{d\bar{\lambda}_0}{dt} = -\frac{2}{n\,a}\,\frac{\partial\mathcal{R}}{\partial a} + \frac{(1 - e^2)^{1/2}\,[1 - (1 - e^2)^{1/2}]}{n\,a^2\,e}\,\frac{\partial\mathcal{R}}{\partial e}$$

$$\quad + \frac{\tan(I/2)}{n\,a^2\,(1 - e^2)^{1/2}}\,\frac{\partial\mathcal{R}}{\partial I}, \tag{B.108}$$

$$\frac{de}{dt} = -\frac{(1 - e^2)^{1/2}}{n\,a^2\,e}\,[1 - (1 - e^2)^{1/2}]\,\frac{\partial\mathcal{R}}{\partial\bar{\lambda}_0} - \frac{(1 - e^2)^{1/2}}{n\,a^2\,e}\,\frac{\partial\mathcal{R}}{\partial\varpi}, \tag{B.109}$$

$$\frac{dI}{dt} = -\frac{\tan(I/2)}{n\,a^2\,(1 - e^2)^{1/2}}\left(\frac{\partial\mathcal{R}}{\partial\bar{\lambda}_0} + \frac{\partial\mathcal{R}}{\partial\varpi}\right) - \frac{(1 - e^2)^{-1/2}}{n\,a^2\,\sin I}\,\frac{\partial\mathcal{R}}{\partial\Omega}, \tag{B.110}$$

$$\frac{d\varpi}{dt} = \frac{(1 - e^2)^{1/2}}{n\,a^2\,e}\,\frac{\partial\mathcal{R}}{\partial e} + \frac{\tan(I/2)}{n\,a^2\,(1 - e^2)^{1/2}}\,\frac{\partial\mathcal{R}}{\partial I}, \tag{B.111}$$

and

$$\frac{d\Omega}{dt} = \frac{1}{n\,a^2\,(1 - e^2)^{1/2}\,\sin I}\,\frac{\partial\mathcal{R}}{\partial I}. \tag{B.112}$$

Equations (B.107)–(B.112), which specify the time evolution of the osculating orbital elements of our planet under the action of the disturbing function, are known collectively as the *Lagrange planetary equations* (Brouwer and Clemence 1961).

In fact, the orbital element $\bar{\lambda}_0$ always appears in the disturbing function in the combination $\bar{\lambda}_0 + \int_0^t n(t')\,dt'$. This combination is known as the *mean longitude* and is denoted $\bar{\lambda}$. It follows that

$$\frac{\partial \mathcal{R}}{\partial \bar{\lambda}_0} = \frac{\partial \mathcal{R}}{\partial \bar{\lambda}}, \tag{B.113}$$

$$\frac{\partial \mathcal{R}}{\partial a} = \frac{\partial \mathcal{R}}{\partial a} + \frac{\partial \mathcal{R}}{\partial \bar{\lambda}} \int_0^t \frac{dn}{da}\,dt'. \tag{B.114}$$

The integral appearing in the previous equation is problematic. Fortunately, it can easily be eliminated by replacing the variable $\bar{\lambda}_0$ by $\bar{\lambda}$. In this case, the Lagrange planetary equations become

$$\frac{da}{dt} = \frac{2}{na}\frac{\partial \mathcal{R}}{\partial \bar{\lambda}}, \tag{B.115}$$

$$\frac{d\bar{\lambda}}{dt} = n - \frac{2}{na}\frac{\partial \mathcal{R}}{\partial a} + \frac{(1-e^2)^{1/2}\,[1-(1-e^2)^{1/2}]}{na^2\,e}\frac{\partial \mathcal{R}}{\partial e}$$
$$\qquad\qquad + \frac{\tan(I/2)}{na^2\,(1-e^2)^{1/2}}\frac{\partial \mathcal{R}}{\partial I}, \tag{B.116}$$

$$\frac{de}{dt} = -\frac{(1-e^2)^{1/2}}{na^2\,e}\,[1-(1-e^2)^{1/2}]\frac{\partial \mathcal{R}}{\partial \bar{\lambda}} - \frac{(1-e^2)^{1/2}}{na^2\,e}\frac{\partial \mathcal{R}}{\partial \varpi}, \tag{B.117}$$

$$\frac{dI}{dt} = -\frac{\tan(I/2)}{na^2\,(1-e^2)^{1/2}}\left(\frac{\partial \mathcal{R}}{\partial \bar{\lambda}} + \frac{\partial \mathcal{R}}{\partial \varpi}\right) - \frac{(1-e^2)^{-1/2}}{na^2\,\sin I}\frac{\partial \mathcal{R}}{\partial \Omega}, \tag{B.118}$$

$$\frac{d\varpi}{dt} = \frac{(1-e^2)^{1/2}}{na^2\,e}\frac{\partial \mathcal{R}}{\partial e} + \frac{\tan(I/2)}{na^2\,(1-e^2)^{1/2}}\frac{\partial \mathcal{R}}{\partial I}, \tag{B.119}$$

and

$$\frac{d\Omega}{dt} = \frac{1}{na^2\,(1-e^2)^{1/2}\,\sin I}\frac{\partial \mathcal{R}}{\partial I}, \tag{B.120}$$

where $\partial/\partial \bar{\lambda}$ is taken at constant a, and $\partial/\partial a$ at constant $\bar{\lambda}$ (Brouwer and Clemence 1961).

B.6 Alternative forms of Lagrange planetary equations

It can be seen, from Equations (B.115)–(B.120), that in the limit of small eccentricity, e, and small inclination, I, certain terms on the right-hand sides of the Lagrange planetary equations become singular. This problem can be alleviated by defining the alternative orbital elements,

$$h = e\,\sin\varpi, \tag{B.121}$$

$$k = e\,\cos\varpi, \tag{B.122}$$

$$p = \sin I\,\sin\Omega, \tag{B.123}$$

and

$$q = \sin I \cos \Omega. \tag{B.124}$$

If we write the Lagrange planetary equations in terms of these new elements, we obtain

$$\frac{da}{dt} = \frac{2}{n\,a} \frac{\partial \mathcal{R}}{\partial \bar{\lambda}}, \tag{B.125}$$

$$\frac{d\bar{\lambda}}{dt} = n - \frac{2}{n\,a} \frac{\partial \mathcal{R}}{\partial a} + \frac{(1-e^2)^{1/2}}{n\,a^2\,[1+(1-e^2)^{1/2}]} \left(h\frac{\partial \mathcal{R}}{\partial h} + k\frac{\partial \mathcal{R}}{\partial k} \right)$$
$$+ \frac{\cos I}{2\,n\,a^2\,\cos^2(I/2)\,(1-e^2)^{1/2}} \left(p\frac{\partial \mathcal{R}}{\partial p} + q\frac{\partial \mathcal{R}}{\partial q} \right), \tag{B.126}$$

$$\frac{dh}{dt} = -\frac{(1-e^2)^{1/2}}{n\,a^2\,[1+(1-e^2)^{1/2}]} h\frac{\partial \mathcal{R}}{\partial \bar{\lambda}} + \frac{(1-e^2)^{1/2}}{n\,a^2} \frac{\partial \mathcal{R}}{\partial k}$$
$$+ \frac{\cos I}{2\,n\,a^2\,\cos^2(I/2)\,(1-e^2)^{1/2}} k\left(p\frac{\partial \mathcal{R}}{\partial p} + q\frac{\partial \mathcal{R}}{\partial q} \right), \tag{B.127}$$

$$\frac{dk}{dt} = -\frac{(1-e^2)^{1/2}}{n\,a^2\,[1+(1-e^2)^{1/2}]} k\frac{\partial \mathcal{R}}{\partial \bar{\lambda}} - \frac{(1-e^2)^{1/2}}{n\,a^2} \frac{\partial \mathcal{R}}{\partial h}$$
$$- \frac{\cos I}{2\,n\,a^2\,\cos^2(I/2)\,(1-e^2)^{1/2}} h\left(p\frac{\partial \mathcal{R}}{\partial p} + q\frac{\partial \mathcal{R}}{\partial q} \right), \tag{B.128}$$

$$\frac{dp}{dt} = -\frac{\cos I}{2\,n\,a^2\,\cos^2(I/2)\,(1-e^2)^{1/2}} p\left(\frac{\partial \mathcal{R}}{\partial \bar{\lambda}} + k\frac{\partial \mathcal{R}}{\partial h} - h\frac{\partial \mathcal{R}}{\partial k} \right)$$
$$+ \frac{\cos I}{n\,a^2\,(1-e^2)^{1/2}} \frac{\partial \mathcal{R}}{\partial q}, \tag{B.129}$$

and

$$\frac{dq}{dt} = -\frac{\cos I}{2\,n\,a^2\,\cos^2(I/2)\,(1-e^2)^{1/2}} q\left(\frac{\partial \mathcal{R}}{\partial \bar{\lambda}} + k\frac{\partial \mathcal{R}}{\partial h} - h\frac{\partial \mathcal{R}}{\partial k} \right)$$
$$- \frac{\cos I}{n\,a^2\,(1-e^2)^{1/2}} \frac{\partial \mathcal{R}}{\partial p}. \tag{B.130}$$

Note that the new equations now contain no singular terms in the limit $e, I \to 0$.

It is sometimes convenient to write the Lagrange planetary equations in terms of the *mean anomaly*, $\mathcal{M} = \bar{\lambda} - \varpi$, and the *argument of the perigee*, $\omega = \varpi - \Omega$, rather than $\bar{\lambda}$ and ω. Making the appropriate substitutions, we see that the equations take the form (Brouwer and Clemence 1961)

$$\frac{da}{dt} = \frac{2}{n\,a} \frac{\partial \mathcal{R}}{\partial \mathcal{M}}, \tag{B.131}$$

$$\frac{d\mathcal{M}}{dt} = n - \frac{2}{n\,a} \frac{\partial \mathcal{R}}{\partial a} - \frac{1-e^2}{n\,a^2\,e} \frac{\partial \mathcal{R}}{\partial e}, \tag{B.132}$$

$$\frac{de}{dt} = \frac{1-e^2}{n\,a^2\,e}\frac{\partial\mathcal{R}}{\partial\mathcal{M}} - \frac{(1-e^2)^{1/2}}{n\,a^2\,e}\frac{\partial\mathcal{R}}{\partial\omega}, \tag{B.133}$$

$$\frac{dI}{dt} = \frac{\cot I}{n\,a^2\,(1-e^2)^{1/2}}\frac{\partial\mathcal{R}}{\partial\omega} - \frac{(1-e^2)^{-1/2}}{n\,a^2\,\sin I}\frac{\partial\mathcal{R}}{\partial\Omega}, \tag{B.134}$$

$$\frac{d\omega}{dt} = \frac{(1-e^2)^{1/2}}{n\,a^2\,e}\frac{\partial\mathcal{R}}{\partial e} - \frac{\cot I}{n\,a^2\,(1-e^2)^{1/2}}\frac{\partial\mathcal{R}}{\partial I}, \tag{B.135}$$

and

$$\frac{d\Omega}{dt} = \frac{(1-e^2)^{-1/2}}{n\,a^2\,\sin I}\frac{\partial\mathcal{R}}{\partial I}. \tag{B.136}$$

Appendix C Expansion of orbital evolution equations

C.1 Introduction

The purpose of this appendix is to derive simplified evolution equations for the osculating orbital elements of a two-planet solar system, starting from the Lagrange planetary equations, Equations (B.125)–(B.130), and exploiting the fact that the planetary masses are all very small compared with the solar mass, as well as the fact that the planetary orbital eccentricities and inclinations (in radians) are small compared with unity. Our approach is mostly based on that of Murray and Dermott 1999.

Let the first planet have position vector $\mathbf{r}$, mass m, and the standard osculating elements a, $\bar{\lambda}_0$, e, I, ϖ, Ω. (See Section 3.12.) It is convenient to define the alternative elements $\bar{\lambda} = \bar{\lambda}_0 + \int_0^t n(t')\,dt'$, $h = e \sin \varpi$, $k = e \cos \varpi$, $p = \sin I \sin \Omega$, and $q = \sin I \cos \Omega$, where $n = (\mu/a^3)^{1/2}$ is the mean orbital angular velocity, $\mu = G(M+m)$, and M is the solar mass. Thus, the osculating elements of the first planet become a, $\bar{\lambda}$, h, k, p, q. Let a', $\bar{\lambda}'$, h', k', p', q' be the corresponding osculating elements of the second planet. Furthermore, let the second planet have position vector $\mathbf{r}'$, mass m', and mean orbital angular velocity $n' = (\mu'/a'^3)^{1/2}$, where $\mu' = G(M + m')$.

C.2 Expansion of Lagrange planetary equations

The first planet's disturbing function can be written in the form [see Equation (9.8)]

$$\mathcal{R} = \frac{\tilde{\mu}'}{a'}\,\mathcal{S}, \tag{C.1}$$

where $\tilde{\mu}' = G\,m'$, and $\mathcal{S}$ is $\mathcal{O}(1)$. Thus, because $\mu = n^2 a^3$, the Lagrange planetary equations, Equations (B.125)–(B.130), applied to the first planet, reduce to

$$\frac{d \ln a}{dt} = 2\,n\,\epsilon'\,\alpha\,\frac{\partial \mathcal{S}}{\partial \bar{\lambda}}, \tag{C.2}$$

$$\frac{d\bar{\lambda}}{dt} = n - 2\,n\,\epsilon'\,\alpha^2\,\frac{\partial \mathcal{S}}{\partial \alpha} + \frac{n\,\epsilon'\,\alpha\,(1-e^2)^{1/2}}{[1 + (1-e^2)^{1/2}]}\left(h\,\frac{\partial \mathcal{S}}{\partial h} + k\,\frac{\partial \mathcal{S}}{\partial k}\right)$$

$$+ \frac{n\,\epsilon'\,\alpha\,\cos I}{2\,\cos^2(I/2)\,(1-e^2)^{1/2}}\left(p\,\frac{\partial \mathcal{S}}{\partial p} + q\,\frac{\partial \mathcal{S}}{\partial q}\right), \tag{C.3}$$

$$\frac{dh}{dt} = -\frac{n\,\epsilon'\alpha\,(1-e^2)^{1/2}}{[1+(1-e^2)^{1/2}]}\,h\,\frac{\partial S}{\partial\bar{\lambda}} + n\,\epsilon'\alpha\,(1-e^2)^{1/2}\,\frac{\partial S}{\partial k}$$

$$+\frac{n\,\epsilon'\alpha\,\cos I}{2\,\cos^2(I/2)\,(1-e^2)^{1/2}}\,k\left(p\,\frac{\partial S}{\partial p} + q\,\frac{\partial S}{\partial q}\right), \tag{C.4}$$

$$\frac{dk}{dt} = -\frac{n\,\epsilon'\alpha\,(1-e^2)^{1/2}}{[1+(1-e^2)^{1/2}]}\,k\,\frac{\partial S}{\partial\bar{\lambda}} - n\,\epsilon'\alpha\,(1-e^2)^{1/2}\,\frac{\partial S}{\partial h}$$

$$-\frac{n\,\epsilon'\alpha\,\cos I}{2\,\cos^2(I/2)\,(1-e^2)^{1/2}}\,h\left(p\,\frac{\partial S}{\partial p} + q\,\frac{\partial S}{\partial q}\right), \tag{C.5}$$

$$\frac{dp}{dt} = -\frac{n\,\epsilon'\alpha\,\cos I}{2\,\cos^2(I/2)\,(1-e^2)^{1/2}}\,p\left(\frac{\partial S}{\partial\bar{\lambda}} + k\,\frac{\partial S}{\partial h} - h\,\frac{\partial S}{\partial k}\right)$$

$$+\frac{n\,\epsilon'\alpha\,\cos I}{(1-e^2)^{1/2}}\,\frac{\partial S}{\partial q}, \tag{C.6}$$

and

$$\frac{dq}{dt} = -\frac{n\,\epsilon'\alpha\,\cos I}{2\,\cos^2(I/2)\,(1-e^2)^{1/2}}\,q\left(\frac{\partial S}{\partial\bar{\lambda}} + k\,\frac{\partial S}{\partial h} - h\,\frac{\partial S}{\partial k}\right)$$

$$-\frac{n\,\epsilon'\alpha\,\cos I}{(1-e^2)^{1/2}}\,\frac{\partial S}{\partial p}, \tag{C.7}$$

where

$$\epsilon' = \frac{\tilde{\mu}'}{\mu} = \frac{m'}{M+m} \tag{C.8}$$

and

$$\alpha = \frac{a}{a'}. \tag{C.9}$$

The Sun is much more massive than any planet in the solar system. It follows that the parameter ϵ' is very small compared with unity. Expansion of Equations (C.2)–(C.7) to first order in ϵ' yields

$$\bar{\lambda}(t) = \bar{\lambda}_0 + n^{(0)}\,t + \bar{\lambda}^{(1)}(t), \tag{C.10}$$

$$a(t) = a^{(0)}\left[1 + \epsilon'\,a^{(1)}(t)\right], \tag{C.11}$$

and

$$n(t) = n^{(0)}\left[1 - (3/2)\,\epsilon'\,a^{(1)}(t)\right], \tag{C.12}$$

where $\bar{\lambda}^{(1)} \sim \mathcal{O}(\epsilon')$, $a^{(1)} \sim \mathcal{O}(1)$, $n^{(0)} = \{\mu/[a^{(0)}]^3\}^{1/2}$, and

$$\frac{d\epsilon' a^{(1)}}{dt} = \epsilon' n^{(0)} \left[2\alpha \frac{\partial \mathcal{S}}{\partial \bar{\lambda}^{(0)}} \right], \tag{C.13}$$

$$\frac{d\bar{\lambda}^{(1)}}{dt} = \epsilon' n^{(0)} \left\{ -\frac{3}{2} a^{(1)} - 2\alpha^2 \frac{\partial \mathcal{S}}{\partial \alpha} + \frac{\alpha(1-e^2)^{1/2}}{[1+(1-e^2)^{1/2}]} \left(h \frac{\partial \mathcal{S}}{\partial h} + k \frac{\partial \mathcal{S}}{\partial k} \right) \right.$$
$$\left. + \frac{\alpha \cos I}{2 \cos^2(I/2)(1-e^2)^{1/2}} \left(p \frac{\partial \mathcal{S}}{\partial p} + q \frac{\partial \mathcal{S}}{\partial q} \right) \right\}, \tag{C.14}$$

$$\frac{dh}{dt} = \epsilon' n^{(0)} \left\{ -\frac{\alpha(1-e^2)^{1/2}}{[1+(1-e^2)^{1/2}]} h \frac{\partial \mathcal{S}}{\partial \bar{\lambda}^{(0)}} + \alpha(1-e^2)^{1/2} \frac{\partial \mathcal{S}}{\partial k} \right.$$
$$\left. + \frac{\alpha \cos I}{2 \cos^2(I/2)(1-e^2)^{1/2}} k \left(p \frac{\partial \mathcal{S}}{\partial p} + q \frac{\partial \mathcal{S}}{\partial q} \right) \right\}, \tag{C.15}$$

$$\frac{dk}{dt} = \epsilon' n^{(0)} \left\{ -\frac{\alpha(1-e^2)^{1/2}}{[1+(1-e^2)^{1/2}]} k \frac{\partial \mathcal{S}}{\partial \bar{\lambda}^{(0)}} - \alpha(1-e^2)^{1/2} \frac{\partial \mathcal{S}}{\partial h} \right.$$
$$\left. - \frac{\alpha \cos I}{2 \cos^2(I/2)(1-e^2)^{1/2}} h \left(p \frac{\partial \mathcal{S}}{\partial p} + q \frac{\partial \mathcal{S}}{\partial q} \right) \right\}, \tag{C.16}$$

$$\frac{dp}{dt} = \epsilon' n^{(0)} \left[-\frac{\alpha \cos I}{2 \cos^2(I/2)(1-e^2)^{1/2}} p \left(\frac{\partial \mathcal{S}}{\partial \bar{\lambda}^{(0)}} + k \frac{\partial \mathcal{S}}{\partial h} - h \frac{\partial \mathcal{S}}{\partial k} \right) \right.$$
$$\left. + \frac{\alpha \cos I}{(1-e^2)^{1/2}} \frac{\partial \mathcal{S}}{\partial q} \right], \tag{C.17}$$

and

$$\frac{dq}{dt} = \epsilon' n^{(0)} \left[-\frac{\alpha \cos I}{2 \cos^2(I/2)(1-e^2)^{1/2}} q \left(\frac{\partial \mathcal{S}}{\partial \bar{\lambda}^{(0)}} + k \frac{\partial \mathcal{S}}{\partial h} - h \frac{\partial \mathcal{S}}{\partial k} \right) \right.$$
$$\left. - \frac{\alpha \cos I}{(1-e^2)^{1/2}} \frac{\partial \mathcal{S}}{\partial p} \right], \tag{C.18}$$

with $\bar{\lambda}^{(0)} = \bar{\lambda}_0 + n^{(0)} t$, and $\alpha = (a/a')^{(0)}$. In the following, for ease of notation, $\bar{\lambda}^{(0)}$, $a^{(0)}$, and $n^{(0)}$ are written simply as $\bar{\lambda}$, a, and n, respectively.

According to Table 3.1, the planets in the solar system all possess orbits whose eccentricities, e, and inclinations, I (in radians), are small compared with unity but large compared with the ratio of any planetary mass to that of the Sun. It follows that

$$\epsilon' \ll e, I \ll 1, \tag{C.19}$$

which is our fundamental ordering of small quantities. Assuming that $I, e', I' \sim \mathcal{O}(e)$, we can perform a secondary expansion in the small parameter e. It turns out that when the normalized disturbing function, $\mathcal{S}$, is expanded to second order in e it takes the general form (see Section C.3)

$$\mathcal{S} = \mathcal{S}_0(\alpha, \bar{\lambda}, \bar{\lambda}') + \mathcal{S}_1(\alpha, \bar{\lambda}, \bar{\lambda}', h, h', k, k') + \mathcal{S}_2(\alpha, \bar{\lambda}, \bar{\lambda}', h, h', k, k', p, p', q, q'), \tag{C.20}$$

where $\mathcal{S}_n$ is $\mathcal{O}(e^n)$. If we expand the right-hand sides of Equations (C.13)–(C.18) to first order in e, we obtain

$$\frac{d\epsilon' a^{(1)}}{dt} = \epsilon' n \left[2\alpha \frac{\partial(\mathcal{S}_0 + \mathcal{S}_1)}{\partial\bar\lambda} \right], \tag{C.21}$$

$$\frac{d\bar\lambda^{(1)}}{dt} = \epsilon' n \left[-\frac{3}{2} a^{(1)} - 2\alpha^2 \frac{\partial(\mathcal{S}_0 + \mathcal{S}_1)}{\partial\alpha} + \alpha \left(h \frac{\partial\mathcal{S}_1}{\partial h} + k \frac{\partial\mathcal{S}_1}{\partial k} \right) \right], \tag{C.22}$$

$$\frac{dh}{dt} = \epsilon' n \left[-\alpha h \frac{\partial\mathcal{S}_0}{\partial\bar\lambda} + \alpha \frac{\partial(\mathcal{S}_1 + \mathcal{S}_2)}{\partial k} \right], \tag{C.23}$$

$$\frac{dk}{dt} = \epsilon' n \left[-\alpha k \frac{\partial\mathcal{S}_0}{\partial\bar\lambda} - \alpha \frac{\partial(\mathcal{S}_1 + \mathcal{S}_2)}{\partial h} \right], \tag{C.24}$$

$$\frac{dp}{dt} = \epsilon' n \left[-\frac{\alpha}{2} p \frac{\partial\mathcal{S}_0}{\partial\bar\lambda} + \alpha \frac{\partial\mathcal{S}_2}{\partial q} \right], \tag{C.25}$$

and

$$\frac{dq}{dt} = \epsilon' n \left[-\frac{\alpha}{2} q \frac{\partial\mathcal{S}_0}{\partial\bar\lambda} - \alpha \frac{\partial\mathcal{S}_2}{\partial p} \right]. \tag{C.26}$$

Note that h, k, p, q are $\mathcal{O}(e)$, whereas α and $\bar\lambda$ are $\mathcal{O}(1)$.

By analogy, writing the second planet's disturbing function as [see Equation (9.9]

$$\mathcal{R}' = \frac{\tilde\mu}{a} \mathcal{S}', \tag{C.27}$$

where $\tilde\mu = G\,m$ and $\mathcal{S}'$ is $\mathcal{O}(1)$, and assuming that $\mathcal{S}'$ takes the form

$$\mathcal{S}' = \mathcal{S}'_0(\alpha, \bar\lambda, \bar\lambda') + \mathcal{S}'_1(\alpha, \bar\lambda, \bar\lambda', h, h', k, k') + \mathcal{S}'_2(\alpha, \bar\lambda, \bar\lambda', h, h', k, k', p, p', q, q'), \tag{C.28}$$

where $\mathcal{S}'_n$ is $\mathcal{O}(e^n)$, we see that the Lagrange planetary equations, applied to the second planet, yield

$$\frac{d\epsilon' a^{(1)'}}{dt} = \epsilon n' \left[2\alpha^{-1} \frac{\partial(\mathcal{S}'_0 + \mathcal{S}'_1)}{\partial\bar\lambda'} \right], \tag{C.29}$$

$$\frac{d\bar\lambda^{(1)'}}{dt} = \epsilon n' \left[-\frac{3}{2} a^{(1)'} + 2 \frac{\partial(\mathcal{S}'_0 + \mathcal{S}'_1)}{\partial\alpha} + \alpha^{-1} \left(h' \frac{\partial\mathcal{S}'_1}{\partial h'} + k' \frac{\partial\mathcal{S}'_1}{\partial k'} \right) \right], \tag{C.30}$$

$$\frac{dh'}{dt} = \epsilon n' \left[-\alpha^{-1} h' \frac{\partial\mathcal{S}'_0}{\partial\bar\lambda'} + \alpha^{-1} \frac{\partial(\mathcal{S}'_1 + \mathcal{S}'_2)}{\partial k'} \right], \tag{C.31}$$

$$\frac{dk'}{dt} = \epsilon n' \left[-\alpha^{-1} k' \frac{\partial\mathcal{S}'_0}{\partial\bar\lambda'} - \alpha^{-1} \frac{\partial(\mathcal{S}'_1 + \mathcal{S}'_2)}{\partial h'} \right], \tag{C.32}$$

$$\frac{dp'}{dt} = \epsilon n' \left[-\frac{\alpha^{-1}}{2} p' \frac{\partial\mathcal{S}'_0}{\partial\bar\lambda'} + \alpha^{-1} \frac{\partial\mathcal{S}'_2}{\partial q'} \right], \tag{C.33}$$

and

$$\frac{dq'}{dt} = \epsilon n' \left[-\frac{\alpha^{-1}}{2} q' \frac{\partial\mathcal{S}'_0}{\partial\bar\lambda'} - \alpha^{-1} \frac{\partial\mathcal{S}'_2}{\partial p'} \right], \tag{C.34}$$

where

$$\epsilon = \frac{\tilde\mu}{\mu} = \frac{m}{M + m'}. \tag{C.35}$$

C.3 Expansion of planetary disturbing functions

Equations (9.8), (9.9), (C.1), and (C.27) give

$$S = S_D + \alpha\,S_E, \tag{C.36}$$

$$S' = \alpha\,S_D + \alpha^{-1}S_I, \tag{C.37}$$

where

$$S_D = \frac{a'}{|\mathbf{r}' - \mathbf{r}|}, \tag{C.38}$$

and

$$S_E = -\left(\frac{r}{a}\right)\left(\frac{a'}{r'}\right)^2 \cos\psi, \tag{C.39}$$

$$S_I = -\left(\frac{r'}{a'}\right)\left(\frac{a}{r}\right)^2 \cos\psi. \tag{C.40}$$

In the preceding equations, ψ is the angle subtended between the directions of $\mathbf{r}$ and $\mathbf{r}'$.
    Now

$$S_D = a'\left[r'^2 - 2\,r'\,r\,\cos\psi + r^2\right]^{-1/2}. \tag{C.41}$$

Let

$$\zeta = \frac{r - a}{a}, \tag{C.42}$$

$$\zeta' = \frac{r' - a'}{a'}, \tag{C.43}$$

$$\delta = \cos\psi - \cos(\vartheta - \vartheta'), \tag{C.44}$$

where $\vartheta = \varpi + \theta$ and $\vartheta' = \varpi' + \theta'$. Here, θ and θ' are the true anomalies of the first and second planets, respectively. We expect ζ and ζ' to both be $\mathcal{O}(e)$ [see Equation (C.54)], and δ to be $\mathcal{O}(e^2)$ [see Equation (C.64)]. We can write

$$S_D = (1 + \zeta')^{-1}\left[1 - 2\,\tilde\alpha\,\cos(\vartheta - \vartheta') + \tilde\alpha^2 - 2\,\tilde\alpha\,\delta\right]^{-1/2}, \tag{C.45}$$

where

$$\tilde\alpha = \left(\frac{1 + \zeta}{1 + \zeta'}\right)\alpha. \tag{C.46}$$

Expanding in e, and retaining terms only up to $\mathcal{O}(e^2)$, we obtain

$$S_D \simeq (1 + \zeta')^{-1}\left[F + (\tilde\alpha - \alpha)\,D\,F + \frac{1}{2}\,(\tilde\alpha - \alpha)^2\,D^2\,F\right] + \delta\,\alpha\,F^3, \tag{C.47}$$

where $D \equiv \partial/\partial\alpha$, and

$$F(\alpha, \vartheta - \vartheta') = \frac{1}{[1 - 2\,\alpha\,\cos(\vartheta - \vartheta') + \alpha^2]^{1/2}}. \tag{C.48}$$

Hence,

$$
\mathcal{S}_D \simeq \left[(1 - \zeta' + \zeta'^2) + (\zeta - \zeta' - 2\zeta\zeta' + 2\zeta'^2)\alpha D \right.
$$
$$
\left. + \frac{1}{2}(\zeta^2 - 2\zeta\zeta' + \zeta'^2)\alpha^2 D^2 \right] F + \delta\alpha F^3. \tag{C.49}
$$

Now, we can expand F and F^3 as Fourier series in $\vartheta - \vartheta'$:

$$
F(\alpha, \vartheta - \vartheta') = \frac{1}{2} \sum_{j=-\infty,\infty} b_{1/2}^{(j)}(\alpha) \cos[j(\vartheta - \vartheta')], \tag{C.50}
$$

$$
F^3(\alpha, \vartheta - \vartheta') = \frac{1}{2} \sum_{j=-\infty,\infty} b_{3/2}^{(j)}(\alpha) \cos[j(\vartheta - \vartheta')], \tag{C.51}
$$

where

$$
b_s^{(j)}(\alpha) = \frac{1}{\pi} \int_0^{2\pi} \frac{\cos(j\psi)\, d\psi}{[1 - 2\alpha\cos\psi + \alpha^2]^s}. \tag{C.52}
$$

Incidentally, the $b_s^{(j)}$ are known as *Laplace coefficients*. Thus,

$$
\mathcal{S}_D \simeq \frac{1}{2} \sum_{j=-\infty,\infty} \left\{ \left[(1 - \zeta' + \zeta'^2) + (\zeta - \zeta' - 2\zeta\zeta' + 2\zeta'^2)\alpha D \right. \right.
$$
$$
\left. \left. + \frac{1}{2}(\zeta^2 - 2\zeta\zeta' + \zeta'^2)\alpha^2 D^2 \right] b_{1/2}^{(j)}(\alpha) + \delta\alpha\, b_{3/2}^{(j)}(\alpha) \right\} \cos[j(\vartheta - \vartheta')], \tag{C.53}
$$

where D now denotes $d/d\alpha$.

Equation (3.86) gives

$$
\zeta \equiv \frac{r}{a} - 1 \simeq -e\cos\mathcal{M} + \frac{e^2}{2}(1 - \cos 2\mathcal{M})
$$
$$
\simeq -e\cos(\bar{\lambda} - \varpi) + \frac{e^2}{2}\left[1 - \cos(2\bar{\lambda} - 2\varpi)\right] \tag{C.54}
$$

to $\mathcal{O}(e^2)$. Here, $\mathcal{M} = \bar{\lambda} - \varpi$ is the first planet's mean anomaly. Obviously, there is an analogous equation for ζ'. Moreover, from Equation (3.85), we have

$$
\sin\theta \simeq \sin\mathcal{M} + e\sin 2\mathcal{M} + \frac{e^2}{8}(9\sin 3\mathcal{M} - 7\sin\mathcal{M}), \tag{C.55}
$$

$$
\cos\theta \simeq \cos\mathcal{M} + e(\cos 2\mathcal{M} - 1) + \frac{e^2}{8}(9\cos 3\mathcal{M} - 9\cos\mathcal{M}). \tag{C.56}
$$

Hence,

$$
\cos(\omega + \theta) \equiv \cos\omega\cos\theta - \sin\omega\sin\theta
$$
$$
\simeq \cos(\omega + \mathcal{M}) + e\left[\cos(\omega + 2\mathcal{M}) - \cos\omega\right]
$$
$$
+ \frac{e^2}{8}\left[-8\cos(\omega + \mathcal{M}) - \cos(\omega - \mathcal{M}) + 9\cos(\omega + 3\mathcal{M})\right], \tag{C.57}
$$

and

$$\sin(\omega + \theta) \equiv \sin\omega \cos\theta + \cos\omega \sin\theta$$

$$\simeq \sin(\omega + \mathcal{M}) + e\,[\sin(\omega + 2\mathcal{M}) - \sin\omega]$$

$$+ \frac{e^2}{8}\,[-8\,\sin(\omega + \mathcal{M}) + \sin(\omega - \mathcal{M}) + 9\,\sin(\omega + 3\mathcal{M})]\,. \quad \text{(C.58)}$$

Thus, Equations (3.72)–(3.74) yield

$$\frac{X}{r} \simeq \cos(\omega + \Omega + \mathcal{M}) + e\,[\cos(\omega + \Omega + 2\mathcal{M}) - \cos(\omega + \Omega)]$$

$$+ \frac{e^2}{8}\,[9\,\cos(\omega + \Omega + 3\mathcal{M}) - \cos(\omega + \Omega - \mathcal{M}) - 8\,\cos(\omega + \Omega + \mathcal{M})]$$

$$+ s^2\,[\cos(\omega - \Omega + \mathcal{M}) - \cos(\omega + \Omega + \mathcal{M})]\,, \quad \text{(C.59)}$$

$$\frac{Y}{r} \simeq \sin(\omega + \Omega + \mathcal{M}) + e\,[\sin(\omega + \Omega + 2\mathcal{M}) - \sin(\omega + \Omega)]$$

$$+ \frac{e^2}{8}\,[9\,\sin(\omega + \Omega + 3\mathcal{M}) - \sin(\omega + \Omega - \mathcal{M}) - 8\,\sin(\omega + \Omega + \mathcal{M})]$$

$$- s^2\,[\sin(\omega - \Omega + \mathcal{M}) + \sin(\omega + \Omega + \mathcal{M})]\,, \quad \text{(C.60)}$$

and

$$\frac{Z}{r} \simeq 2\,s\,\sin(\omega + \mathcal{M}) + 2\,e\,s\,[\sin(\omega + 2\mathcal{M}) - \sin\omega)]\,, \quad \text{(C.61)}$$

where $s \equiv \sin(I/2)$ is assumed to be $\mathcal{O}(e)$. Here, X, Y, Z are the Cartesian components of **r**. There are, of course, completely analogous expressions for the Cartesian components of **r**′.

Now,

$$\cos\psi = \frac{X}{r}\frac{X'}{r'} + \frac{Y}{r}\frac{Y'}{r'} + \frac{Z}{r}\frac{Z'}{r'}\,, \quad \text{(C.62)}$$

so

$$\cos\psi \simeq (1 - e^2 - e'^2 - s^2 - s'^2)\,\cos(\bar{\lambda} - \bar{\lambda}') + e\,e'\,\cos(2\,\bar{\lambda} - 2\,\bar{\lambda}' - \varpi + \varpi')$$

$$+ e\,e'\,\cos(\varpi - \varpi') + 2\,s\,s'\,\cos(\bar{\lambda} - \bar{\lambda}' - \Omega + \Omega')$$

$$+ e\,\cos(2\,\bar{\lambda} - \bar{\lambda}' - \varpi) - e\,\cos(\bar{\lambda}' - \varpi) + e'\,\cos(\bar{\lambda} - 2\,\bar{\lambda}' + \varpi')$$

$$- e'\,\cos(\bar{\lambda} - \varpi') + \frac{9\,e^2}{8}\,\cos(3\,\bar{\lambda} - \bar{\lambda}' - 2\,\varpi) - \frac{e^2}{8}\,\cos(\bar{\lambda} + \bar{\lambda}' - 2\,\varpi)$$

$$+ \frac{9\,e'^2}{8}\,\cos(\bar{\lambda} - 3\,\bar{\lambda}' + 2\,\varpi') - \frac{e'^2}{8}\,\cos(\bar{\lambda} + \bar{\lambda}' - 2\,\varpi')$$

$$- e\,e'\,\cos(2\,\bar{\lambda} - \varpi - \varpi') - e\,e'\,\cos(2\,\bar{\lambda}' - \varpi - \varpi')$$

$$+ s^2\,\cos(\bar{\lambda} + \bar{\lambda}' - 2\,\Omega) + s'^2\,\cos(\bar{\lambda} + \bar{\lambda}' - 2\,\Omega')$$

$$- 2\,s\,s'\,\cos(\bar{\lambda} + \bar{\lambda}' - \Omega - \Omega')\,. \quad \text{(C.63)}$$

It is easily demonstrated that $\cos(\vartheta - \vartheta')$ represents the value taken by $\cos\psi$ when $s = s' = 0$. Hence, from Equations (C.44) and (C.63),

$$
\begin{aligned}
\delta \simeq\ & s^2 \left[\cos(\bar\lambda + \bar\lambda' - 2\,\Omega) - \cos(\bar\lambda - \bar\lambda')\right] \\
& + 2\, s\, s' \left[\cos(\bar\lambda - \bar\lambda' - \Omega + \Omega') - \cos(\bar\lambda + \bar\lambda' - \Omega - \Omega')\right] \\
& + s'^{\,2} \left[\cos(\bar\lambda + \bar\lambda' - 2\,\Omega') - \cos(\bar\lambda - \bar\lambda')\right].
\end{aligned}
\tag{C.64}
$$

Now,

$$
\cos[j(\vartheta - \vartheta')] \equiv \cos(j\,\vartheta)\,\cos(j\,\vartheta') + \sin(j\,\vartheta)\,\sin(j\,\vartheta').
\tag{C.65}
$$

However, from Equation (3.85),

$$
\begin{aligned}
\cos(j\,\vartheta) \equiv\ & \cos[j\,(\varpi + \theta)] \\
\simeq\ & (1 - j^2\,e^2)\,\cos(j\,\bar\lambda) + e^2\left(\frac{j^2}{2} - \frac{5\,j}{8}\right)\cos[(2 - j)\,\bar\lambda - 2\,\varpi] \\
& + e^2\left(\frac{j^2}{2} + \frac{5\,j}{8}\right)\cos[(2 + j)\,\bar\lambda - 2\,\varpi] \\
& - j\,e\,\cos[(1 - j)\,\bar\lambda - \varpi] + j\,e\,\cos[(1 + j)\,\bar\lambda - \varpi)],
\end{aligned}
\tag{C.66}
$$

because $\mathcal{M} = \bar\lambda - \varpi$. Likewise,

$$
\begin{aligned}
\sin(j\,\vartheta) \equiv\ & \sin[j\,(\varpi + \theta)] \\
\simeq\ & (1 - j^2\,e^2)\,\sin(j\,\bar\lambda) + e^2\left(\frac{5\,j}{8} - \frac{j^2}{2}\right)\sin[(2 - j)\,\bar\lambda - 2\,\varpi] \\
& + e^2\left(\frac{5\,j}{8} + \frac{j^2}{2}\right)\sin[(2 + j)\,\bar\lambda - 2\,\varpi] \\
& + j\,e\,\sin[(1 - j)\,\bar\lambda - \varpi] + j\,e\,\sin[(1 + j)\,\bar\lambda - \varpi].
\end{aligned}
\tag{C.67}
$$

Hence, we obtain

$$
\begin{aligned}
\cos[j(\vartheta - \vartheta')] \simeq\ & (1 - j^2\,e^2 - j^2\,e'^{\,2})\,\cos[j\,(\bar\lambda - \bar\lambda')] \\
& + e^2\left(\frac{5\,j}{8} + \frac{j^2}{2}\right)\cos[(2 + j)\,\bar\lambda - j\,\bar\lambda' - 2\,\varpi] \\
& + e^2\left(\frac{j^2}{2} - \frac{5\,j}{8}\right)\cos[(2 - j)\,\bar\lambda + j\,\bar\lambda' - 2\,\varpi] \\
& + j\,e\,\cos[(1 + j)\,\bar\lambda - j\,\bar\lambda' - \varpi] - j\,e\,\cos[(1 - j)\,\bar\lambda + j\,\bar\lambda' - \varpi] \\
& + e'^{\,2}\left(\frac{j^2}{2} - \frac{5\,j}{8}\right)\cos[j\,\bar\lambda + (2 - j)\,\bar\lambda' - 2\,\varpi'] \\
& + e'^{\,2}\left(\frac{5\,j}{8} + \frac{j^2}{2}\right)\cos[j\,\bar\lambda - (2 + j)\,\bar\lambda' + 2\,\varpi'] \\
& - j\,e'\,\cos[j\,\bar\lambda + (1 - j)\,\bar\lambda' - \varpi'] + j\,e'\,\cos[j\,\bar\lambda - (1 + j)\,\bar\lambda' + \varpi'] \\
& - j^2\,e\,e'\,\cos[(1 + j)\,\bar\lambda + (1 - j)\,\bar\lambda' - \varpi - \varpi']
\end{aligned}
$$

$$- j^2 \, e \, e' \, \cos[(1-j)\,\bar{\lambda} + (1+j)\,\bar{\lambda}' - \varpi - \varpi']$$

$$+ j^2 \, e \, e' \, \cos[(1+j)\,\bar{\lambda} - (1+j)\,\bar{\lambda}' - \varpi + \varpi']$$

$$+ j^2 \, e \, e' \, \cos[(1-j)\,\bar{\lambda} - (1-j)\,\bar{\lambda}' - \varpi + \varpi']. \tag{C.68}$$

Equations (C.53), (C.54), (C.64), and (C.68) yield

$$\mathcal{S}_D = \sum_{j=-\infty,\infty} \mathcal{S}^{(j)}, \tag{C.69}$$

where

$$
\begin{aligned}
\mathcal{S}^{(j)} \simeq & \left[\frac{b_{1/2}^{(j)}}{2} + \frac{1}{8}\left(e^2 + e'^2\right)\left(-4\,j^2 + 2\,\alpha\,D + \alpha^2\,D^2\right) b_{1/2}^{(j)} \right. \\
& \left. - \frac{\alpha}{4}\left(s^2 + s'^2\right)\left(b_{3/2}^{(j-1)} + b_{3/2}^{(j+1)}\right) \right] \cos[j\,(\bar{\lambda}' - \bar{\lambda})] \\
& + \frac{e\,e'}{4}\left(2 + 6\,j + 4\,j^2 - 2\,\alpha\,D - \alpha^2\,D^2\right) b_{1/2}^{(j+1)} \cos(j\,\bar{\lambda}' - j\,\bar{\lambda} + \varpi' - \varpi) \\
& + s\,s'\,\alpha\,b_{3/2}^{(j+1)} \cos(j\,\bar{\lambda}' - j\,\bar{\lambda} + \Omega' - \Omega) \\
& + \frac{e}{2}\,(-2\,j - \alpha\,D)\,b_{1/2}^{(j)} \cos[j\,\bar{\lambda}' + (1-j)\,\bar{\lambda} - \varpi] \\
& + \frac{e'}{2}\,(-1 + 2\,j + \alpha\,D)\,b_{1/2}^{(j-1)} \cos[j\,\bar{\lambda}' + (1-j)\,\bar{\lambda} - \varpi'] \\
& + \frac{e^2}{8}\left(-5\,j + 4\,j^2 - 2\,\alpha\,D + 4\,j\,\alpha\,D + \alpha^2\,D^2\right) b_{1/2}^{(j)} \cos[j\,\bar{\lambda}' + (2-j)\,\bar{\lambda} - 2\,\varpi] \\
& + \frac{e\,e'}{4}\left(-2 + 6\,j - 4\,j^2 + 2\,\alpha\,D - 4\,j\,\alpha\,D - \alpha^2\,D^2\right) b_{1/2}^{(j-1)} \cos[j\,\bar{\lambda}' + (2-j)\,\bar{\lambda} - \varpi' - \varpi] \\
& + \frac{e'^2}{8}\left(2 - 7\,j + 4\,j^2 - 2\,\alpha\,D + 4\,j\,\alpha\,D + \alpha^2\,D^2\right) b_{1/2}^{(j-2)} \cos[j\,\bar{\lambda}' + (2-j)\,\bar{\lambda} - 2\,\varpi'] \\
& + \frac{s^2}{2}\,\alpha\,b_{3/2}^{(j-1)} \cos[j\,\bar{\lambda}' + (2-j)\,\bar{\lambda} - 2\,\Omega] \\
& - s\,s'\,\alpha\,b_{3/2}^{(j-1)} \cos[j\,\bar{\lambda}' + (2-j)\,\bar{\lambda} - \Omega' - \Omega] \\
& + \frac{s'^2}{2}\,\alpha\,b_{3/2}^{(j-1)} \cos[j\,\bar{\lambda}' + (2-j)\,\bar{\lambda} - 2\,\Omega']. \tag{C.70}
\end{aligned}
$$

Likewise, Equations (C.39), (C.40), (C.54), and (C.63) give

$$
\begin{aligned}
\mathcal{S}_E \simeq & \left(-1 + \frac{e^2}{2} + \frac{e'^2}{2} + s^2 + s'^2\right) \cos(\bar{\lambda}' - \bar{\lambda}) \\
& - e\,e'\,\cos(2\,\bar{\lambda}' - 2\,\bar{\lambda} - \varpi' + \varpi) - 2\,s\,s'\,\cos(\bar{\lambda}' - \bar{\lambda} - \Omega' + \Omega) \\
& - \frac{e}{2}\,\cos(\bar{\lambda}' - 2\,\bar{\lambda} + \varpi) + \frac{3\,e}{2}\,\cos(\bar{\lambda}' - \varpi) - 2\,e'\,\cos(2\,\bar{\lambda}' - \bar{\lambda} - \varpi') \\
& - \frac{3\,e^2}{8}\,\cos(\bar{\lambda}' - 3\,\bar{\lambda} + 2\,\varpi) - \frac{e^2}{8}\,\cos(\bar{\lambda}' + \bar{\lambda} - 2\,\varpi) \\
& + 3\,e\,e'\,\cos(2\,\bar{\lambda} - \varpi' - \varpi) - \frac{e'^2}{8}\,\cos(\bar{\lambda}' + \bar{\lambda} - 2\,\varpi') \\
& - \frac{27\,e'^2}{8}\,\cos(3\,\bar{\lambda}' - \bar{\lambda} - 2\,\varpi') - s^2\,\cos(\bar{\lambda}' + \bar{\lambda} - 2\,\Omega) \\
& + 2\,s\,s'\,\cos(\bar{\lambda}' + \bar{\lambda} - \Omega' - \Omega) - s'^2\,\cos(\bar{\lambda}' + \bar{\lambda} - 2\,\Omega'), \tag{C.71}
\end{aligned}
$$

and

$$
\begin{aligned}
\mathcal{S}_1 \simeq & \left(-1 + \frac{e^2}{2} + \frac{e'^2}{2} + s^2 + s'^2\right)\cos(\bar{\lambda}' - \bar{\lambda}) \\
& - e\,e'\,\cos(2\,\bar{\lambda}' - 2\,\bar{\lambda} - \varpi' + \varpi) - 2\,s\,s'\,\cos(\bar{\lambda}' - \bar{\lambda} - \Omega' + \Omega) \\
& - 2\,e\,\cos(\bar{\lambda}' - 2\,\bar{\lambda} + \varpi) + \frac{3\,e'}{2}\,\cos(\bar{\lambda} - \varpi') - \frac{e'}{2}\,\cos(2\,\bar{\lambda}' - \bar{\lambda} - \varpi') \\
& - \frac{27\,e^2}{8}\,\cos(\bar{\lambda}' - 3\,\bar{\lambda} + 2\,\varpi) - \frac{e^2}{8}\,\cos(\bar{\lambda}' + \bar{\lambda} - 2\,\varpi) \\
& + 3\,e\,e'\,\cos(2\,\bar{\lambda} - \varpi' - \varpi) - \frac{e'^2}{8}\,\cos(\bar{\lambda}' + \bar{\lambda} - 2\,\varpi') \\
& - \frac{3\,e'^2}{8}\,\cos(3\,\bar{\lambda}' - \bar{\lambda} - 2\,\varpi') - s^2\,\cos(\bar{\lambda}' + \bar{\lambda} - 2\,\Omega) \\
& + 2\,s\,s'\,\cos(\bar{\lambda}' + \bar{\lambda} - \Omega' - \Omega) - s'^2\,\cos(\bar{\lambda}' + \bar{\lambda} - 2\,\Omega').
\end{aligned}
\tag{C.72}
$$

We can distinguish two different types of term that appear in our expansion of the disturbing functions. *Periodic terms* vary *sinusoidally* in time as our two planets orbit the Sun (i.e., they depend on the mean ecliptic longitudes, $\bar{\lambda}$ and $\bar{\lambda}'$), whereas *secular terms* remain *constant* in time (i.e., they do not depend on $\bar{\lambda}$ or $\bar{\lambda}'$). We expect the periodic terms to give rise to relatively short-period (i.e., of order a typical orbital period) oscillations in the osculating orbital elements of our planets. On the other hand, we expect the secular terms to produce an initially linear increase in these elements with time. Because such an increase can become significant over a long period of time, even if the rate of increase is small, it is necessary to evaluate the secular terms in the disturbing function to higher order than the periodic terms. Hence, in the following, we shall evaluate periodic terms to $\mathcal{O}(e)$ and secular terms to $\mathcal{O}(e^2)$.

Making use of Equations (C.20), (C.36), and (C.69)–(C.71), as well as the definitions (B.121)–(B.124) (with $p \simeq I \sin\Omega$ and $q \simeq I \cos\omega$), we see that the order-by-order expansion (in e) of the first planet's disturbing function becomes $\mathcal{S} = \mathcal{S}_0 + \mathcal{S}_1 + \mathcal{S}_2$, where

$$
\mathcal{S}_0 = \frac{1}{2}\sum_{j=-\infty,\infty} b_{1/2}^{(j)}\,\cos[\,j\,(\bar{\lambda} - \bar{\lambda}')] - \alpha\,\cos(\bar{\lambda} - \bar{\lambda}'),
\tag{C.73}
$$

$$
\begin{aligned}
\mathcal{S}_1 = \frac{1}{2}\sum_{j=-\infty,\infty} & \Big\{ k\,(-2\,j - \alpha\,D)\,b_{1/2}^{(j)}\,\cos[(1-j)\,\bar{\lambda} + j\,\bar{\lambda}'] \\
& + h\,(-2\,j - \alpha\,D)\,b_{1/2}^{(j)}\,\sin[(1-j)\,\bar{\lambda} + j\,\bar{\lambda}'] \\
& + k'\,(-1 + 2\,j + \alpha\,D)\,b_{1/2}^{(j-1)}\,\cos[(1-j)\,\bar{\lambda} + j\,\bar{\lambda}'] \\
& + h'\,(-1 + 2\,j + \alpha\,D)\,b_{1/2}^{(j-1)}\,\sin[(1-j)\,\bar{\lambda} + j\,\bar{\lambda}']\Big\} \\
& + \frac{\alpha}{2}\Big\{-k\,\cos(2\,\bar{\lambda} - \bar{\lambda}') - h\,\sin(2\,\bar{\lambda} - \bar{\lambda}') + 3\,k\,\cos\bar{\lambda}' + 3\,h\,\sin\bar{\lambda}' \\
& - 4\,k'\,\cos(\bar{\lambda} - 2\,\bar{\lambda}') + 4\,h'\,\sin(\bar{\lambda} - 2\,\bar{\lambda}')\Big\},
\end{aligned}
\tag{C.74}
$$

and

$$
\mathcal{S}_2 = \frac{1}{8}(h^2 + k^2 + h'^2 + k'^2)(2\alpha D + \alpha^2 D^2)b_{1/2}^{(0)} - \frac{1}{8}(p^2 + q^2 + p'^2 + q'^2)\alpha b_{3/2}^{(1)}
$$

$$
+ \frac{1}{4}(kk' + hh')(2 - 2\alpha D - \alpha^2 D^2)b_{1/2}^{(1)} + \frac{1}{4}(pp' + qq')\alpha b_{3/2}^{(1)}. \tag{C.75}
$$

Likewise, the order by order expansion of the second planet's disturbing function becomes $\mathcal{S}' = \mathcal{S}'_0 + \mathcal{S}'_1 + \mathcal{S}'_2$, where

$$
\mathcal{S}'_0 = \frac{\alpha}{2} \sum_{j=-\infty,\infty} b_{1/2}^{(j)} \cos[j(\bar{\lambda}' - \bar{\lambda})] - \alpha^{-1} \cos(\bar{\lambda}' - \bar{\lambda}), \tag{C.76}
$$

$$
\mathcal{S}'_1 = \frac{\alpha}{2} \sum_{j=-\infty,\infty} \left\{ k(-2j - \alpha D)b_{1/2}^{(j)} \cos[j\bar{\lambda}' + (1-j)\bar{\lambda}] \right.
$$

$$
+ h(-2j - \alpha D)b_{1/2}^{(j)} \sin[j\bar{\lambda}' + (1-j)\bar{\lambda}]
$$

$$
+ k'(-1 + 2j + \alpha D)b_{1/2}^{(j-1)} \cos[j\bar{\lambda}' + (1-j)\bar{\lambda}]
$$

$$
\left. + h'(-1 + 2j + \alpha D)b_{1/2}^{(j-1)} \sin[j\bar{\lambda}' + (1-j)\bar{\lambda}] \right\}
$$

$$
+ \frac{\alpha^{-1}}{2} \left\{ -k' \cos(2\bar{\lambda}' - \bar{\lambda}) - h' \sin(2\bar{\lambda}' - \bar{\lambda}) + 3k' \cos\bar{\lambda} + 3h' \sin\bar{\lambda} \right.
$$

$$
\left. - 4k \cos(\bar{\lambda}' - 2\bar{\lambda}) + 4h \sin(\bar{\lambda}' - 2\bar{\lambda}) \right\}, \tag{C.77}
$$

and

$$
\mathcal{S}'_2 = \frac{1}{8}(h^2 + k^2 + h'^2 + k'^2)\alpha(2\alpha D + \alpha^2 D^2)b_{1/2}^{(0)}
$$

$$
- \frac{1}{8}(p^2 + q^2 + p'^2 + q'^2)\alpha^2 b_{3/2}^{(1)}
$$

$$
+ \frac{1}{4}(kk' + hh')\alpha(2 - 2\alpha D - \alpha^2 D^2)b_{1/2}^{(1)}
$$

$$
+ \frac{1}{4}(pp' + qq')\alpha^2 b_{3/2}^{(1)}. \tag{C.78}
$$

Bibliography

Abramowitz M. (ed.), and Stegun, I.A. (ed.). 1965. *Handbook of Methematical Functions: With Formulas, Graphs, and Mathematical Tables*. Dover. Chapter 22.

Adams, J.C. 1900. *Lectures on the Lunar Theory*. Cambridge University Press.

Barbour, J.B. 2001. *The Discovery of Dynamics: A Study from a Machian Point of View of the Discovery and the Structure of Dynamical Theories*. Oxford University Press.

Bate, R.R., Mueller, D.D., and White, J.E. 1977. *Fundamentals of Astrodynamics*. Dover.

Bertotti, B., Farinella, P., and Vokrouhlický, D. 2003. *The Physics of the Solar System: Dynamics and Evolution, Space Physics, and Spacetime Structure*. Kluwer Academic Publishers.

Black, G.J., Nicholson, P.D., and Thomas, P.C. 1995. *Hyperion: Rotational Dynamics*. Icarus **117**, 149.

Brouwer, D., and Clemence, G.M. 1961. *Methods of Celestial Mechanics*. Academic Press.

Brouwer, D., and van Woerkom, A.J.J. 1950. *The Secular Variations of the Orbital Elements of the Principal Planets*. Astronomical Papers of the American Ephemeris **13**, 81.

Brown, E.W. 1896. *An Introductory Treatise on the Lunar Theory*. Cambridge University Press.

Chandrasekhar, S. 1969. *Ellipsoidal Figures of Equilibrium*. Yale University Press.

Chapront, J., Chapront-Touzé, M., and Francou, G. 2002. *A New Determination of Lunar Orbital Parameters, Precession Constant, and Tidal Acceleration from LLR Measurements*. Astronomy and Astrophysics **387**, 700.

Chapront-Touzé, M., and Chapront, J. 1988. *A Semi-Analytic Lunar Ephemeris Adequate for Historical Times*. Astronomy and Astrophysics **190**, 342.

Cook, A. 1988. *The Motion of the Moon*. Adam Hilger.

Cotter, C.H. 1968. *A History of Nautical Astronomy*. The Bodley Head.

Danby, J.M.A. 1992. *Fundamentals of Celestial Mechanics*, 2nd Edition, Revised and Enlarged. Willmann-Bell.

Delaunay, C. 1867. *Théorie du Mouvement de la Lune*, two volumes. Mallet-Bachelier.

de Pater, I., and Lissauer, J.J. 2010. *Planetary Sciences*, 2nd Edition. Cambridge University Press.

Evans, J. 1998. *The History and Practice of Ancient Astronomy*. Oxford University Press.

Fowles, G.R., and Cassiday G.L. 2005. *Analytical Mechanics*, 7th Edition. Brooks/Cole–Thomson Learning.

Godfray, H. 1853. *An Elementary Treatise on the Lunar Theory*. Macmillan.

Goldstein, H., Poole, C., and Safko. J. 2001. *Classical Mechanics*, 3rd Edition. Addison-Wesley.

Gradshteyn, I.S., and Ryzhik, I.M. 1980a. *Tables of Integrals, Series, and Products*, Corrected and Enlarged Edition, tr. A. Jeffrey. Academic Press. Section 8.411.

Gradshteyn, I.S., and Ryzhik, I.M. 1980b. *ibid.* Section 8.440.

Gradshteyn, I.S., and Ryzhik, I.M. 1980c. *ibid.* Chapters 13–15.

Hagihara, Y. 1971. *Celestial Mechanics*, Volume II, Part 1, *Perturbation Theory*. MIT Press.

Heath, T.L. 1991. *Greek Astronomy*. Dover.

Hoffleit, D., and Warren, W.H. Jr. 1991. *The Bright Star Catalogue*, 5th Revised Edition. Astronomical Data Center, NSSDC/ADC.

Jin, W., and Li, J. 1996. *Determination of Some Physical Parameters of the Moon with Lunar Laser Ranging Data*. Earth, Moon, and Planets **73**, 259.

Konopliv, A.S., Binder, A.B., Hood, L.L., et al. 1998. *Improved Gravity Field of the Moon from Lunar Prospector*. Science **281**, 1476.

Lamb, H. 1923. *Dynamics*, 2nd Edition. Cambridge University Press.

Laskar, J. 1990. *The Chaotic Motion of the Solar System: A Numerical Estimate of the Size of the Chaotic Zones*. Icarus **88**, 266.

Love, A.E.H. 2011. *A Treatise on the Mathematical Theory of Elasticity*, 4th Edition. Dover.

Margot, L.J., Peale, S.J., Jurgens, R.F., et al. 2007. *Large Longitude Libration of Mercury Reveals a Molten Core*. Science **316**, 710.

Meeus, J. 2005. *Astronomical Algorithms*, 2nd Edition. Willmann-Bell.

Moulton, F.R. 1914. *An Introduction to Celestial Mechanics*, 2nd Revised Edition. Macmillan.

Murray, C.D., and Dermott, S.F. 1999. *Solar System Dynamics*. Cambridge University Press.

Pannekoek, A. 2001. *A History of Astronomy*. Dover.

Plummer, H.C. 1960. *An Introductory Treatise on Dynamical Astronomy*. Dover.

Press, W.H., Teukolsky, S.A., Vetterling, W.T., and Flannery, B.P. 1992. *Numerical Recipes in C: The Art of Scientific Computing*, 2nd Edition. Cambridge University Press. Chapter 11.

Rindler, W. 1977. *Essential Relativity: Special, General, and Cosmological*, 2nd Edition. Springer.

Roy, A.E. 2005. *Orbital Motion*, 4th Edition. Taylor & Francis.

Siedelmann, P., Archinal, B.A., A'Hearn, M.F., et al. 2007. *Report of the IAU/IAG Working Group on Cartographic Coordinates and Rotational Elements: 2006*. Celestial Mechanics and Dynamical Astronomy **90**, 155.

Smart, E.H. 1951. *Advanced Dynamics*, Volume I. Macmillan.

Smart, W.M. 1977. *Textbook on Spherical Astronomy*, 6th Edition. Cambridge University Press.

Standish, E.M., and Williams, J.G. 1992. *Orbital Ephemeris of the Sun, Moon, and Planets*, in *Explanatory Supplement to the Astronomical Almanac*, ed. P.K. Seidelmann. University Science Books. Table 8.10.3.

Stephenson, F.R., and Morrison, L.V. 1995. *Long-Term Fluctuations in the Earth's Rotation: 700 BC to AD 1990*. Philosophical Transactions of the Royal Society A **351**, 165.

Stewart, M.G. 2005. *Precession of the Perihelion of Mercury's Orbit*. American Journal of Physics **73**, 730.

Strogatz, S.H. 2001. *Nonlinear Dynamics and Chaos: With Applications to Physics, Biology, Chemistry, and Engineering*. Westview.

Taton, R., and Wilson, C. (eds.) 1995. *The General History of Astronomy*, Volume 2: *Planetary Astronomy from the Renaissance to the Rise of Astrophysics*, Part B: *The Eighteenth and Nineteenth Centuries*. Cambridge University Press.

Thomas, P.C., Black, G.J., and Nicholson, P.D. 1995. *Hyperion: Rotation, Shape, and Geology from Voyager Images*. Icarus **117**, 128.

Thornton, S.T., and Marion, J.B. 2004. *Classical Dynamics of Particles and Systems*, 5th Edition. Brooks/Cole—Thomson Learning.

Williams, J.G., and Dickey, J.O. 2003. *Lunar Geophyics, Geodesy, and Dynamics*, in *13th International Workshop on Laser Ranging*, eds. R. Noomen, S. Klosko, C. Noll, and M. Pearlman (NASA/CP-2003-212248). p. 75.

Willner, K., Oberst, J., Hussmann, H., et al. 2010. *Phobos Control-Point Network, Rotation, and Shape*. Earth and Planetary Science **294**, 541.

Yoder, C.F. 1995. *Astrometric and Geodetic Properties of Earth and the Solar System*, in *Global Earth Physics: A Handbook of Physical Constants*, ed. T. Ahrens. American Geophysical Union.

Zhang, C.Z. 1992. *Love Numbers of the Moon and of the Terrestrial Planets*. Earth, Moon, and Planets **56**, 193.

Index

BV - #0003 - 181022 - C0 - 253/177/16 [18] - CB - 9781107023819 - Gloss Lamination